AF516019

BIOTECHNOLOGY
Cracking New Pastures

BIOTECHNOLOGY
Cracking New Pastures

Edited By

C P MALIK
AMAN VERMA
School of Life Sciences,
Jaipur National University
Jaipur

MD PUBLICATIONS PVT LTD
NEW DELHI
www.mdppl.com

Published by :

MD Publications Pvt Ltd
"MD House", 11, Darya Ganj
New Delhi-110 002
Phone : +91-11-41563325 (Customer Service)
+91-11-41562846 (Editorial)
E-mail (orders): order@mdppl.com
Website : www.mdppl.com

ISBN : 978-81-7533-233-1

Published and Printed by Mr. Pranav Gupta on behalf of **MD Publications Pvt Ltd** at Times Press, New Delhi.

Contents

PREFACE

The term 'BIOTECHNOLOGY' was coined in 1919 by Hungarian engineer Karl Ereky. Since then, biotechnology is one of the key technologies in benefiting the human race. It must be seen in the context of systems biology, synthetic biology, bioinformatics and nanotechnology, whose convergence will drive new products and technologies in the future. It is a disruptive technology whose potential is not well appreciated.

Breathtaking progress has been made in biotechnology in the past 50 years, especially in the past 20 years. The global biotechnology industry is now at the initiation of a technology curve whose growing potential seems limitless.

Biotechnology is the next driver of innovation and economic growth. This cutting edge technology is already beginning to usher in complex, rapidly emerging and far-reaching new changes in several areas, particularly food and nutrition, security, healthcare and environmental sustainability. It has sparked several discoveries. Developments in this area have radically altered our concept of healthcare with the arrival of numerous products that were unthinkable 20 years ago. Such is the pace of biotechnology that it can be anticipated in the coming years that even greater developments will occur. Thanks to the projects like Human Genome Project. The cost of this global collaborative project ran approximately around $3 billion. In 2006, the full genome of, James Watson was published in full, but the project cost just $300 million. There are companies offering to sequence personnel gene map for only $1,000 or so. It is predicted that the individual genome map would soon be available for $100 or around. This is one instance of the tremendous progress biotechnology is experiencing through some cutting edge technological tools. Such projects will open up opportunities for treatment of diseases at the individual basis. On top of this several of these developments owe their advances on computer software, electronic instrumentation, imaging, and several other enabling technologies.

The fundamentals of biotechnology remain, as always, same *i.e.* production of goods and services that are needed and can be provided with safety and reasonable cost. Keeping in view the remarkable advance being made in this growing discipline, we have attempted to highlight compendium of some of these enabling technologies and advances in a volume entitled *Biotechnology: Cracking New Pastures.*

Biotechnology has revolutionized the concepts in agriculture, food and industrial feed stocks in the recent past. Techniques to enhance agricultural productivity (green biotechnology) raise value added products and healthcare systems for better environments. Biotechnology has ushered tremendous changes in agriculture, medicine and cell biology.

India has enormous opportunities to exploit the potential of this new biology and this is attributable to several factors *e.g.* enormous genetic diversity in plants, animals, microbes;

strong coordinated system; and expertise in mathematical and computational areas. In fact genomics has opened new horizons for gene mining and allele mining, which are highly significant in various research areas especially in private breeding companies.

The present book *Biotechnology: Cracking New Pastures* discusses selected topics to provide the reader a sound knowledge of selected areas in different guises, ranging from Lectins, RNAi, Bacterial biosorption of heavy metals, Increasing yield potential of grain legumes, Science of Omics, Endophytic fungi; advances in DNA transfer to prions, Chemistry and Functions of NO, etc. Each topic is written by an expert and offers the most comprehensive and up-to-date information on pertinent topic. Most of these topics are sought after by researchers and students at the graduate as well as postgraduate levels. Each chapter discusses the current status.

The strength of the book is lavishly used images, and extensive literature citation in each chapter. Certain to become the standard reference for biotechnologists, molecular biologists, breeders, applied biologists, a must for teachers and students engaged in teaching and research in physiology and plant sciences, the book is the definitive source for those who are keen to remain updated with the recent advances in biotechnology. To give a snapshot of state-of-art research in the related area where things move faster than anywhere else is next to impossible, so it is inevitable that by the time the book goes into print, some developments will have superseded, those described here. Umpteen authors and co-authors have worked on this project, and although we tried to find a more or less uniform style, the authors with their unique views and values are still recognizable.

We would like to thank all those who have contributed to the making of this book, not the least those who contributed *via* frank discussions. We sincerely apologize unreservedly for any lapses, all of which are ours.

Finally, we would like to thank our families (CP Malik, his spouse Surakhsha; Aman, her husband Kuldeep Singh, daughter Anushka Singh) for their continued forbearance, support, encouragement and love throughout the compilation of this book.

Finally, we thank Mr. Pranav, our publisher, and Pragati madam for believing in us and extending their support.

C P Malik
Aman Verma

LIST OF CONTRIBUTORS

Akhilesh Kumar
National Research Centre on Plant Biotechnology (NRCPB), IARI, New Delhi

Aman Verma
School of Life Sciences, Jaipur National University, Jaipur-302 025

B. K. Garg
Central Arid Zone Research Institute, Jodhpur - 342 003

Chitra Wadhwani
School of Life Sciences, Jaipur National University, Jaipur- 302 025

C. P. Malik
School of Life Sciences, Jaipur National University, Jaipur-302 025

H Wani Shabir
Department of Plant Breeding and Genetics, Punjab Agricultural University, Ludhiana-141 004

Jin Ho Joo
Department of Biological Environment, Kangwon National University, Chuncheon-200 701, Korea

Kartikeya Tiwar
School of Life Sciences, Jaipur National University, Jaipur- 302 025

K.R. Koundal
National Research Centre on Plant Biotechnology (NRCPB), IARI, New Delhi

Manjit S Gill
Department of Plant Breeding and Genetics, Punjab Agricultural University, Ludhiana-141 004

Mohammad Humayun Kabir
School of Biotechnology, Kangwon National University, Chuncheon-200 701, Korea.

Mukesh K Upadhya
School of Life Sciences, Jaipur National University, Jaipur- 302 025

Pushp Sharma
Department of Plant Breeding and Genetics, Punjab Agricultural University, Ludhiana-141 004

Rekha Kansal
National Research Centre on Plant Biotechnology (NRCPB), IARI, New Delhi

Ritu Mahajan
School of Biotechnology, University of Jammu, Jammu

Sang Eun Oh
Department of Biological Environment, Kangwon

Sarita Bhuradia
School of Life Sciences, Jaipur National University, Jaipur- 302 025

Varsha Chaturvedi
School of Life Sciences, Jaipur National University, Jaipur- 302 025

Yaksha Singh
School of Life Sciences, Jaipur National University, Jaipur- 302 025

1

LECTINS: THEIR PROPOSED BIOLOGICAL FUNCTIONS

REKHA KANSAL, AKHILESH KUMAR, K.R. KOUNDAL AND V.K.GUPTA

INTRODUCTION

Farmers still rely on the use of pesticides for crop protection against insect pests. However, the continuous use of chemical pesticides has added to environmental problems and has led to the development of resistance in many pests. This calls for alternative approaches for crop protection that lead to reduction of inputs in energy and chemicals and also will not generate harmful effects like pesticide residues (Jouanin *et al.* 1998). Consequently, plant genetic engineering may be exploited for development of insect resistant plants by insertion and expression of genes encoding entomopathogenic proteins. The use of transgenic plants carrying δ-endotoxin encoding genes from *Bacillus thuringiensis* is already in practice. More emphasis is now being laid towards the use of plant derived genes such as those encoding enzyme inhibitors and lectins that specifically inhibit insect pests. Lectins are currently receiving the maximum interest as insecticidal agents against sap sucking homopterans including aphids, leaf and plant hoppers.

The plant lectins are a heterogenous group of carbohydrate binding proteins which differ in their molecular structure and carbohydrate binding specificity (Barre *et al.* 2001). They are found in seeds and other tissues such as leaves, bark, stems, rhizomes, bulbs, tubers, *etc.* (Peumans and Van Damme, 1995; 1998). Because of the broad distribution and ease of isolation, plant lectins are the largest and the best characterized group among lectins. Rarely, lectins that differ in their sugar specificity have been found in the same plant (Table 1), especially in legumes and the Gramineae (Lis and Sharon, 1986 a, b).

Lectins are composed of several polypeptides which are encoded by related and linked genes of lectin genes family. Lectins play an important role in seed maturation, cell-wall assembly, defense mechanism and rhizobial nodulation of legume roots (Diaz *et al.* 1989). The lectins share extensive sequence homology and three-dimensional structural similarity but differ in carbohydrate specificity (Rouge *et al.* 1991; Sharon and Lis, 1990). Many of them have been purified and characterized with respect to their molecular structure, sugar binding specificity and biological role. More than 50 of them have been sequenced, mostly by protein sequencing methods (Peumans and Van Damme, 1999; Sharon and Lis, 1990; Van Damme *et al.* 1992, 1998). Yim *et al.* (2001) suggested the use of novel plant lectin libraries to discriminate in a highly specific manner among a variety of cell types. This technology might be very useful in a number of different applications requiring a high level of specificity in cell identification. Uses of plant lectins in bioscience and biomedicine have been reviewed by Pusztai *et al.* (2008).

Lectins other than those of the legumes also combine with carbohydrates, primarily by a network of hydrogen bonds and hydrophobic interactions. However, different amino acids are often employed for the same purpose, even for combination with the same ligand. In fact, essentially all side chains capable of forming hydrogen bonds with sugars do so in one lectin-carbohydrate complex or another (Table 1). Although carboxylic and carboxyamino acids are most common, basic and hydroxyamino acids, as well as aromatic ones, also participate in ligand binding. Because most saccharides are uncharged, ionic (charge-charge) interactions are not often found in lectin-saccharide complexes (Sharon and Lis, 2002).

HISTORY

Plant lectins were originally detected by the medical student Stillmark (1888), in the University of Dorpat/Estonia when he was working on his dissertation thesis on castor beans (*Ricinus communis* L.). At this time he described the presence of a toxic proteinaceous factor in extracts from castor beans, which also agglutinated red blood cells. To indicate the source in which it had been found the name ricin was given. Thus, historically, this fact is considered the true starting point of the research on plant lectins (Rudiger and Gabius, 1998). Soon after Stillmark's discovery, ricin and the related toxin from jequirity beans (*Abrus precatorius*, Leguminosae) played a fundamental role as model antigen in the pioneering studies of Paul Ehrlich. The first legume lectin isolated was concanavalin A from jack bean (*Canavalia ensiformis*) seeds. The prominent biochemist James B. Summer, well known for demonstrating that an enzyme (urease) is nothing but a protein, succeeded in purifying also the agglutinating principle from these seeds by crystallization and called Con A is able to interact with red blood cells as well as with starch, glycogen and mucins and these interaction can be prevented by low molecular weight carbohydrates such as sucrose.

DEFINITION

The term lectin (derived from a latin verb "legre" means "to select") was coined by Boyd many years later to emphasize the ability of some hemagglutinins to discriminate blood cells within the ABO blood group system (Boyd and Shapleigh, 1954). Presently, the term lectin is prevalent over that of hemagglutinin and is broadly one non-catalytic domain, which binds reversibly to a specific mono- or oligosaccharide (Peumans and Van Damme, 1998). Alternatively, lectins are also defined as proteins or glycoproteins of non-immune origin with one or more binding sites per subunit, which can reversibly bind to specific sugar segments through hydrogen bonds and Van der Waals interactions (Lis and Sharon, 1998). Both these definitions are very comprehensive and include non-agglutinating and some enzyme proteins (Van Damme *et al.* 1998, 1997) such as the class I plant chitinase, considered as a lectin because its single carbohydrate-binding domain is structurally distinct of its catalytic site and does not directly participate in the chitinolytic activity of this enzyme (Peumans and Van Damme, 1998). Obviously, chitinases cannot be called as agglutinins or phytohemagglutinins since they do not possess agglutination and/or glycoconjugate precipitation properties, as they are not multivalent carbohydrate-binding proteins. Another class of plant proteins which are also considered as lectins, according to the above definitions, is the type 2 RIPs (ribosome-inactivating proteins). Type-2-RIPs are the lectins that catalytically inactivate ribosomes or eukaryotes and therefore irreversibly shut down protein synthesis. They bind galactose, N-acetyl galactosamine and N-acetyl neuraminic acid.

OCCURRENCE

Generally, the richest sources for most lectins are the seeds or storage organ of plants. Apart from seeds in most plants studied so far there are few examples which show lectin presence.

Root : *Urtica, Phytolacca, Sambucus, Calystegia*

Tuber or Bulb : *Solanum, Galanthus, Scilla, Allium*

Bark : *Sambucus, Sophora, Robinia, Cytisus*

Leaves : *Aloe, Lactuca, Vicia unijuga, Viscum album*

Lectins are found in the protein bodies inside cells which proliferate at endoplasmic reticulum (ER) and transported *via* Golgi apparatus and originate by subdividing the vacuole. In addition to this intracellular site lectins have also been found in the cytoplasm and in intracellular space. The amount of lectin can vary markedly from one plant species to other. The data compiled on the basis of purification from various plants is given in Table 2.

BIOSYNTHESIS

The biosynthesis of most plant lectins studied so far proceeds *via* the secretory pathway. In brief, the lectin is synthesized at ribosomes attached to the ER, enters the lumen, and is further transported through the Golgi apparatus. In contrast to secretory proteins in a narrow sense, which moves to the cell membrane by vesicular transport and then leave the cell by exocytosis, lectins (other protein as typical storage proteins) end up in the vacuole.

TABLE 1 : Different lectins combining with the same ligand

Ligand	*Lectin*	*H-bonding residues*
Mannose	Con A	Asn, Asp
	GNL	Asn, Asp, Gln, Tyr
	MBP-A	Asn, Glu
Galactose	ACA	Asn, His, Thr
	ECorL	Asn, Asp
	Galectin 1	Arg, Asn, His, Glu

Con A, concanavalin A; GNL, *Galanthus nivalis* lectin; MBP-A, mannose binding protein A; ACA, *Amaranthus caudatus* agglutinin; ECorL, *Erythtrina corallodendron* lectin

TABLE 2 : Legume lectins: their carbohydrate specificity and amount in seeds

S.No.	*Plant Species*	*Carbohydrate specificity (Monosaccharides)*	*mg lectin per 100 g seeds*
1	*Canavalia ensiformis*	Man/Glc	2100
2	*Ricinus communis*	Gal	1400
3	*Vicia cracca*	Man/Glc	1400
4	*Phaseolus vulgaris*	No monosaccharide known	1200
5	*Glycine max.*	GalNAc/Gal	300
6	*Arachis hypogaea*	Gal	190
7	*Pisum sativum*	Man/Glc	140
8	*Dolichos biflorus*	Gal/Nac	110
9	*Lens culinaris*	Man/Glc	60
10	*Bauhinia purpurea*	GalNAc	28

Man, Mannose ; Glc, Glucose ; Gal, Galactose ; Gal N Ac, N-acetylgalactosamine

On the way from the site of synthesis to its final destination, a lectin is subject to a series of covalent modifications that are common for proteins on this route: the N-terminal signal sequence of about 20-30 amino acids responsible for initiating transmembrane passage is split off co-translationally as soon as sizeable segment of the newly synthesized protein has entered the lumen of the ER. The lectins are generally synthesised as a precursor, which subsequently, after an initial removal of an N-terminal signal peptide, is post-translationally processed into a mature protein. Depending on each specific lectin, the different precursors may or may not possess carbohydrate binding activity. In about two third of the molecules, the original N- and C-termini are then ligated, resulting in a circularly permuted protein (Loris *et al.* 1998).

CLASSIFICATION

Plant lectins are subdivided into five groups, according to the monosaccharide for which they exhibit the highest affinity: D-mannose/D-glucose, D-galactose/N-acetyl-D-galactosamine, N-acetyl-D-glucosamine, L-fucose and N-acetylneuraminic acid (Goldstein *et al.* 1997). Thus depending on the specificity toward a given monosaccharide, the lectin will selectively bind to one of the above sugars which are typical constituents of eukaryotic cell surfaces (Lis and Sharon, 1998).

Considering the overall structure of plant lectins, they are subdivided into four major classes:

1- Merolectins	:	Single carbohydrate-binding domain;
2- Hololectins	:	Di- or multivalent carbohydrate-binding sites;
3-Chimerolectins	:	Consisting of one or more carbohydrate-binding domain(s) plus an additional catalytic or another biological activity dependent on a distinct domain other than the carbohydrate-binding site;
4- Superlectins	:	At least two carbohydrate-binding domains but differ from the hololectins because their sites are able to recognize structurally unrelated sugars.

Although plant lectins are considered a very complex and heterogeneous group of proteins from the biochemical/physicochemical viewpoint yet sequence, structural information and molecular cloning of lectin genes enable subdivision of plant lectins in seven families of structurally and evolutionary related proteins: legume lectins; chitin binding lectins; type 2- RIPs and related lectins; monocot mannose-binding lectins; jacalin-related lectins; amaranthin lectin family and cucurbit phloem lectins (Murdock and Shade, 2002). With regard to their localization, plant lectins are generally most abundant in the seeds but they are also found in different vegetative tissues such as roots, leaves, bark, flowers, bulbs and rhizomes (Van Damme *et al.* 2000).

At present the most accepted classification is based on protein sequences which groups lectins into seven distinct families (Peumans and Van Damme, 1999) *viz.* the legume lectins, monocot mannose–binding lectins, jacalins, chitin-binding lectins, type 2-RIPs, amaranthins, and cucurbitaceae phloem lectins (Table 3). Among these, legume lectins are extensively studied lectin proteins, which are present in several hundreds of species of legumes. A database named Lectindb has been developed for plant lectins (Chandra *et al.* 2006).

LEGUME LECTINS

Although, among all the grain crops, legumes (also known as pulses) rank third after cereals and oilseeds in world production, but constitute an important dietary constituent for humans and animals daily diet. Grain legumes provide about a third of all dietary protein nitrogen and a third of processed vegetable oil for human consumption. Seeds of grain legumes contain at least 20-40% of protein (Wang *et al.* 2003) and in India it is the major source of protein, which people take in their daily diet. Thus, India stands top in the world for legume production (Table 4).

An important additional reason for the interest in legume lectins is their structural similarity to lectins from other sources, such as the galectins of animals. Typically, the members of this protein family consist of two or four subunits (or protomers), either identical or slightly different, each with a single, small carbohydrate-combining site with the same specificity.

TABLE 3 : Types of lectins and their source

S. No.	*Plant lectin type*	*Source*
1	Legume lectins	*Arachis hypogaea* (Peanut), *Canavalia ensiformis* (Jackbean-Con A), *Dolichos biflorus* lectin (DBL), *Lens culinaris* (lentil lectin), *Pisum sativum* (Pea lectin), *Glycine max* (soybean agglutinin), *Phaseolus vulgaris (PHA-L)* and *Dolichos lablab (Fril)*
2	Monocot mannose-binding (Bulb) lectins	*Galanthus nivalis* agglutinin (Snowdrop lectin) and *Allium sativum* (Garlic bulb lectin),
3	Cereal or Chitin-binding lectins	*Triticum vulgaris (WGA)*
4	Jacalin lectins	*Artocarpus integrifolia* (Jacalin), *Helianthus tuberosus* lectin (Heltuba)
5	Type 2-RIPs	*Ricinus communis* (Ricin B)
6	Amaranthin lectins	*Amaranthus caudatus* (ACA)
7	Cucurbitaceae phloem lectins	*Lagenaria siceraria, Cucumis melo*

TABLE 4 : Statistics of common Indian legume crops

Legume species (% of world production)	*Centre of origin*	*Production (MMT)*		*Major Producers*	*India's rank in production*
		World	*Developing Country*	*(%)*	
Peanut (*Arachis hypogaea*)	South America	34.07	32.32	China (44)	Second
Common bean (*Phaseolus vulgaris*)	Central America	18.33	15.75	Brazil (17)	Second
Chickpea (*Cicer arietinum*)	Turkey	7.80	7.41	India (86)	First
Cowpea (*Vigna unguiculata*)	Ethiopia	3.68	3.64	Nigeria (65)	First
Pigeon pea (*Cajanas cajan*)	India	2.99	2.99	India (81)	First
Lentil (*Lens culinaris*)	Near East	2.94	2.94	India (33)	First
Green pea (*Pisum sativum*)	Peninsula	9.06	6.28	India (42)	First

(Source: FAO stat 2002)

They also contain one ion each of tightly attached calcium and manganese per subunit, which are required for carbohydrate binding. The subunits are of 25-30 kDa and are commonly composed of a single polypeptide chain of 250 amino acids. Most, but not all, are glycosylated and carry up to three asparagine-linked oligosaccharides per subunit (Imberty *et al.* 2000). The primary structures of the lectins of this family are remarkably homologous, with 20% of invariant amino acids and close to 20% of similar ones. The conserved amino acids include several of those that participate in hydrogen bonding or hydrophobic interactions with the monosaccharide held in the combining site and almost all residues that coordinate the metal ions. Currently, over 100 three-dimensional structures of more than 20 legume lectins, without as well as with various carbohydrate ligands, have been elucidated by high-resolution X-ray crystallography (Bettler *et al.* 2001). The legume lectin subunits, all of which are nearly super imposable, are in the shape of a dome. They consist largely of anti parallel sheets and possess a topology related to the jelly roll fold that is known as "the lectin fold" (Srinivasan *et al.* 1996). The same fold has been observed in a number of other proteins from diverse sources, including the mammalian galectins. The carbohydrate-binding sites of the legume lectins, similarly to those of other lectins, are in the form of shallow depressions on the surface of the protein. They are located at the top of each subunit and are accessible not just to monosaccharides but to oligosaccharides as well. The Ca^{2+} and Mn^{2+} are 4.25 Å apart and are in close proximity (9-13 Å) to the carbohydrate-binding site. The sites appear to be preformed because few conformational changes occur upon carbohydrate binding.

STRUCTURE

The first legume lectin structure to be solved was of concanavalin A (con A) (Baker, 1975; Hardman and Ainsworth, 1972). The structure belonged to 'jelly-roll motif' commonly found in viral coat proteins. It has a nearly flat six-stranded rear β-sheet and a curved seven-stranded front β-sheet. A short five membered β-sheet on top holds the two larger sheets together. In addition, a number of loops connecting the strands are also present. Two metal ions, calcium and a transition metal, found in all the legume lectin structures are essential for the carbohydrate binding. The metal ions are bound to four protein atoms and two water molecules. Till date more than 100 legume lectins have been isolated and out of them 50 have been sequenced. Among these, more than 20 legume lectins 3-D structure have been solved. Legume lectins consist of two or four subunits of 25-30 kDa, each with one Ca^{2+} and one Mn^{2+} required for activity. Their tertiary structures are super imposable; quaternary structures are different, possess diverse carbohydrate specificities and have a single combining site per subunit (Sharon and Lis, 2002).

One interesting aspect about legume lectin structures is their different quaternary association, independent of carbohydrate specificity. It is observed that small sequence dependent variations in tertiary structure lead to large changes in quaternary association (Manoj and Suguna, 2001). The dimeric structure found in the crystals of ConA, Pea lectin, Lentil lectin and Lathyrus lectin is termed as 'canonical dimer'. This dimer is formed by side-by-side arrangement of two monomers in such a way that the two rear β sheets form a contiguous 12-stranded β-sheet. However, this type of canonical dimer is missing in the crystal structures of *EcoRI, GSIV* and Peanut lectins. The dimers of these lectins posses back to back arrangement ('hand shake dimer') of the subunits (Srinivas *et al.* 2001).

All legume lectins have similar carbohydrate-binding sites (Loris *et al.* 1998). The four loops A, B, C, D, shown in red, purple, yellow and blue color in Fig. 1 associated with the concave face of the seven-stranded curved β-sheet at the top front side of the subunit form the binding site (Sharma and Surolia, 1997). The amino acid residues which bind metal ions (Ca^{2+} and Mn^{2+}) are highly conserved and the residues which constitute the sugar binding site are less conserved but similar in properties (Lis and Sharan, 1998).

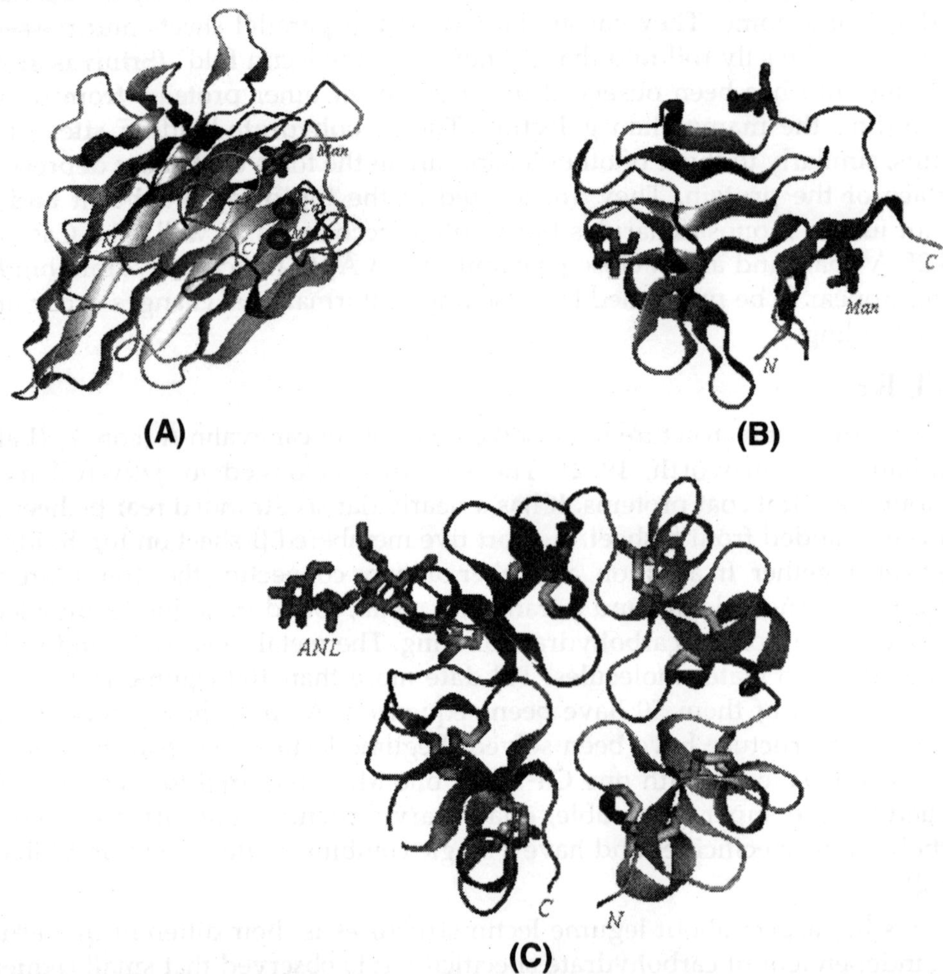

Fig.1. Crystal structures of subunits of, (A) Concanavalin A bound to mannose (Man), Ca and Mn are calcium and manganese ions, (B) *Galanthus nivalis* agglutinin (GNA) bound to mannose at three sites, (C) Wheat germ agglutinin bound to a-N-Acetyl Neuraminyl Lactose (ANL), disulfide bonds are shown in yellow.

The conserved residues Asp and Gly (or Arg) are from loops A and B, whereas Asn and the hydrophobic residues (Phe, Tyr, Trp or Leu) are from loop C. The size of the backbone of this loop is the primary determinant of carbohydrate specificity in legume lectins. The overall conformational changes observed upon sugar binding are small in legume lectins. The aligned sequences of legume lectins show 4 to 7 gaps in the binding loop D, admitting variation in the loop size (Sharma and Surolia, 1997). This partly explains the broad specificity of legume lectins compared to other lectin families.

ANTI-NUTRITIONAL PROPERTIES

A few lectins such as the lectin found in red kidney beans (the PHA lectin), are, however, known to damage the structure of the small intestine, or affect the pancreas and production of digestive enzymes when eaten by humans or animals. This is why kidney beans are always soaked and boiled before use, a process, which effectively destroys the lectin, making beans fit for human consumption. One of the most nutritionally important features of plant lectins is their ability to survive digestion by the gastrointestinal tract of consumers. This allows the lectins to bind to membrane glycosyl groups of the cells lining the digestive tract. As a result of this interaction a series of harmful local and systemic reactions are triggered placing this class of molecules as nutritative and/or toxic substances. Locally, they can affect the turnover and loss of gut epithelial cells, damage the luminal membranes of the epithelium, interfere with nutrient digestion and absorption, stimulate shifts in the bacterial flora and modulate the immune state of the digestive tract. Thus, they are also detrimental to numerous insect pests of crop plants although less is presently known about their insecticidal mechanism of action (Iika and Oliveira, 2004).

In addition to feeding studies made with transgenic potato themselves, several experiments included the addition of isolated purified lectins (Con A and GNA) to raw and cooked potato both in amounts similar to that in the transgenic potato and in amounts greatly in excess of those in the transgenic plant. One long term feeding study and several short-term studies were made with potato tuber expressing GNA. These showed, as expected, no discernable effect on growth. No comparable experiments were made with transgenic plants expressing the ConA in a concentration some 5000-times that found tuber expressing the ConA gene. Addition of isolated ConA at this concentration did cause a small depression in growth rate. However, lectins from different sources have different properties, and thus one lectin may be mitogenic, while another lectin is not. With the exception of several well-studied lectins, little is known about their precise effects when fed as part of the diet on the cells of the gut and immune system of mammals. Given the interest in the possible future introduction of lectin genes to provide pest resistance in crops, it is important that suitable tools are developed to evaluate the potential effect of lectins on animals, including any possible effect on the immune system.

PHYSIOLOGICAL FUNCTIONS

The wide distribution of lectins in all tissues of plants and their ubiquitous presence in the plant kingdom suggest important roles for these proteins. Despite their long history, the true physiological role of legume lectins is still not well understood. Sharon (2007) gave an overview of plant lectins including their physiological roles. Many hypotheses have been formulated in the course of the years but, at present, no physiological function for any legume lectin has been established with certainty. The difficulty in assigning a precise role arises from several of their features. The defining characteristic of all lectins is their carbohydrate-binding ability. This activity has been preserved during evolution, suggesting that it is essential for exploitation of their function. Nevertheless, some lectins possess additional activities that may be non-secondary in determining their following physiological role.

(*i*) In signal transduction

Several legume lectins, for instance, present adenine-combining sites able to bind active forms of the cytokinins, a major class of plant hormones. (Gegg *et al.* 1992) The specificity of adenine-binding sites has been maintained during evolution despite variations in carbohydrate specificity and tissue distribution. Therefore, it may be possible that in such cases proteins have a primary role unrelated to carbohydrates, *e.g.* hormone-binding, and that the carbohydrate-combining activity plays a regulatory or transducing function. In general, a single legume plant can contain a variety of lectins that may have evolved by gene duplication and become specialized for different roles in the plant (Etzler, 1992) Furthermore, despite being concentrated in plant seeds, lectins are also present in different tissues, where their function probably requires lower concentrations. Lectin plays a role in jasmonate/ethylene-responsive defense reaction. Plant lectins have not traditionally been thought to be inducible proteins but it has been found that lectin gene expression is induced by jasmonate and ethylene, consistent with defense role for this gene. Thus, in other words we can say that lectin is involved in endogenous developmental cues or external stresses (Lyou *et al.* 2009).

(*ii*) As plant antibodies

Early investigators noted the similarities of lectins to antibodies and asparagines that lectins might function as plant antibodies (Bird, 1959). One of the early events in the defence response of legumes is the production of phytoalexins (Dixon, 1995), whose synthesis is stimulated by the release of elicitors, many of which are oligosaccharides that arise from the breakdown of plant or pathogen cell wall components (Cote and Hahn, 1994). Therefore, lectins could participate in this step of the defense process either acting as receptors for elicitors or the elicitors in structures required by their receptors. It has been shown that the levels of *Dolichos biflorus* stem and leaf lectin, DB58, increase upon plant wounding (Etzler, 1998).

(*iii*) In symbiosis

Again, based on their carbohydrate-binding ability, lectins have also been thought to be involved in the establishment of symbiosis between nitrogen-fixing bacteria and plants (De Hoff *et al.* 2009; Etzler, 1986 and Sharon and Lis, 1989). Legumes are able to associate specifically and form symbiosis with soil bacteria of the rhizobia family, a phenomenon that makes them independent from soil nitrogen supplies. The idea that lectins are responsible for this association was advanced over 20 years ago and it has been the subject of much controversy. It has been suggested that root lectin recognized by bacterial receptor molecules is an important determinant of host plant specificity in *Rhizobium* legume symbiosis (Bohlool and Schmidt, 1974). Rhizobium adheres to the surface of differentiated root cells and is then internalized into the root hair to form nitrogen-fixing nodules. This symbiosis is specific in that certain species of Rhizobium can only associate with a particular species of legumes. It has been shown that bacteria that nodulate beans are agglutinated by bean extracts. Also, erythrocytes that can be agglutinated by the bean lectin bind to root hairs. Infact, nitrogen-fixing symbiosis is a multistep process that requires the formation of the root nodule, followed by the adhesion of the bacteria to the roots with the help of lectin (Mylona *et al.* 1995).

(*iv*) As toxins

It has been shown that plant lectins possess cytotoxic, fungitoxic, anti-insect and anti-nematode properties either *in vitro* or *in vivo* and are toxic to higher animals (Oka *et al.* 1997; Oliveira *et al.* 1994 and Ripoll *et al.* 2003); for example, the toxic effects of PHA, the lectin from *Phaseolus vulgaris*, on monogastric animals are well documented (Pusztai, 1991; Martinez-Aragon *et al.* 1995), and the expression in transgenic plants of genes encoding for the lectins GNA (from *Galanthus nivalis*; snowdrop), PSA (*Pisum sativum*; pea), WGA (*Triticum vulgare*; wheatgerm), ConA (*Canavalia ensiformis*; jack bean), AIA (*Artocarpus integrifolia*; jack fruit), OSA (*Oryza sativa*; rice) and UDA (*Urtica dioica*; stinging nettle) resulted in added protection against insect pests and/or phytopathogenic nematodes or fungi (Hilder *et al.* 1995; Hilder and Boulter, 1999; Powell, 2001; Ripoll *et al.* 2003). One of the most important features of plant lectins, compatible with the proposed defensive function, is the remarkably high resistance to proteolysis and stability over a large range of pH, even when they are out of their natural environment.

(*v*) In embryogenesis

Evidences exist for their involvement in embryogenesis, development and reproduction. Two lectin genes MsLEC1 and MsLEC2 from alfalfa cloned in antisense orientation and expressed in alfalfa lead to severe embryogenic, developmental and reproductive abnormalities (Brill *et al.* 2001).

(*vi*) As clinical tool

The lectins, proteins which characteristically bind with carbohydrate moieties, have been extensively studied in many biochemical and structural aspects in order to understand the molecular basis behind this non-catalytic event. At the same time, their clinical and agricultural potentials have been growing fast day-by-day. Over the last few decades, lectins have become a topic of interest to a large number of researchers owing to their potentially exploitable biological properties including antitumor, immunomodulatory and anti-insect, antifungal, antibacterial, anti-HIV, and mitogenic activities. Because of their sugar binding properties, lectins have been extensively studied and used as molecular tools for the study of carbohydrate architecture and dynamics on the cell surface, and have been exploited for such practical applications as distinguishing between normal and malignant cells, purification of glycoconjugate, and coating of drugs to enhance their gastrointestinal tract absorption. Although lectins, mainly those from legume plants, have been investigated for biological properties and plenty of studies concerning their physiological functions have been done in last decade (De Mejia *et al.* 2003; Frédérich *et al.* 2009; Islam *et al.* 2009; Rhodes, 1999). Kaur *et al.* (2005) isolated two mannose/glucose specific seed lectins from *Parkia biglandulosa* and *Parkia roxburghii* and characterized these for various physicochemical properties. *Parkia biglandulosa* lectin was found to be T-cell mitogen as revealed by IL-2 bioassay. These lectins showed anti-proliferative effect on two murine macrophage cancer cell lines *i.e.* P 388DI (50%) and J774 (70%). Liu *et. al.* (2009) have briefly presented the advances of plant lectins in elucidating their complex anti-cancer mechanisms implicated in apoptosis and autophagy. In addition, they have discussed the pre-clinical and clinical studies of plant lectins for their potential therapeutic applications.

(*vii*) In growth process

Probably, interaction with the hormone system determines the role of lectin in growth and development. Maximal lectin content was found in rapidly growing adventitious roots and leaf base of wheat seedlings. A possible mechanism is the involvement of lectins in inhibition of auxin-dependent catabolism of cell wall polysaccharides and growth by stretching. In the case of compatible pollination in petunia and onion plants, an increase in the content of lectins in the pistil sporophyte tissue was observed along with changing their carbohydrate specificity (Babosha, 2008).

(*viii*) As entomopathogenic protein

One of the possible physiological functions that have emerged is the defensive role of these carbohydrate-binding proteins against phytopathogenic microorganisms, phytophagous insects and plant-eating animals (Chrispeels and Raikhel, 1991; Gatehouse *et al.* 1995, 1996, 1997; Harper *et al.* 1995; Van Damme, 2008). Several lectin genes have been exploited for developing insect-resistant transgenic crops (Table 5). An essential feature of any active defense agent is the ability to specifically recognize the pathogen. Based on their carbohydrate-binding specificity and also considering their abundance in plant seeds and bark, lectins seem to possess all the necessary characteristics to exploit this function (Datta *et al.* 2000; Koundal *et al.* 2003; Qureshi *et al.* 2003, 2006; Sharma *et al.* 2004).

At present strategy employed by most of the worker is not only to use single genes, but also to use gene combinations whose products are targeted at different biochemical and physiological processes with in the insect. The availability of suitable promoters and effective insecticidal genes means that the application of transgenic technology can now be extended to the control of almost any order of insect in near future (Chakraborti *et al.* 2009; Datta *et al.* 2005; Hossain *et al.* 2006; Saha *et al.* 2006).

TABLE 5 : Applications of lectin genes in transgenic development

Transgenic plant	*Lectin gene used*	*Target insect*
Tobacco	Pea lectin	*Heliothis virescens*
	Snowdrop lectin (GNA)	*Myzus persicae* *Aulacorthum solani*
	Allium sativum leaf agglutinin (ASAL)	*Myzus persicae*
Potato	Snowdrop lectin (GNA)	*Myzus persicae* *Lacanobia oleracea*
Tomato	Snowdrop lectin (GNA)	*Lacanobia oleracea*
Rice	*Allium sativum* leaf agglutinin (ASAL)	*Nilaparvata lugens* *Nephotettix virescens*
Pea	Bean α-amylase inhibitor	*Callosobruchus malculatus* *Bruchus pisorum*
	Wheat germ agglutinin	*Callosobruchus maculatus*
Chickpea	*Allium sativum* leaf agglutinin (ASAL)	*Aphis craccivora*
Mustard	*Allium sativum* leaf agglutinin (ASAL)	*Lipaphis erysimi*
Sugarcane	Snowdrop lectin (GNA)	*Antitrogus consanguineus*
Cotton	Snowdrop lectin (GNA)	*Heliothis virescens*

Legume lectins represent the largest and most thoroughly studied family of these sugar-binding proteins. Over 100 members of this family are known, almost all isolated from seeds of the plants (Sharon and Lis, 2002). Because they are readily obtainable in purified form and exhibit an amazing variety of sugar specificities, these lectins are eminently suitable for tackling the question of how proteins bind carbohydrates. Lectin genes of plant origin have an added advantage, as when transferred, they are correctly transcribed, translated and processed in a plant host. Most of the lectin genes are devoid of introns and therefore, both genomic as well as cDNA clones can be mobilized for developing transgenic plants. Owing to the popularity of lectins as a tool in biochemistry, cell biology, medicine and plant biotechnology, an overview of lectins was presented in this chapter.

REFERENCES

Babosha AV (2008) Inducible lectins and plant resistance to pathogens and abiotic stress *Biochemistry* (Moscow) **73(7)**: 812-825

Baker JW Reeke GN Wang JL Cunningham BA and Edelman GM (1975) The covalent and three-dimensional structure of concanavalin A *J Biol Chem* **250**: 1513-1524

Barre A Bourne Y Van Damme EJM Peumans WJ and Rouge P (2001) Mannose binding plant lectins: Differential structural scaffolds for a common recognition process *Biochimie* **83**: 645-651

Bettler E Loris R and Imberty A (2001) 3D lectin databank www.cermav.cnrs.fr/databank.lectine

Bird GWG and Sang V (1959) Hemagglutinins in seeds *British Med Bull* **15**: 165

Bohlool BB and Schmidt EL (1974) Lectins: A possible basis for specificity in the *Rhizobium*-Legume root nodule symbiosis *Science* **185**: 269-271

Boyd WC and Shapleigh E (1954) Specific precipitating activity of plant agglutinins (lectins) *Science* **119**: 419

Brill LM Evans CJ and Hirsch AM (2001) Expression of MsLEC1 and MsLEC2 antisense genes in alfalfa plant lines causes severe embryogenic developmental and reproductive abnormalities *Plant J* **25**: 453-461

Chakraborti D Sarkar A Mondal HA and Das S (2009) Tissue specific expression of potential insecticidal, *Allium sativum* leaf agglutinin (ASAL) in important pulse crop chickpea (*Cicer arietinum*) to resist the phloem feeding *Aphis craccivora Transgenic Res* **18(4)**: 529-544

Chandra NR Kumar N Jeyakani J Singh DD Gowda SB Prathima MN (2006) Lectindb: a plant lectin database *Glycobiology* **16:** 938–946

Chrispeels MJ and Raikhel NV (1991) Lectins, lectin genes and their role in plant defense *Plant Cell* **3**: 1-9

Cote F and Hahn MG (1994) Oligosaccharins: structures and signal transduction *Plant Mol Biol* **26:** 1379-1411

Datta I Saha P Majumdar P Sarkar A Chakraborti D Banerjee S and Das S (2005) The efficacy of a novel insecticidal protein *Allium sativum* leaf lectin (ASAL) against homopteran insects monitored in transgenic tobacco *Plant Biotechnol J* **3(6)**: 601-611

Datta S Kansal R and Koundal KR (2000) Construction and screening of cowpea (*Vigna unguiculata* L.) cDNA library for lectin genes *J Plant Biochem Biotechnol* **6**: 67-71

De Hoff PL Brill LM and Hirsch AM (2009) Plant lectins: the ties that bind in root symbiosis and plant defense *Mol Genet Genomics* **282(1)**: 1-15

De Mejia EG Bradford T and Hasler C (2003) The anticarcinogenic potential of soybean lectin and lunasin *Nutr Rev* **61(7)**: 239-246

Diaz C Melchers LS Hooykaas PJJ Lugtenberg BJJJ and Kijne JW (1989) Root lectins as determinant of host plant specificity in the *Rhizobium*-legume symbiosis *Nature* **338**: 579-581

Dixon RA and Paiva NL (1995) Stress-induced phenylpropanoid metabolism *Plant Cell* **7**: 1085-1097

Etzler ME (1986) In: *The Lectins: Properties, Functions and Applications in Biology and Medicine*; Liener IE Sharon N and Goldstein IJ (eds.) Academic Press Orlando pp 371

Etzler ME (1998) From structure to activity: New insights into the functions of legume lectins *Trends Glycosci Glycotechnol* **10:** 247–55

Etzler ME (1992) Plant lectins: Molecular biology, synthesis and function In:*Glycoconjugates: Composition, Structure and Function;* Allen HJ and Kisailus EC (eds.) Marcel Dekker New York pp 521-539

Frédérich M Marcowycz A Cieckiewicz E Mégalizzi V Angenot L and Kiss R (2009) *In vitro* anticancer potential of tree extracts from the Walloon region forest *Planta Med* [Epub ahead of print]

Gatehouse AMR Down RE Powell KS Sauvion N Rahbe Y Newell CA Merryweather A Hamilton WDO and Gatehouse JA (1996) Transgenic potato plants with resistance to the peach potato aphid *Myzus persicae Entomol Exp Appl* **79:** 295-307

Gatehouse JA Gatehouse AMR and Fitches E (1997) Effects of snowdrop lectin (GNA) delivered *via* artificial diet and transgenic plants on the development of tomato moth (*Lacanobia oleracea*) larvae in laboratory and glasshouse trials *J Insect Physiol* **43(8)**: 727-739

Gatehouse AMR Powell KS Peumans WJ Van Damme EJ and Gatehouse JA (1995) Insecticidal properties of plant lectins: Their potential in plant protection In: *Lectins: Biomedical Perspectives*; Pusztai A and Bardocz S (eds.) Taylor and Francis London pp 35–38

Gegg CV Roberts DD Segel IH and Etzler ME (1992) Characterization of the adenine binding sites of two *Dolichos biflorus* lectins *Biochemistry* **31(30)**: 6938-6942

Goldstein IJ Winter HC and Poretz RD (1997) Plant lectins: Tools for the study of complex carbohydrates In: *Glycoproteins II*; Montreuil J Vliegenthart JFG and Schachter H (eds.) Elsevier Amsterdam pp 403–474

Hardman KD and Ainsworth CF (1972) Structure of concanavalinA at 2.4-A° resolution *Biochemistry* **11**: 4910-4919

Harper SM Crenshaw RW Mullins MA and Privalle LS (1995) Lectin binding to insect brush border membranes *J Econ Entomol* **88(6)**: 1197–1202

Hilder VA Powell KS Gatehouse AMR Gatehouse JA Gatehouse LN Shi Y Hamilton WD Merryweather A Newell CA and Timans JC (1995) Expression of snowdrop lectin in transgenic tobacco plants results in added protection against aphids *Transgenic Res* **4:** 18–25

Hilder VA and Boulter D (1999) Genetic engineering of crop plants for insect resistance a critical review *Crop Protection* **18:** 177-191

Hossain MA Maiti MK Basu A Sen S Ghosh AK and Sen SK (2006) Transgenic expression of onion leaf lectin gene in Indian mustard offers protection against aphid colonization *Crop Sci* **46:** 2022-2032

Iika M and Oliveira VT (2004) Antinutritional properties of plant lectins *Toxicon* **44 (4):** 385-403

Imberty A Gautier C Lescart J P´erez S Wyns L and Loris R (2000) An unusual carbohydrate-binding site revealed by the structures of two *Maackia amurensis* lectins complexed with sialic acid-containing oligosaccharides *J Biol Chem* **275:** 17541–48

Islam B Khan SN Naeem A Sharma V Khan AU (2009) Novel effect of plant lectins on the inhibition of *Streptococcus mutans* biofilm formation on saliva-coated surface *J Appl Microbiol* **106(5)**: 1682-1689

Jouanin L Bonade-Bottino M Girard C Morrot G and Giband M (1998) Transgenic plants for insect resistance *Plant Sci* **131**: 1-11

Koundal KR Dash PK and Kansal R (2003) Plant defense proteins: Mechanism and potential of pest control through genetic manipulation *J Plant Biol* **30(2)**: 211-227

Lis H and Sharon N (1998) Lectins: Carbohydrate-specific proteins that mediate cellular recognition *Chem Rev* **98**: 637-674

Lis H and Sharon N (1986a) Lectins as molecules and as tools *Annu Rev Biochem* **55**: 35-67

Lis H and Sharon N (1986b) In: *The Lectins: properties, functions and applications in biology and medicine*; Liener IE Sharon N and Goldstein IJ (eds.) Academic press New York

Liu B Bian H and Bao J (2009) Plant lectins: Potential antineoplastic drugs from bench to clinic *Cancer Lett* doi:10.1016/j.canlet.2009.05.013

Loris R Hamelryck T Bouckaert J and Wyns L (1998) Legume lectin structure *Biochim Biophys Acta* **1383**: 9-36

Lyou SH Park HJ Jung C Sohn HB Lee G Kim CH Kim M Choi YD and Cheong JJ (2009) The Arabidopsis At LEC gene encoding a lectin like protein is up-regulated by multiple stimuli including developmental signal, wounding, jasmonate, ethylene, and chitin elicitor *Mol Cells* **27**: 75-81

Manoj N and Suguna K (2001) Signature of quaternary structure in the sequences of legume lectins *Protein Engg* **14**: 735-745

Martinez-Aragon A Cavallé C Frühbeck G Tosar A Santidrian S Stewart JC Rubio L and Pusztai A (1995) Identification and biological activity of lectins of different subunit composition isolated from *Phaseolus vulgaris* L var. athropurpurea *J Food Sci Technol* **68**: 375-381

Murdock LL and Shade RE (2002) Lectins and protease inhibitors as plant defenses against insects *J Agric Food Chem* **50**: 6605- 6611

Mylona P Mylona D Pawlowski and Bisseling T (1995) Symbiotic nitrogen fixation *Plant Cell* **7**: 869-885

Navjot K Jatinder S Kamboj SS Agrewala JN and Kaur M (2005) Two novel lectins from *Parkia biglandulosa* and *Parkia roxburghii*: isolation, physicochemical characterization, mitogenicity and anti- proliferative activity *Protein Peptide Lett* **12(6)**: 589-595

Oka Y Chet I and Spiegel Y (1997) Accumulation of lectins in cereal roots invaded by the cereal cyst nematode *Heterodera avenae Physiol Mol Plant Pathol* **51**: 333-345

Oliveira SM Dias-Baruffi SMO Thomaz LM Beltramini and Roque-Barreira MC (1994) Neutrophil migration-inducing lectin from *Artocarpus integrifolia J Immunol* **153**: 1798–1807

Peumans WJ and Van Damme EJM (1995) Lectins as plant defense proteins *Plant Physiol* **109**: 347-352

Peumans WJ and Van Damme EJM (1998) Plant lectins: specific tools for the identification, isolation, and characterization of O-linked glycans *Crit Rev Biochem Mol Biol* **33(3)**: 209-258

Peumans WJ and Van Damme EJM (1999) Seed lectins In: Seed Proteins Shewry PR and Casey R (eds.) Kluwer Academic Publishers The Netherlands 657-683

Powell KS (2001) Antimetabolic effects of plant lectins towards nymphal stages of the planthoppers *Tarophagous proserpina* and *Nilaparvata lugens Entomologia Experimentalis et Applicata* **99**: 71–77

Pusztai A Watt WB and Stewart JC (1991) A comprehensive scheme for the isolation of trypsin inhibitors and the agglutinin from soybean seeds *J Agri Food Chem* **39**: 862-866

Pusztai A Bardocz S and Ewen SW (2008) Uses of plant lectins in bioscience and biomedicine *Front Biosci* **13**: 1130-1140

Qureshi IA Dash PK Kansal R Srivastav PS and Koundal KR (2003) Isolation and characterization of a lectin from seeds of *Cicer arietinum Indian J Plant Physiol* **8**: 625-629

Qureshi IA Dash PK Kansal R Srivastav PS and Koundal KR (2006) Purification and characterization of an N-acetyl-D-galactosamine specific lectin from seeds of chickpea (*Cicer arietinum* L.) *Phytochem Anal* **17(5)**: 350-356

Rhodes JM (1999) Beans means lectins *Gut* **44(5)**: 593-594

Ripoll C Favery B Lecomte P Van Damme E Peumans W and Abad P (2003) Evaluation of the ability of lectin from snowdrop (*Galanthus nivalis*) to protect plants against root-knot nematodes *Plant Sci* **164 (4)**: 517-523

Rouge P Cambillu C and Bourne Y (1991) The three-dimensional structure of legume lectins In: Lectin Reviews Kilpatrick DC Van Driessche E and BogHansen TC (eds.) Vol.1 Sigma Chemical Co. St.Louis USA pp.143-159

Rudiger H and Gabius HJ (2001) Plant lectins: Occurrence, biochemistry, functions and applications *Glycoconj J* **18**: 589-613

Saha P Dasgupta I and Das S (2006) A novel approach for developing resistance in rice against phloem limited viruses by antagonizing the phloem feeding hemipteran vectors *Plant Mol Biol* **62(4-5)**: 735-752

Sharma M Sahni R Kansal R and Koundal KR (2004) Transformation of oil seed mustard *Brassica juncea* (L) cv Pusa Jai Kissan with snowdrop lectin gene *Indian J Biotechnol* **3**: 97-102

Sharma V and Surolia A (1997) Analyses of carbohydrate recognition by legume lectins: size of the combining site loops and their primary activity *J Mol Biol* **267**: 433-445

Sharon N and Lis H (1989) Lectins as cell recognition molecules *Science* **246**: 227-234

Sharon N and Lis H (1990) Legume lectins a large family of homologous proteins *The FASEB J* **4**: 3198-3208

Sharon N and Lis H (2002) How protein bind carbohydrates: Lesson from Legume lectins *J Agric Food Chem* **50(22)**: 6586-6591

Sharon N (2007) Lectins: carbohydrate-specific reagents and biological recognition molecules *J Biol Chem* **282(5)**: 2753-2764

Srinivas VR Reddy GB Ahmad N Swaminathan CP Mitra N and Surolia A (2001) Legume lectin family, the 'natural mutants of the quaternary state', provide insights into the relationship between protein stability and oligomerization *Biochim Biophys Acta* **1527**: 102–111

Srinivasan N Rufino SD Pepys MB Wood SP and Blundell TL (1996) A superfamily of proteins with the lectin fold Chemtracts *Biochem Mol Biol* **6**: 149-164

Van Damme EJ Smeets K Torrekens S Van Leuven F and Peumans WJ (1992) Isolation and characterization of alliinase cDNA clones from garlic (*Allium sativum* L.) and related species *Eur J Biochem* **209(2)**: 751-757

Van Damme EJ Van Leuven F and Peumans WJ (1997) Isolation, characterization and molecular cloning of the bark lectins from *Maackia amurensis Glycoconj J* **14(4)**: 449-456

Van Damme EJM Astoul CH Barre A Rougém P and Peumans WJ (2000) Cloning and characterization of a monocot mannose binding lectin from *Crocus vernus* (family Iridaceae) *Eur J Biochem* **267**: 5067-5077

Van Damme JM Smeets K Torrekens S Van Leuven F and Peumans WJ (1994) Characterization and molecular cloning of mannose-binding lectins from the Orchidaceae species *Listera ovata, Epipactis helleborine* and *Cymbidium* hybrid *Eur J Biochem* **221 (2)**: 769-77

Van Damme EJM Peumans WJ Barre A and Rouge P (1998) Plant lectins: a composite of several distinct families of structurally and evolutionarily related proteins with diverse biological roles *Crit Rev Plant Sci* **17**: 575-692

Van Damme EJM (2008) Plant lectins as part of the plant defense system against insects In: *Induced Plant Resistance to Herbivory*; Schaller A (ed.) Springer Science and Business Media pp 285-303

Wang W Peumans WJ Rougé P Rossi C Proost P Chen J and Van Damme EJ (2003) Leaves of the Lamiaceae species *Glechoma hederacea* (ground ivy) contain a lectin that is structurally and evolutionary related to the legume lectins *Plant J* **33(2)**: 293-304

Yim M Ono T and Irimura T (2001) Mutated plant lectin library useful to identify different cells *PNAS* **98(5)**: 2222-2225

2

PHYSIOLOGICAL ASPECTS OF DROUGHT TOLERANCE IN ARID LEGUMES

B. K. GARG

Clusterbean, moth bean, cowpea and horse gram grouped together as arid legumes are known to possess different mechanisms of drought resistance to survive and reproduce under drought-prone environments. The development of short duration and early maturing varieties that enable the crop to escape severe soil moisture deficits is of paramount importance in these legumes. Most of these legumes display several mechanisms of dehydration avoidance to maintain turgor potential for normal metabolic functioning under water deficits. These comprise maximization of water uptake through deep and extensive root systems and reduction of water loss by appropriate shoot attributes and osmotic adjustment for continued photosynthesis and growth at low water potentials. Significant genotypic differences exist for different physiological attributes and drought resistance traits which need to be better understood for selection and development of drought tolerant genotypes with higher yield potential under water limited conditions. These legumes are generally most sensitive to water stress at the flowering stage and genotypic differences at this critical stage are most important. Maintenance of higher leaf turgidity, net photosynthetic rate and stomatal conductance at similar stress levels, alongwith least alterations in leaf metabolites in moth bean was also reflected in its higher yield stability under water stress as compared to clusterbean and other legumes. In addition to drought tolerance characteristics, various improved cultural practices, optimum plant nutrition and adoption of drought ameliorative measures such as thiourea application must be considered to reduce drought impact and to obtain sustainable crop yields under arid environments. Arid legumes comprising cluster bean, moth bean, cow pea and horse gram are the integrated components of dryland farming, being the main stay of pulse and forage production in arid and semi-arid regions of the country. Their average productivity is, however, very low as they are confined and concentrated in arid areas having poor fertility soils and scanty and erratic rainfall leading to frequent droughts. Therefore, it is of paramount importance to identify and develop drought tolerant genotypes based on sound physiological mechanisms that influence the performance of these legumes under water deficits (Garg, 2003; Garg and Burman, 2002 a, b). Furthermore, appropriate crop management and improved cultural practices are equally important to reduce the hazards of drought for ensuring stable yields and increased water use efficiency under drought-prone environments of arid regions (Arnon, 1992; Turner, 1997 and Garg, 2002). The available information on physiological aspects of drought tolerance in these legume crops and some of the approaches for yield improvement under drought is discussed here.

DROUGHT RESISTANCE

Drought resistance refers to the ability of a crop to produce its economic product with minimum loss in water-deficit environment relative to the water constraint free management. According to Turner (1986) the ability of a crop to produce satisfactory in areas subjected to water deficits has been termed as drought resistance. Basically three mechanisms namely drought escape, drought avoidance and drought tolerance are involved in drought resistance (Levitt, 1980) while Turner (1986) suggested that they could be classified into three major categories, explained as follows :

(1) Drought escape

It is the ability of a plant to complete its life cycle before acute soil and plant water deficit develop. This is an important strategy of matching phenological development with the period of soil moisture availability to minimise the impact of drought where the growing season is short and terminal drought predominates. This mechanism involves rapid phenological development (early flowering and maturity), developmental plasticity (variation in duration of growth period depending on the extent of water stress) and mobilization of parenthesis assimilates to grain.

(2) Drought avoidance

It is the ability of plants to maintain relativity high tissue water potential inspite a shortage of soil moisture. Drought avoidance is achieved by maintenance of turgor through increased rooting depth, efficient root systems and internal hydraulic conductance and by reduction of water loss through reduced stomatal conductance, reduced absorption of solar radiation by leaf rolling and reduces evaporation surface *i.e.* leaf area reduction, cluster bean and moth bean survive under drought conditions by doing a balancing act between maintenance of turgor and reduction of water loss.

(3) Drought tolerance

It is the capacity of plants to withstand water deficit with low tissue water potential. Tolerance to dehydration is considered to arrive at the molecular level and depends upon membrane structure and enzyme activity. Basically it is due to the ability of the cells to maintain integrity of the cell membranes and prevent denaturation of proteins.

However, crop plants including arid legumes use more than one mechanism at a time to resist drought.

DROUGHT TOLERANCE ASSESSMENT

Drought is by for the most important environmental stress in agriculture and many efforts have been made to improve crop productivity under water limiting conditions. While natural selection has favoured mechanisms for adaptations and survival, breeding activity has directed selection towards increasing the economic yield of cultivated species. Meanwhile, fundamental research has provided significant gains in the understanding of the physiological and molecular responses of plants to water deficits but there is still a large gap between yield in optimal and stress conditions. A number of physiological studies have identified some traits for which presence or expression is associated with plant adaptability to drought-prone environments. However, a critical aspect in all studies dedicated to drought tolerance is the assessment of the degree of drought tolerance of different genotypes. In many studies the identification of tolerant and susceptible cultivars is based on few physiological measures related to drought response. The difficulty in identifying physiological parameters as reliable indicator of yield in dry conditions has suggested that yield performance over a range of environments should be used as the main indicator for drought tolerance (Voltas *et al.* 2005). Several indices were proposed to describe yield performance of a given genotype under stress and nonstress conditions or in comparison with the average yield or the yield of a superior genotype. Clearly there are only few studies on this line in the arid legumes (Garg, 2008). Some authors have expressed the yielding capacity with regard to an environment-related physiological trait, such as canopy temperature (Idso *et al.* 1981) or water potential (Karamanos and Papatheohari, 1999). Rizza *et al.* (2004) proposed on integrated "Water stress Index" (WSI) based on a simple soil water balance and the integrated reduction of plant transpiration relative to potentials transpiration. They related yield potential and adaptability of cultivars under water stress to the intercept and slope of the linear regression of yield versus WSI.

PHYSIOLOGICAL BASES FOR YIELD UNDER DROUGHT

The water content and the water potential of plant tissues are the most important physiologically relevant integrators of droughts effects on plant (Jones, 2007). Once a drop in water potential develops, responses of a wide range of physiological processes are induced, either directly or by plant hormones signalling changes in water status. Physiological traits relevant for the responses to water deficits and/or modified by water stress span a wide range of vital processes (Table 1). As a consequence, it can be expected that there is no single response pattern that is highly correlated with yield under all drought environments. Furthermore, different crop developmental stagers show different sensitivity to water stress which has also been found to be the case in arid legumes (Garg *et al.* 1998, 2005). Several studies have addressed yield under drought stress as a function of single physiological trait, in attempts to understand which metabolic processes or morpho-physiological traits are crucial in ensuring high yield performance under wide range of environments. But most of these studies relate to cereals crops like wheat, rice, maize and sorghum. Some attempts have been made in our laboratory to short list few of physiological traits contributing to high yield of arid legumes under stressful condition of Indian arid zone (Garg, 2003, 2008). However, it will be pertinent to mention here some of the single physiological parameters associated to yield in cereals which include-

(*i*) Stomatal conductance in wheat cultivars (Fisher *et al.* 1998)

(*ii*) Stay green in maize, rice and sorghum (Campos *et al.* 2004)

(*iii*) Osmotic adjustment (Morgan, 1986; Moinuddin *et al.* 2005)

(*iv*) Carbon isotope discrimination in C3 species (Farquhar and Richards, 1984).

TABLE 1 : Physiological traits relevant for response to drought condition (Adapted from Cattivelli *et al.* 2008)

Plant traits	*Modulation under stress*	*Effect relevant to yield*
Stomatal conductance	Decreases under stress	Less water consumption
Photosynthetic capacity	Reduction under stress	Calvin cycle enzymes, light reactions, biomass accumulation
Timing of phenological phases barley delayed in rice	Advanced flowering in wheat, duration, reduced grain number	Early/late flowering, maturity and growth
Starch availability	Inhibition of photosynthetic activity	Abortion, reduced grain number
Stem reserve utilization	Remobilization and compensation for reduced current photosynthesis	Higher remobilization for grain filling, high yield
Stay green	Resistance to premature senescence	Delayed senescence, post-flowering, drought resistance
Single plant leaf area	Reduced under stress	Plant size and related productivity
Rooting depth	Reduced total mass but increased root/shoot ratio	Higher/lower tapping of soil moisture
Osmotic adjustment	Slow response to plant water potential	Accumulation of solutes/ions, increased membrane stability
Antioxidative stress	Acclimatization of defence system	Protection against active oxygen species
Accumulation of stress-related proteins	Accumulated under stress	Protection of cellular structure and protein activities

However, these parameters have not received adequate attention in assessment of drought tolerance in arid legumes.

MOLECULAR MARKERS FOR DROUGHT TOLERANCE TRAITS

Many efforts have been dedicated in recent years to understand the genetic basis of physiological traits conferring advantages in dry environments, while less attention has been given to understanding the high yield stability in dry and wet environments. Single genes such as those controlling flowering time, ear type, plant height and osmotic adjustment may have important roles in adaptation to drought-prone environments. During the last 10 years, the application of QTL analysis provided very good opportunities to identify chromosomes regions controlling variations in almost all the physiological, morphological and developmental changes observed during plant growth in water-limiting conditions. Attention has particularly been paid to -

(1) Genetic variation of the osmotic adjustment.

(2) Genetic basis of stay green phenotype.

(3) The ability of roots to exploit deep soil moisture to meet ET demands.

(4) Reduction of leaf area to limit water use.

(5) Isotopic discrimination.

(6) Non-stomatal water loss from the leaves *e.g.* through the cuticle.

(7) Rate of leaf elongation in response to soil moisture and ET.

(8) Stress induces LEA proteins.

(9) Stomatal sensitivity to ABA.

(10) Overproduction of free proline.

However, most of their efforts have been made in crops like wheat, rice, maize, sorghum and some temperate legumes and at present there is no specific information on genetic basis of drought tolerance in arid legumes. As drought is a very complex phenomenon, use of above biotechnological tools for increasing efficiency of selection for drought tolerance in arid legumes is also advocated.

DROUGHT TOLERANCE IN ARID LEGUMES

In arid and semi-arid regions because of low and erratic rainfall, sowing is generally delayed until there is sufficient soil moisture for stand establishment. Thus, the most important characteristics that enable crop plants to yield satisfactorily are the matching of phenological development with soil moisture supply. In regions with a characteristics terminal drought, extension of crop production in drier regions has been accompanied by the development of short duration varieties that enable the crop to escape severe soil water deficits. In cluster bean and moth bean seed yields were significantly more in early flowering/ maturing genotypes under low rainfall conditions (Garg *et al.* 2003, 2004a and Vyas *et al.* 2003). Significant genotypic differences were observed in crop performance and various physiological attributes among early and late genotypes of moth bean under extreme dry conditions in the field (Vyas *et al.* 2003 a). Early flowering (30-31 days), genotypes such as RMO-40 and RMO-257 consumed soil moisture at a slower rate during vegetative stage and conserved it for the critical flowering stage. In contrast late flowering (45-50 days) genotypes consumed water at a faster rate during the vegetative stage. Consequently early flowering genotypes produced higher seed yield and exhibited greater WUE than late types. This was attributed to more efficient partitioning of photosynthates by early flowering genotypes. Observations on plant water status, NPK concentrations as well as Nitrogen Harvest Index (NHI) and Nitrogen Use Efficiency (NUE) also showed superiority of early genotypes over late genotypes under field drought conditions (Table 2).

TABLE 2 : Seed yield, nitrogen uptake, nitrogen harvest index and nitrogen use efficiency of moth bean genotypes under field drought condition (Vyas *et al.* 2003)

Genotypes	*Seed yield* (*kg ha^{-1}*)	*N uptake* (*kg ha^{-1}*)	*NHI* (*%*)	*NUE* (*kg seed kg^{-1}N*)
		Early flowering		
RMO-40	723	16.04	35.2	45.1
RMD-117	495	11.77	37.8	42.1
RMO-224	644	14.88	46.8	43.3
RMO-225	481	11.05	50.9	43.5
RMO-257	443	9.73	50.9	45.5
Mean	557	12.69	44.3	43.9
		Late Flowering		
Jadia	408	17.88	32.4	22.8
FMO-1023-1	447	18.74	0.5	29.2
FMO-1096-1	385	1.72	27.9	35.9
Jawala	486	15.62	37.0	31.0
JMM-DBS-1	356	11.52	27.8	3.9
FMO-1118-1	443	16.68	31.1	26.6
IPCMO-880	345	12.27	31.2	28.1
FMO-1460-1	244	12.13	24.7	20.1
JMM-259	297	12.20	22.4	24.3
Mean	390	14.20	29.6	27.7

Likewise field studies undertaken with diverse early and late maturing genotypes of cluster bean under rainfed conditions for two consecutive years (1993- low rainfall, 1994- high rainfall) revealed significant genotypic differences in seed yield, WUE, nutrient uptake and leaf metabolism (Garg *et al.* 2003). The seed yield of early genotypes (RGC-936 and Suvidha) was significantly higher during low rainfall year while late genotypes (Maru guar and HG-75) out yielded early types during high rainfall year. WUE was 56.8% higher in early as compared to late genotypes during drought year. This was accompanied by higher NHI and NUE in early maturing genotypes indicating a better utilization of nitrogen (Table 3). Studies have further indicated that early genotypes of both these legumes maintain higher plant water status and experience less metabolic changes than late genotypes under water stress (Garg, 2003; Garg *et al.* 2001). For instance genotype CZM-32E of moth bean displayed higher drought tolerance capacity than cv. Maru moth due to maintenance of higher ψ plant and RWC, higher photosynthetic rate, higher nitrate reductase activity and better carbohydrate metabolism under different intensities of water stress (Table 4). Water stress had maximum adverse effect on seed yield at the flowering stage in genotypes Maru guar and FS-277 but at the pod development stage in cluster bean genotypes Suvidha, HFG-182 and JGC-19 (Garg *et al.* 1998). Different physiological and biochemical parameters also displayed higher metabolic derangements underwater stress at the flowering stage in all the genotypes. In moth bean early flowering genotypes (RMO-40 and CZM-99) consistently displayed less metabolic alterations than late genotypes (Maru moth and CAZRI moth-1) at vegetative and flowering stages but flowering stages more are sensitive to water stress. The maintenance of higher photosynthetic rate and better metabolic efficiency in early genotypes under water stress led to significantly less reduction in their seed yield and DMP as compared to late flowering genotypes (Garg *et al.* 2004a). Results of various studies on cluster bean, mung bean and moth bean indicated stress mediated changes in various leaf metabolites and enzyme activities which modulated the growth and yield of the legumes (Garg and Burman, 2002 a, b). It was found that under increasing water stress, normal GS-GOGAT system becomes less effective but GDH activity increases many fold indicating a shift in the pathway of ammonia assimilation which is an adaptive mechanism to detoxify the accumulated ammonia under acute water stress (Vyas *et al.* 1996; Garg *et al.* 1998). Evaluation of relative drought tolerance of arid legumes at vegetative and flowering stages revealed that moth bean displayed higher drought tolerance than cluster bean, mung bean and cowpea (Garg *et al.* 2005). Mung bean at the vegetative stage and cowpea at the flowering stage recorded maximum reduction in seed yield due to more alterations in various physiological and biochemical parameters. Maintenance of high leaf turgidity, net photosynthetic rate and stomatal conductance at almost similar stress levels (ψ plant) alongwith less alteration in leaf metabolites such as total chlorophyll, starch, reducing sugars, soluble protein and free amino acids in moth bean as compared to cluster bean and other legumes was reflected in its yield stability under water stress (Table 5). Further, higher plant water potential, leaf turgiscence and high photosynthetic rate on restoration of water supply to stressed plants showed high capacity of recovery in all the legumes, but it is more in case of moth bean. Osmotic adjustment is the active accumulation of solutes within the plant tissues in response to lowering of soil water potential (Morgan, 1983). The degree of osmotic adjustment varies with species and genotypic variation within a species has been demonstrated in a number of crops including several legumes (Turner, 1997). Osmotic adjustment has been shown to enable photosynthesis to be maintained at low leaf water potential, to delay leaf senescence, to improve floral survival and to improve yield under water limited conditions. A wide variety of organic solutes in plant tissues during water stress contribute to osmotic adjustment. Stress induced accumulation of free proline, free amino acids and reducing sugars has been observed in cluster bean and moth bean (Vyas *et al.* 1996; Garg *et al.* 1998, 2001, 2005) and other corps. However, cluster bean and pearl millet accumulate high levels of free proline as compared to moth bean and sesame plant under water deficits (Table 6). Over production of proline and glycine betaine which acts as osmo protectants and their correlation with stress tolerance has been well emphasised (Kishore *et al.* 1995). These studies indicated the possibility of genetic manipulation of proline over-production in drought resistant plants and control of osmoregulation in others. Most of the adaptations to drought have some limitations. A genotype of short duration usually yields less, compared to that of normal duration under better moisture availability (Garg, 2003). The mechanisms that confer drought resistance by reducing water loss (such as stomatal closure and reduced leaf area) generally result in reduced assimilation of carbon dioxide *i.e.* reduced photosynthetic efficiency.

TABLE 3 : Seed yield, harvest index, WUE and NUE of early and late genotypes of cluster bean under low and high rainfall condition (Garg *et al.* 2003)

Year	*Rainfall (mm)*	*Seed yield ($kg\ ha^{-1}$)*		*HI (%)*		*WUE ($kg\ ha^{-1}\ mm^{-1}$)*		*NUE ($kg\ seed\ kg^{-1}\ N$)*	
		Early	*Late*	*Early*	*Late*	*Early*	*Late*	*Early*	*Late*
1993	206	995	625	33.2	19.7	466	2.987	17.2	12.2
1994	279	660	800	116	12.7	1.83	2.38	11.7	10.5

Osmotic adjustment increases drought resistance by maintaining plant vigour but the increased solute concentration responsible for osmotic adjustment may have additional energy requirements at the cost of plant growth (Turner, 1986, 1997).

TABLE 4 : Water stress effects on water potential, net photosynthetic rate and starch in moth bean genotypes (Garg *et al.* 2001)

Water stress	*ψ plant(MPa)*			*Net photosynthetic rate (μ mol m-2s^{-1})*			*Starch ($mg\ g^{-1}\ dw$)*		
(d)	*V1*	*V2*	*V3*	*V1*	*V2*	*V3*	*V1*	*V2*	*V3*
0	- 0.60	- 0.70	- 0.70	8.0	8.4	11.8	131	127	146
3	- 0.90	- 0.95	- 0.90	4.2	3.5	9.4	77	120	128
6	- 1.25	- 1.20	- 1.10	2.7	2.6	5.2	60	73	77
9	- 1.50	- 1.35	- 1.30	1.5	1.7	4.2	55	57	71

YIELD IMPROVEMENT UNDER DROUGHT

Crop yields under drought can be improved by laying emphasis on integrated traits like seedling establishment, early growth vigour, developmental plasticity, leaf area maintenance, osmotic adjustment, high transpiration efficiency and mobilization of pre-anthesis stored reserves. In addition better crop management including appropriate planting dates, row spacing and plant populations, optimum fertilizer use and adoption of drought ameliorative measures may greatly help in alleviation of drought effects. In recent years several attempts have been made to minimise the detrimental effects of drought in cluster bean and moth bean which are briefly outlined here.

TABLE 5: Influence of drought at flowering stage on comparative performance, water status, photosynthesis and leaf metabolism of cluster bean, moth bean, mung bean and cowpea

(Garg *et al.* 2005)

Parameter	*Treatment*	*Cluster bean*	*Moth bean*	*Mung bean*	*Cowpea*
Seed yield	C	7.60	4.13	7.64	10.03
(g plant^{-1})	D	5.10	3.17	4.545	5.35
ψ plant	C	0.87	0.83	0.80	0.80
(-MPa)	D	1.90	1.93	1.87	1.77
RWC (%)	C	94.6	95.6	83.3	90.4
	D	65.4	95.6	83.3	90.4
Net photosynthesis	C	9.10	6.50	6.50	5.50
(μmol m^{-2} s^{-1})	D	4.48	4.80	2.51	2.61
Chlorophyll	C	9.79	9.96	9.77	9.97
(mg g^{-1} dw)	D	7.93	8.87	7.22	7.55
Reducing sugars	C	17.12	26.61	20.68	20.95
(mg g^{-1} dw)	D	33.04	27.984	22.75	24.68
Starch	C	18.06	219.2	308.4	210.1
(mg g^{-1} dw)	D	142.4	179.3	190.6	147.2
Soluble protein	C	89.6	61.8	74.1	75.0
(mg g^{-1} dw)	D	85.6	60.7	71.9	69.5

C-Control Plants; D-Droughted Plants

In cluster bean and moth bean seed yields were significantly more in early flowering/maturing genotypes under low rainfall situation. However, late genotypes provided higher yields under good rainfall conditions.

TABLE 6 : Influence of drought intensity on accumulation of free proline in certain arid zone crop plants (Garg, 2003)

Water stress levels	*Proline concentration (mg g^{-1}dw)*			
	Pearl millet	*Cluster bean*	*Moth bean*	*Sesame*
Control	0.75	0.47	0.38	0.45
Mild	3.26	3.91	0.71	0.62
Moderate	12.31	8.53	1.71	1.15
Severe	15.34	12.91	1.82	1.15

TABLE 7: Yield stability in moth bean and cluster bean through varietal intercrops (Kathju *et al.* 2003; Vyas *et al.* 2003)

Planting pattern	*Seed yield ($kg\ ha^{-1}$)*		*WUE ($kg\ ha^{-1}\ mm^{-1}$)*		*Net Photosynthesis ($\mu\ mol\ m^{-2}\ s^{-1}$)*	
	1995	*1996*	*1995*	*1996*	*1995*	*1996*
			MOTH BEAN			
Early (RMO-40)	744	825	2.66	2.26	11.3	19.2
Late (Maru moth)	528	1011	1.88	2.80	9.7	20.5
E + L	755	1188	2.72	3.32	12.3	22.0
			CLUSTER BEAN			
Early(Subidha)	833	986	2.89	2.66	22.7	26.8
Late (Maru guar)	717	1122	2.54	2.93	21.0	26.0
E + L	928	1172	3.20	3.16	23.6	27.8

1995: Low rainfall year (267 mm); 1996: High rainfall year (424 mm)

For stability of yield, planting of early and late genotypes of these legumes in 1:1 ration was found most appropriate under low as well as higher rainfall conditions (Table 7). The improved performance of mixed crop, particularly under low rainfall situation could be attributed to more efficient use of water, low initial canopy development and higher photosynthetic rates (Kathju *et al.* 2003; Vyas *et al.* 2003b). Hall (1988) indicated that varietal intercrops having alternate rows of early erect and medium spreading cowpea varieties improved the stability and productivity of cowpea under varying rainfall conditions of semi-arid Africa. Optimum plant population or row spacing is one of the major management variables under dry land conditions. In water deficit regions large canopy growth may be disadvantageous as it may exhaust the available soil moisture more rapidly from the root zone during drought.

TABLE 8 : Influence of thiourea of cluster bean (Garg *et al.* 2006)

Treatment	*Seed yield ($kg\ ha^{-1}$)*	*DMP ($kg\ ha^{-1}$)*	*Net Photosynthesis ($\mu mol\ m^{-2}\ s^{-1}$)*	*NRA ($mg\ NO_2\ g^{-1}\ dwh^{-1}$)*
Control	295	1172	10.3	348
Seed Treatment (Thiourea)	382	1409	13.1	512
Seed treatment + spray (Thiourea)	410	1512	13.5	581

Studies with cluster bean genotypes have shown that 40-45 cm row spacing was significantly better for achieving higher seed yields, WUE and NUE under arid conditions (Garg *et al.* 2002). In another study moth bean genotypes planted at 30 × 15 cm provided significantly higher seed yields, DMP and higher WUE as compared to 45 × 10 cm spacing (Burman *et al.* 2002). Addressing the soil fertility constraints can also alleviate drought stress effects on crop productivity in arid environments (Lahiri, 1980). The supply of phosphorus is particularly critical as declining water availability causes reduced phosphorus availability and absorption by moth bean and cluster bean under dryland conditions. Phosphorus application significantly ameliorated the detrimental effects of drought on water status, rates of net photosynthesis, contents of chlorophyll, starch, soluble protein and nitrate reductase activity in moth bean genotypes (Garg *et al.* 2004b). Consequently seed yield was significantly improved despite drought condition. Similarly application of potassium under soil moisture stress has been found to play a major role through its favourable influence on plant water relations, physiological processes like net photosynthesis and stomatal conductance and seed yields (Mengel and Kirkby, 1980). Maintenance of adequate potassium in the soil significantly improved plant water status, net photosynthesis, and yield of cluster bean under drought imposed at various developmental stages (Vyas *et al.* 2001). Similarly seed treatment with thiourea (500 ppm) followed by foliar sprays (1000 ppm) at vegetative and flowering stages significantly mitigated the detrimental effects of field drought in cluster bean through enhancement of photosynthetic efficiency and nitrogen metabolism (Garg *et al.* 2006, Table 8). Likewise, kinetin (5 ppm either as pre-sowing seed treatment or as foliar spray or both (at 25 and 40 DAS) significantly improved the growth, DMP, seed yield and WUE of cluster bean under field drought conditions (Burman *et al.* 2003). Combined application of phosphorus and thiourea has been found to act synergistically and improved growth and yield of cluster bean under simulated drought condition (Burman *et al.* 2004). This has been found to be mediated due to their favourable and additive effects on photosynthetic efficiency, leaf area development and nitrogen metabolism. The usefulness of thiourea application has also been demonstrated at farmer's fields (Garg and Burman, 2006) where seed treatment with 500 ppm thiourea followed by a single spray of thiourea (1000 ppm) at the pre-flowering stage significantly improved the growth and seed yield of cluster bean and moth bean (Table 9).

TABLE 9 : Influence of thiourea application on seed yield and DMP of cluster bean and moth bean genotypes at farmer's field (Garg and Burman, 2006)

Legumes	*Seed yield (kg ha^{-1})*		*Dry matter production (kg ha^{-1})*	
	Control	*Thiourea*	*Control*	*Thiourea*
		Moth bean		
RMO-40	507	569	1218	1320
CZM-99	450	521	1125	1254
		Cluster bean		
RGC-936	405	467	1416	1602
RGC-1003	438	543	1377	1994

It has been concluded that thiourea treatment can significantly enhance the productivity of both these legumes under arid conditions (Garg, 2007). Improvement in performance of a crop under water shortage can arise both from genetic changes to the crop that enable it to perform better in drought-prone environments and from agronomic changes that enable it to perform better under these constraints. It is evident that physiological manipulation along with good crop management is required for the most efficient use of limited rain water by different arid legumes. Thus, to face drought stress and to achieve sufficient and sustainable seed yield in the future, requires a multidisciplinary approach based on plant genomics, physiology, agronomy and modelling.

REFERENCES

Arnon I (1992) *Agriculture in Drylands - Principles and Practice* Elesiver, Amesterdam

Blum A (1988) Improving wheat grain filling under stress by stem reserved mobilization *Euphytica* **100:** 77-83

Burman U Garg BK and Kathju S (2002) Effect of spacing on seed yield and physiological traits in moth bean (*Vigna aconitifolia*) *Indian Journal of Agricultural Sciences* **72:** 76-79

Burman U Garg BK and Kathju S (2003) Influence of kinetin on photosynthesis,nitrogen metabolism and yield of cluster bean under moisture deficit conditions *Indian Journal of Plant Physiology* **8:** 287-291

Burman U Garg BK and Kathju S (2004) Interactive effect of thiourea and Garg BK and Burman U 2006 Yield improvement of cluster bean and moth bean by thiourea application under arid conditions *Journal of Arid Legumes* **3:** 51-54 phosphorus under simulated water stress *Biologia Plantarum* **48:** 61-65

Campos H Coper M Habben JE EDmeades GO and Schussler JR (2004) Improving drought tolerance in maize: a view from industry *Field Crop Research* **90:** 19-34

Cattivelli L Rizza F Badeck FW Mazzucotelh E Mastrangelu AM Francia E Mar C Tondelli A and Stanca AM (2008) Drought tolerance improvement in crop plants: An integrated view from breeding to genomics *Field Crops Research* **105:** 1-4

Farquhar GD and Richards RA (1984) Isotopic composition of plant carbon correlates with water-use efficiency of wheat genotypes *Australian Journal of Plant Physiology* **11:** 39-52

Fisher RA Rees D Sayree KD Lu ZM and Candon AG (1998) Wheat yield progress associated with higher stomatal conductance and photosynthetic rate and cooler canopies *Crop Science* **38:** 1467-1475

Garg BK (2002) Mechanisms of drought tolerance and strategies for yield improvement under drought *Current Agriculture* **26:** 15-21

Garg BK (2003) Physiological aspects of abiotic stress tolerance in arid legumes (in) *Advances in Arid Legumes Research* pp 347-354. A Hentry D Kumar and NB Singh (Eds) Scientific Publishers, Jodhpur

Garg BK (2007) Role of thiourea in improving the productivity or arid legumes *Journal of Arid Legumes* **4:** 1-5

Garg BK (2008) Physiology of drought tolerance of cluster bean and moth bean *Journal of Arid Legumes* **5:** 8-14

Garg and Burman U 2002a Moth bean physiology (in) *Moth bean in India* pp 87-102 D Kumar and NB Singh (Eds) Scientific Publishers, Jodhpur Garg BK and Burman U 2002b Physiology (in) *Guar in India* pp 63-83 D

Garg BK and Burman U 2006 Yield improvement of cluster bean and moth bean by thiourea application under arid condition *Journal of Arid Legumes* **3:** 51-54

Garg BK Burman and Kathju S (2004a) Effect of water stress on moth bean (*Vigna aconitifolia*) genotypes *Indian Journal of Plant Physiology* **9:** 29-35

Garg BK Burman U and Kathhju S (2004b) The influence of phosphorus nutrition the physiological response of moth bean genotypes of drought *Journal of Plant Nutrition and Soil Science* **167:** 503-508

Garg BK Kathju S and Burman U (2001) Influence of water stress on water relations, photosynthetic rate and nitrogen metabolism of moth bean genotypes *Biologia Plantarum* **44:** 289-292

Garg BK Burman U and Kathju S (2005) Comparative water relations, photosynthesis and nitrogen metabolism of arid legumes under water stress *Journal of Plant Biology* **32:** 83-93

Garg BK Burman U and Kathju S (2006) Influence of thiourea on photosynthesis, nitrogen metabolism and yield of cluster bean (*Cyamopsis tetragonoloba*(L.) Taub.) under rainfed conditions of Indian arid zone *Plant Growth Regulation* **48:** 237-245

Garg BK Kathju S Vyas SP and Lahiri AN (1998) Influence of water deficit stress at various growths stages on some enzymes of nitrogen metabolism and yield in cluster bean genotypes *Indian Journal of Plant Physiology* **3:** 214-218

Garg BK Kathju S Vyas SP and Lahiri AN (2002) Influence of spacings on performance, water use, nutrient uptake and nitrogen use efficiency of cluster bean under diverse rainfall conditions (in) *Human Impact on Desert Environment* pp 294-304 P Narain S Kathju Amal Kar and MP Singh (Eds) Scientific Publishers Jodhpur

Garg BK Kathju S Vyas SP and Lahiri AN (2003) Relative performance, evapotranspiration, water and nitrogen use efficiency and nitrogen metabolism of cluster bean genotypes under arid environment (in) *Advances in Arid Legumes Research* pp 212-220 A Hentry D Kumar and NB Singh (Eds) Scientific Publishers, Jodhpur

Idso SB Reginado R Reicosky D and Katfield J (1981) Determining soil induced plant water potential depression in alfalfa by means of infrared thermometer *Agronomy Journal* **73:** 826-830

Jones HG (2007) Monitoring plant and soil water status established and novel methods revisited and their relevance to studies of drought tolerance *Journal of Experimental Botany* **58:** 119-130

Karamanos AJ and Papatheohari AY (1999) Assessment of drought resistance of crop genotypes by means of Water Potential Index *Crop Science* **39:** 1792-1797

Kathju S Garg BK Vyas SP and Lahiri AN (2003) Sustainable production of moth bean through genotype management under arid environments *Journal of Arid Environments* **53:** 137-143

Kishore PBR Hong-Zong L Verma PAS and Miiao CH (1995) Over production of DELTA-pyrroline S-carbohylase synthetase increases proline production and confers osmo tolerance in transgenic plants *Plant Physiology* 108 1387-1394 D Kumar and NB Singh (Eds) Scientific Publishers Jodhpur

Lahiri AN 1980 Interaction of water stress and mineral nutrition on growth and yield (in) *Adaptation of Plants to Water and High Temperature Stress* pp 341-362 NC Turner and PJ Kramer (Eds) John Wiley, New York

Levitt J (1980) *Responses of Plants to Environmental Stress* Vol 2 Academic Press New York

Mengel K and Kirkby EA (1980) Potassium in crop production *Advances in Agronomy* **33:** 59-110

Moinuddin A Fisher RA Sayre KD and Reynolds MP (2005) Osmotic adjustment in wheat in relation to grain yield under water-deficit environments *Agronomy Journal* **97:** 162-1071

Morgan JM (1984) Osmoregulation and water stress in higher plants *Annual Review of Plant Physiology* **35:** 292-319

Rizza F Badeck FW Cattivelli I and Staca AM (2004) Use of a water stress index to identify barely genotypes adapted to rainfed and irrigated conditions *Crop Science* **44:** 2127-2137

Turner NC (1986) Adaptations to water deficits *Australian Journal of Plant Physiology* **13:** 175-190

Turner NC (1997) Further Progress in crop water relations *Advances in Agronomy* **39:** 1-51

Voltas J Lopez Coircoles H and Borras G (2005) Use of bipolar analysis and factorial regression for the investigation of superior genotypes in multi-environment trails *European Journal of Agronomy* **22:** 309-324

Vyas SP Kathju S Garg BK and Lahiri AN (1996) Activities of nitrate reductase and ammonia assimilating enzymes of moth bean under water stress *Science and Culture* **62:** 213-214

Vyas SP Garg BK Kathju S and Lahiri AN (2001) Influence of potassium on water relations, photosynthesis, nitrogen metabolism and yield of cluster bean under soil moisture deficit stress *Indian Journal of Plant Physiology* **6:** 30-37

Vyas SP Garg BK Kathju S and Lahiri AN (2003a) Relative drought tolerance of moth bean genotypes (in) *Advances in Arid Legumes Research* pp 394-397 A Henry D Kumar and NB Singh (Eds) Scientific Publishers, Jodhpur

Vyas SP Garg BK Kathju S and Lahiri AN (2003b) Genotypic management of cluster bean for sustainable production under arid environment *Indian Journal of Pulse Research* **16:** 104-107

3

RNA INTERFERENCE: ITS CONCEPT AND APPLICATION IN CROP PLANTS

GULZAR S SANGHERA, SHABIR H WANI, MANJIT S GILL, PREM L KASHYAP AND SATBIR S GOSAL

INTRODUCTION

Traditional plant breeding has been tremendously successful in improving the nutritional value of food and feed (Davies 2003). However this process is time-consuming and the limited genetic resources of most; crops have left little room for continued improvement by these means. There are many reasons for the limited genetic resources available for breeding (Hoisington *et al.* 1999). Two of the most important ones are:

(*i*) Loss of gene pools occurring during the domestication and breeding of crop plants (Lee 1998); and

(*ii*) Many of the natural gene traits that may be beneficial in one plant tissue such as seeds and fruits may be deleterious in other plant tissues such as vegetative tissues (Frankard *et al.* 1992 and Zhu and Galili 2003).

Over the past few decades breeding possibilities have been broadened by genetic engineering and gene transfer technologies as well as by gene mapping and identification of the genome sequences of model plants and crops. Directed efforts are now underway to use genetic engineering of metabolic pathways to alter plant nutrient levels (Galili *et al.* 2002). These efforts depend on a detailed understanding of plant metabolic pathways and constituent enzymes as well as on regulatory genes that control these enzymes. Modern technologies such as trancriptomics, proteomics and metabolomics are now proving to be important in understanding plant metabolic pathways and the role of key genes associated with their regulation. This can facilitate new insights into the complex metabolite neighborhoods that give rise to a given phenotype and may allow discovery of new target genes to modify a given pathway. Such genes can then be subject to new metabolic engineering efforts and applications.

The major strategy for transgenic enhancement of plant nutrient levels involves increasing the levels of expression of anabolic biosynthetic genes. Unfortunately the efficacy of this approach for crop engineering has been restricted by two main obstructions. First the introduction of extra gene copies can have the non-intuitive effect of decreasing expression from the introduced and homologous endogenous loci a phenomenon of gene silencing known as "co-suppression" (Napoli *et al.* 1990). Second feedback metabolic loops tend to maintain homeostatic nutrient levels. For example plants might respond to increased nutrient production induced by overexpression of biosynthetic genes by activating a degradation pathway that either negates nutrient accumulation or converts nutrients into undesirable metabolites (Karchi *et al.* 1994).

During the last decade our knowledge repertoire of RNA-mediated functions has hugely increased with the discovery of small non-coding RNAs which play a central part in a process called RNA silencing. Ironically the very important phenomenon of co-suppression has recently been recognized as a manifestation of RNA interference (RNAi) an endogenous pathway for negative post-transcriptional regulation. RNAi has revolutionized the possibilities for creating custom "knock-downs" of gene activity. RNAi operates in both plants and animals and uses double stranded RNA (dsRNA) as a trigger that targets homologous mRNAs for degradation or inhibiting its transcription or translation (De Bakker *et al.* 2002; Almeida and Allshire 2005 and Wadhwani *et al.* 2008) whereby susceptible genes can be silenced. This RNA-mediated gene control technology has provided new platforms for developing eco-friendly molecular tools for crop improvement by suppressing the genes responsible for various stresses and improving novel traits in plants including disease resistance and will be a promising future therapeutic agent to combat plant invaders. It has emerged as a method of choice for gene targeting in fungi (Nakayashiki 2005), viruses (Baulcombe 2004), bacteria (Escobar *et al.* 2001) and plants (Brodersen and Voinnet 2006) as it allows the study of the function of hundreds of thousands of genes to be tested (Godge *et al.* 2008). It can silence a gene throughout an organism or in specific tissues (Shahinul Islam *et al.* 2005) offer the versatility to partially silence or completely turn off genes work in both cultured cells and whole organisms and can selectively silence genes at particular stages of the organism's life cycle (Milhavet *et al.* 2003). Methods that introduce dsRNA into plant and animal cells have been enormously successful for decreasing cognate gene expression *in vivo* (Hannon 2002; Shi 2003 and Vaucheret *et al.* 2001). Due to all these elegant and unique features of RNAi our this chapter specifically focuses on the concept of RNAi current knowledge of its pathways and induction in plants and explore the possibilities for the applications of this technology in the improvement of plants with special reference to plant nutritional value and disease management.

Abbreviations: ABA abscisic acid; **ACMV** *African cassava mosaic virus;* **AGO** Argonaute; ***AtRdRP*** Arabidopsis RNA-dependent RNA polymerase ; **bp** base pair; **CaLCuV** *Cabbage leaf curl virus;* ***Chl*** chalcone synthase; ***CMV*** *Cucumber mosaic virus;* **CP**

coat protein; ***DCL*** dicer-like protein; **dpi** days post infection; **dsRNA** double-stranded RNA ; **ds-siRNAs** double-stranded small interfering RNAs; **eIF2C** elongation factor 2C; ***GFP*** green fluorescent protein; ***HC-Pro*** helper-component proteinase; **HIV** *Human immunodeficiency virus*; **hpRNAs** hairpin RNA; **kDa** kilo Dalton; **LISW** laser-induced stress wave; **miRISC** micro RNA-induced silencing complex; **miRNAs** microRNAs ; **mRNA** messenger RNA; **MSP** major sperm protein; **MSV** *Maize streak virus*; **MYMIV** *Mungbean yellow mosaic India virus*; **nt** nucleotide; **PME** pectin methylesterase; ***Pol*IV** RNA *Polymerase* IV; **pri-tasiRNAs** primary transacting siRNAs; **PTGS** posttranscriptional gene silencing; ***PVX*** *Potato virus X*; **RDR** RNA-dependent RNA polymerase; **RDR** RNA-dependent RNA polymerase; **RISC** RNA-induced silencing complex; **RITS** RNA-induced transcriptional silencing complex; **RLC** RISC loading complex; **RMD** RNA-mediated defense; **RNAi** RNA interference; **rRNA** ribosomal RNA; **SARS-Cov** severe acute respiratory syndrome-associated coronavirus; **SDE** silencing defective; **SGS** suppressor of gene silencing; **siRISC** small interfering RNA-induced silencing complex; **siRNAs** Small interfering RNAs; **ssRNA** single-stranded RNA; **ta-siRNAs** transacting siRNAs; **T-DNA** transfer DNA; **TGMV** *Tomato golden mosaic virus*; **TGS** Transcriptional gene silencing; **TMV** *Tobacco mosaic virus*; **tRNA** transfer ribonucleic acid; **TRV** *Tobacco rattle virus*; **VIGS** virus-induced gene silencing

1. RNAi: THE CONCEPT

'RNA interference' refers collectively to diverse RNA-based processes that all result in sequence-specific inhibition of gene expression either at the transcription mRNA stability or translational levels. It has most likely evolved as a mechanism for cells to eliminate foreign genes. The unifying features of this phenomena are the production of small RNAs 21-26 nucleotides (nt) that act as specific determinants for down-regulating gene expression (Waterhouse *et al.* 2001; Hannon 2002; Pickford and Cogoni 2003) and the requirement for one or more members of the Argonaute family of proteins (Hammond *et al.* 2000). RNAi operates by triggering the action of dsRNA intermediates which are processed into RNA duplexes of 21-24 nucleotides by a ribonuclease III-like enzyme called Dicer (Fire *et al.* 1998; Bernstein *et al.* 2001 and Wesley *et al.* 2001). Once produced these small RNA molecules or short interfering RNAs (siRNAs) are incorporated in a multi-subunit complex called RNA induced silencing complex or RISC (Hammond *et al.* 2000; Tang *et al.* 2003). RISC is formed by a siRNA and an endonuclease among other components. The siRNAs within RISC acts as a guide to target the degradation of complementary messenger RNAs (mRNAs) (Hammond *et al.* 2000 and Tang *et al.* 2003). The host genome codifies for small RNAs called miRNAs that are responsible for endogenous gene silencing. The dsRNAs triggering gene silencing can originate from several sources such as expression of endogenous or transgenic antisense sequences expression of inverted repeated sequences or RNA synthesis during viral replication (Voinnet 2005) (Fig 1). When dsRNA molecules produced during viral replication trigger gene silencing the process is called virus-induced gene silencing (VIGS) (Lu *et al.* 2003). One interesting feature of RNA silencing in plants is that once it is triggered in a certain cell a mobile signal is produced and spread through the whole

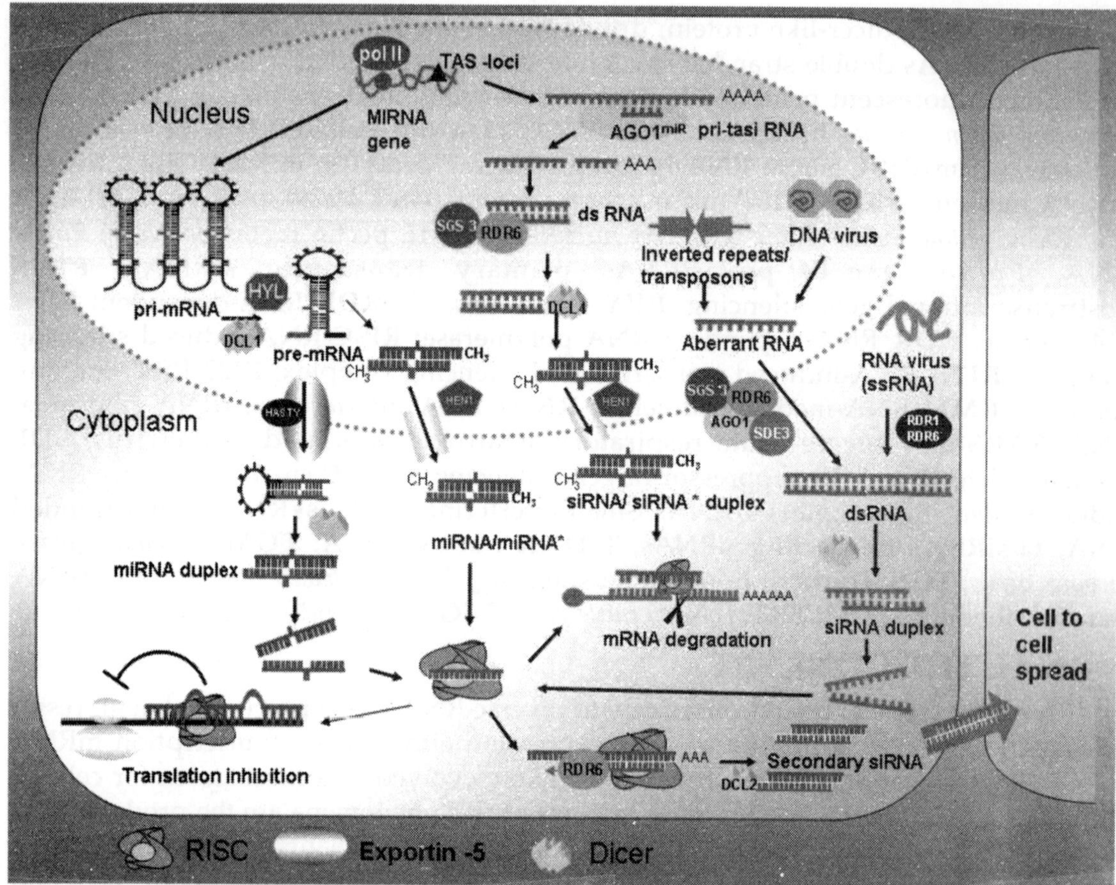

Fig. 1: Different RNAi pathways in Plants. 1).miRNA and tasi-RNA biogenesis regulating expression of MIRNA gene and TAS loci. With the help of DCL1 and HYL1,pri-miRNA is processed to short lived miRNA: miRNA* duplex in plants and are methylated by HEN1 within the nucleus. Thereafter, the miRNA is exported to the cytoplasm by HASTY and after maturation, methylated miRNA is incorporated into a RISC. In this complex, the miRNA is capable of targeting complementary RNAs for cleavage by AGO1, and also for translational repression. Similarly in second phase, trans-acting small interfering RNAs (ta-siRNA) duplexes can be set by miRNA-directed cleavage of the *TAS* transcript. One strand from each siRNA duplex is stably incorporated into RISC, and the other is degraded. siRNAs in RISCs guide cleavage of complementary RNAs. 2) Heterochromatic siRNAs, endogenous siRNAs/siRNAs from transgenes or viral RNA or DNA are generated through similar or partially overlapping pathways. Long double-stranded RNA, generated through the action of RNA-dependent RNA polymerases (RDRs), and iteratively processed by Dicer-like (DCL) proteins to yield multiple siRNA duplexes. One strand from each siRNA duplex is stably incorporated into a RISC, and the other is degraded. siRNAs in RISCs guide cleavage of complementary RNAs. Pol IV is involved in heterochromatic siRNA production in plants, either transcribing the genomic DNA to produce the single-stranded RNA or transcribing the double-stranded RNA to amplify the single-stranded RNA.

plant causing the entire plant to be silenced (Dunoyer *et al.* 2007). After triggering RNA silencing the mobile signaling molecules can be relay-amplified by synthesis of dsRNAs on the primary cleavage of product templates or by their cleavage into secondary siRNAs. This amplification leads to the transitory nature of silencing reaction that may spread along the mRNA though initiated by a locally targeted single siRNA (Klahre *et al.* 2002) and spreads in both the 5′ and 3′ directions (Tang *et al.* 2003). This bi-directional transition further have been witnessed by a process where both the 5′ and 3′ cleavage products of the initial target RNA act as aberrant mRNAs to trigger dsRNA synthesis (Herr *et al.* 2006) and induce secondary silencing reactions (Fig 2). This silencing process is also enhanced by the enzymatic activity of the RISC complex mediating multiple turnover reactions (Hutvagner and Zamore 2002 and Tang *et al.* 2003). Furthermore production of the secondary siRNAs leads to enrichment of silencing *via* its spread from the first activated cell to neighboring cells and systemically through the system (Himber *et al.* 2003). The cell-to-cell spread can be mediated as passive spread of the small RNAs

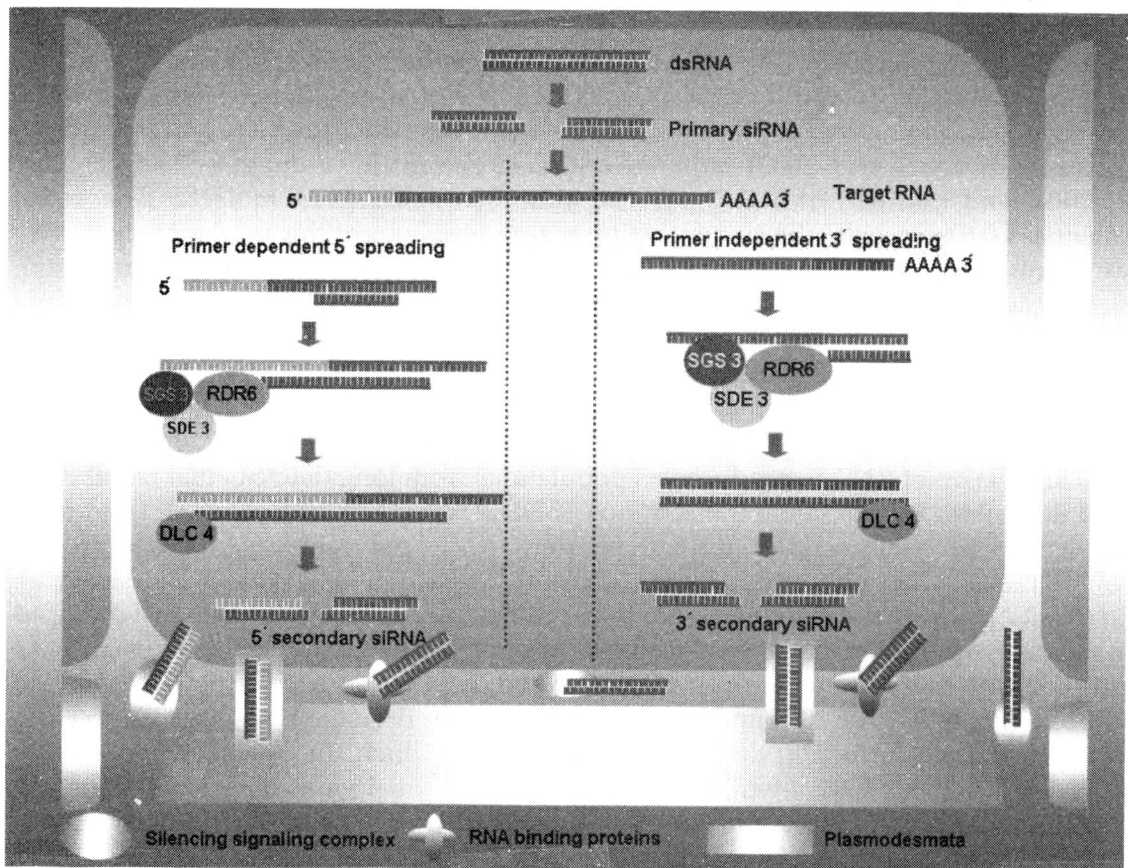

Fig. 2: Model depicting bi-directional mobilization of silencing signal in plant cell. In transitive RNA silencing the mobile signaling molecules relay-amplified by synthesis of dsRNAs into secondary siRNAs by a process where both the 5' and 3' cleavage products of the initial target RNA act as aberrant mRNAs to trigger dsRNA synthesis. siRNAs is produced as shown in scheme (1)showing primer dependent 5' spreading. Further dsRNA synthesis according to the scheme depicted in (2) primer independent 3' spreading. DCL4 appears as putatively involved in secondary siRNA biogenesis that induce secondary silencing reactions. The cell-to-cell spread can be mediated as passive spread of the small RNAs via plasmodesmata or by the silencing signal complex.

via plasmodesmata or by the silencing signal complex which is between 27 and 54 kDa (Kobayashi and Zambryski 2007). The systemic spread in phloem is mediated by the 24 nt siRNAs (Himber *et al.* 2003) unloading of the systemic signal appears to be mediated *via* plasmodesmata since it does not spread into meristematic cells (Voinnet *et al.* 2005). The discovery of RNA-binding protein (PSRP1) in the phloem and its ability to bind 25 nt ssRNA species add further to the argument that siRNAs (24-26 nt) are the key components for systemic silencing signal. The extent of cell-to-cell movement is dependent on the levels of siRNAs produced at the site of silencing initiation but is independent of the presence of siRNA target transcripts in either source or recipient cells (Li and Ding 2006).

2. KEY PLAYERS OF RNAi

RNAi is regulated by different RNA molecules like siRNAs and miRNAs that provide specificity to guide various activities of RNAi machinery including RNA cleavage (Elbashir *et al.* 2001; Llave *et al.* 2002; Hily and Liu 2007), translational repression (Doench *et al.* 2003) and methylation of chromatin (Volpe *et al.* 2002; Chan *et al.* 2004) and outlined in the Fig 1. The aforementioned RNA molecules have the potential to trigger gene silencing by acting as nuts and bolts of this machinery. dsRNAs are precursors of siRNAs whereas ssRNAs with stem-loop structures serve as precursors of miRNAs (pre-miRNAs). However both these RNA molecules are products of the single protein Dicer (Li and Ding 2006). These small RNA species can be divided into various classes which are briefly described in following sections.

2.1 SMALL INTERFERING RNA (siRNA)

This species of RNA is produced as populations from long dsRNAs that result from read through or bi-directional transcription of DNA repeats or transposon loci and from the action of host-encoded RNA-dependent RNA polymerases that synthesize complementary strands from cellular RNAs (Mourrain *et al.* 2000 and Dalmay *et al.* 2000). siRNAs are also generated from transgenic pan-handled transcripts that are used to provide experimental RNAi (Waterhouse and Helliwell 2003). Endogenous siRNAs either direct the endonucleolytic cleavage of homologous transcripts (transacting siRNAs) (Peragine *et al.* 2004 and Vazquez *et al.* 2004) or promote DNA methylation and heterochromatin formation at the genetic loci from which they originate (*cis*-acting siRNAs) (Xie *et al.* 2004) often resulting in transcriptional gene silencing (TGS). *Cis*-acting siRNAs are produced in the nucleus by DCL3 (dicer-like protein 3) whereas *trans*-acting siRNAs require DCL1 for their biogenesis (Xie *et al.* 2004).

2.2 MICRO RNA (miRNA)

It constitutes another class of endogenous small RNA. These molecules are excised by DCL1 from nuclear and non-coding precursor transcripts of approximately 70-200 nts in length which acquire a partial stem-loop structure (Dunoyer and Voinnet 2005). Mature miRNAs are cytoplasmic and direct the cleavage or translational repression of mRNAs that carry discrete complementary target sites (Bartel and Bartel 2003 and Bartel

2004). These miRNA sequences reside between protein coding genes or within introns (Bartel and Bartel 2003).

The first plant miRNA targets to be identified were a series of evolutionarily conserved transcription factors that control important developmental fates (Rhoades *et al.* 2002) but further advancement in the work indicates that miRNAs regulate many other biological processes (Jones-Rhoades and Bartel 2004; Sunkar and Zhu 2004). Moreover gene inversion or duplication events can generate species-specific miRNAs that probably contribute to the ability of plants to adapt to their environment (Allen *et al.* 2004 and Voinnet 2004).

2.3 TRANSACTING siRNA (tasiRNA)

tasiRNA is a class of plant endogenous sRNA which evolved from non-coding single-stranded transcripts called the pri-tasiRNAs. These are converted into dsRNA by RDR6–SGS3 giving rise to siRNAs produced as discrete species in a specific 21-nt phase (Peragine *et al.* 2004; Vazquez *et al.* 2004; Brodersen and Voinnet 2006). Much like plant miRNAs mature tasiRNAs guide cleavage and degradation of homologous cellular transcripts. The tasiRNA generating loci (TAS1–3) have been identified in *Arabidopsis* (Allen *et al.* 2005) but they probably also exist in other plant species and organisms that contain RNA dependent RNA polymerase (RDRs) such as *C. elegans* or *Neurospora crassa*. tasiRNA production involves an interesting mix of miRNA action and the siRNA biogenesis machinery (Brodersen and Voinnet 2006). Pri-tasiRNAs contain a binding site for a miRNA that guides cleavage at a defined point. The initial miRNA-guided cut has two important consequences. First it triggers RDR6-mediated transitivity on the pri-tasiRNA cleavage products allowing dsRNA production either 50 or 30 of the cleavage site (Allen *et al.* 2005) and secondly it provides a well defined dsRNA terminus which is crucial for the accuracy of a phased dicing reaction performed by DCL4 which produces mature tasiRNAs.

2.4 RNA DEPENDENT RNA POLYMERASE (RdRP)

As silencing is triggered by dsRNA conversion of ssRNA sequences to ds form is required for silencing of endogenous transcripts. In *Arabidopsis* this RdRP function is mediated by proteins encoded by the SILENCING DEFECTIVE1 (SDE1)/ SUPPRESSOR OF GENE SILENCING2 (SGS2) RdRP genes now referred to as RdRP6 (Xie *et al.* 2004) and the SDE3 RNA helicase gene (Mourrain *et al.* 2000; Dalmay *et al.* 2000b 2001 and Tang *et al.* 2003). These gene functions are not needed for virus-induced PTGS (Dalmay *et al.* 2000 and Voinnet 2001) or for PTGS of transgenes that are transcribed into sense-antisense RNAs forming dsRNA structures (Waterhouse *et al.* 2001 and Beclin *et al.* 2002). But host-encoded and pathogenesis-inducible RdRP identified as NtRdRP1 in tobacco (Xie *et al.* 2001) and as *AtRdRP1* in *Arabidopsis* (Yu *et al.* 2003) are known to enhance the silencing-based defense reaction against different RNA viruses. This suggests that the dsRNA accumulation may be the rate-limiting step in the silencing-based defense reaction.

One of the silencing pathways leading to DNA methylation is mediated by RNA *Polymerase* IV (Pol IV) RdRP2 and DCL3. In this pathway Pol IV and RdRP2 would synthesize dsRNA which is then be cleaved by DCL3 to generate siRNA. In one other pathway Pol IV acts together with AGO4 and one of the RdRP to produce specific siRNAs. These two silencing pathways are independent in the leaves but interdependent in the flowers (Herr *et al.* 2005). Synthesis of other small RNAs requires activity of specific RdRPs. At least tasiRNAs production also depends on RdRP6/SGS3 proteins (Allen *et al.* 2005).

2.5 DICER

Dicer is the key enzyme initiating the RNA-silencing process. It is a dsRNA specific Ribonuclease III-like endonuclease which cleaves the target dsRNAs into fragments of 21-24 nts leaving 3′-hydroxyl and 5′-phosphate ends and 2 nt 3′ overhangs at the termini of the duplex (Bernstein *et al.* 2001; Lau *et al.* 2001 see also Wadhwani *et al.* 2008). Dicer appears to interact with the RISC complex (see RISC assembly and Fig 4) and after cleavage the dsRNA fragments disassociate from dicer and become associated with the RISC. Only one strand of the miRNA fragments (the strand complementary to the target sequence) but either strand of the short lived ds-siRNAs are incorporated into the RISC where these RNAs function as sequence specific tags and target the silencing function to the homologous RNAs (Dunoyer *et al.* 2004).

Arabidopsis and rice genomes encode at least for four different dicer-like proteins (DCL1-DCL4) (Schauer *et al.* 2002). DCL1 produces miRNAs (Bartel 2004 and Chen *et al.* 2005) DCL3 produces 24 nt long siRNAs involved in DNA methylation and in heterochromatin formation (Xie *et al.* 2004) and DCL2 cleaves natural antisense transcripts into siRNAs (Deleris *et al.* 2006). DCL4 generates siRNAs of 21 nt which mediate viral RNA silencing (Dunoyer *et al.* 2005; Gasciolli *et al.* 2005; Herr 2005; Qi *et al.* 2005 and Xie *et al.* 2005). If DCL4 is not functional then DCL2 and DCL3 produce 22 and 24 nt-long siRNAs respectively from viral sequences but only siRNAs produced either by DCL4 or by DCL2 can mediate antiviral silencing. These dicers can restrict virus accumulation into inoculated leaves and DCL4 produce the silencing signal which can inhibit the virus spread because viral siRNAs are produced by DCL4 and DCL2 partially antagonizes the production of miRNAs by DCL1 at least in leaf and fully substitutes for DCL4 to produce viral siRNAs when DCL4 is inhibited by viruses (Bouche *et al.* 2006 and Dunoyer *et al.* 2007).

2.6 RNA INDUCED SILENCING COMPLEX (RISC)

In order to guide suppression of homologous targets the small (s) RNA duplex must become single-stranded which needs a less stable strand with the 5′ phosphate end of the helix that is incorporated as a guide RNA into an effector complex containing an Argonaute (AGO) protein (Schwarz *et al.* 2003 and Herr 2005). Two types of effector complexes *viz.* cytoplasmic and nuclear complexes have been described for induction of RNAi. Cytoplasmic complex known as (RISC) (Hammond *et al.* 2000) can mediate both mRNA cleavage and translational inhibition (Sigova *et al.* 2004). RISC is multi-subunit large assemblies of 250-500 kDa associated with the single stranded fragments of the si- or miRNA (Nykänen *et al.* 2001 and Omarov *et al.* 2007). However the nuclear complex known as RNA-induced Transcriptional Silencing (RITS) complex was first described in fission yeast for repressing DNA expression (Verdel *et al.* 2004). Both these complexes contain (AGO) proteins which are specific to each effector complex *e.g.* RITS have AGO4 and RISC involves AGO1 which are homolog of the translation initiation factor eIF2C (Hammond *et al.* 2000; Hall 2005 and Song and Joshua-Tor 2006).

2.7 ARGONAUTE PROTEINS

Argonaute proteins are the direct binding partners of siRNAs and form the core of RISC. These are about 100 kDa highly basic proteins comprising PAZ and PIWI domains. The PAZ domain which occurs also in the Dicer enzymes may mediate protein-protein interactions and facilitate binding with the Dicer complex. PAZ and PIWI domains are responsible for 3-2 nt overhang recognition and endonucleolytic activities respectively (Hammond *et al.* 2000; Hall 2005; Song and Joshua-Tor 2006). Due to their basic character the argonaute proteins bind RNAs (*e.g.* siRNAs) and guide them to functional complexes (Tang *et al.* 2003; Bartel and Bartel 2003; Dugas and Bartel 2004; Lecellier and Voinnet 2004; Matzke *et al.* 2004 and Vaucheret *et al.* 2004). It is now well known that AGO1 which has slicer activity is an essential component of RISCs and cleaves the target mRNAs which are homologous to the miRNA or siRNA sequences in the complex (Vaucheret *et al.* 2004; Baumberger and Baulcombe 2005; Qi *et al.* 2005; Ronemus *et al.* 2006) and AGO4 is essential for DNA and histone methylation in *Arabidopsis* (Fagard *et al.* 2000; Morel *et al.* 2002; Zilberman *et al.* 2003 and Irvine *et al.* 2006). In the absence of small interfering RNAs (siRNAs) NRDE-3 resides in the cytoplasm. NRDE-3 binds siRNAs generated by RNA-dependent RNA polymerases acting on messenger RNA templates in the cytoplasm and redistributes to the nucleus. Nuclear redistribution of NRDE-3 requires a functional nuclear localization signal for nuclear RNAi and results in NRDE-3 association with nuclear-localized nascent transcripts (Guang *et al.* 2008). Thus specific Argonaute proteins can transport specific classes of small regulatory RNAs to distinct cellular compartments to regulate gene expression.

2.8 OTHER PLAYERS OF RNAi

Hua enhancer 1 (HEN1) a dsRNA methylase (Park *et al.* 2002; Boutet *et al.* 2003 and Xie *et al.* 2004) is required for miRNA accumulation in *Arabidopsis* and for the methylation of miRNA duplexes (Yu *et al.* 2005). HEN1 can also methylate the 3′-OH of siRNAs with less efficiency when they are in duplexes with a 2 nt overhang and preferably of 23 nt. HEN1 is also involved in the production of siRNAs (Akbergenov *et al.* 2006 and Yang *et al.* 2006). HYPONESTIC LEAF 1 (HYL1) a dsRNA binding protein affects miRNA but not siRNA accumulation (Anantharaman *et al.* 2002; Han *et al.* 2004; Vazquez *et al.* 2004 and Wu *et al.* 2007). HYL1 function is also required for hormonal (auxin, ABA and cytokinin) responses (Lu and Federoff 2000). It is possible that other cellular proteins are also affecting the silencing processes. For instance enhanced pectin methylesterase (PME) can enhance the degradation of the viral RNAs and the RNA silencing mechanism (Dorokhov *et al.* 2006). Still some other proteins *e.g.* dsRNA binding proteins (Han *et al.* 2004; Vazquez *et al.* 2004 and Kurihara *et al.* 2006) SDE3

(Dalmay *et al.* 2001) and SGS3 (Mourrain *et al.* 2000) may operate with different DCL proteins in different RNA silencing pathways (Herr *et al.* 2006).

3. GENERAL APPROACHES TO SUPPRESS THE EXPRESSION OF UNDESIRABLE GENES

To reduce the levels of undesirable gene products two general approaches are commonly used: recessive gene disruption and dominant gene silencing. In gene disruption approaches the target sequence is mutated to eliminate either expression or function whereas dominant gene silencing methods induce either the destruction of the gene transcript or the inhibition of transcription. The advantages of the dominant gene silencing methodologies over the gene disruption approach are 2-fold. First dominant gene silencing is easier to manipulate genetically and to screen for subsequent transgenic plants. Second dominant gene silencing can be manipulated in a spatial and temporal manner using specific promoters. Among the dominant gene silencing approaches dsRNA triggered RNAi is the most powerful method (Smith *et al.* 2000); it is the most efficient in terms of the extent of gene silencing and the resulting silencing is almost as complete as in a gene knockout approach. It appears that dsRNA triggered RNAi directly bypasses the requirement for dsRNA synthesis via RDR which is likely the rate-limiting step in the plant RNAi pathway.

4. RNAi SILENCING TECHNOLOGIES IN PLANTS

Although dsRNA-triggered gene silencing was discovered in plants *via* crossing parents that overexpress sense and antisense RNA by (Waterhouse *et al.* 1998) and in animals (Fire *et al.* 1998), the first to apply this technology to plants was Chuang and Meyerowitz 2000. In this case dsRNAs were produced and triggered efficient silencing of flower identity genes using inverted repeats. This vector-based RNAi technology was further improved using an intron instead of a fragment of DNA as the linker by Peter Waterhouse and colleagues (Smith *et al.* 2000). These RNAi vectors were specifically designed to generate long dsRNA species with the same sequence as the target genes. Similarly vectors designed to express hairpin RNAs (hpRNAs) were also successfully applied to silence the corresponding target genes (Wesley *et al.* 2003 and Wadhwani *et al.* 2008). Since constitutive expression of dsRNA or hairpin RNA of some genes often leads to unexpected adverse effects on plant growth and development chemically inducible RNAi silencing vectors were developed to enable temporal and spatial control of gene silencing (Chen *et al.* 2003 and Guo *et al.* 2003). The application of such chemically inducible silencing systems to the study of plant functional genomics is significant but the large-scale use of chemicals for plant improvement is impractical and harmful to the environment. Tissue- or organ-specific control of gene silencing might be a better choice in developing plant RNAi technologies.

In plants three RNAi pathways occur that seem to be involved in a variety of regulatory and immune functions. The first pathway regulates gene expression by microRNAs (miRNAs) (Jones-Rhoades *et al.* 2006) and *trans*-acting smRNAs (ta-smRNAs) (Bartel 2005 and Vaucheret 2005). Both predominantly direct cleavage of near-perfect complementary target mRNAs (Vaucheret *et al.* 2004) but suppression of translation has also been observed in plants (Chen 2004). The working of miRNA based RNAi pathway (Fig 1) emerged after the expression of a specific gene "MIRNA" which is predominantly found within genomic segments previously known as intergenic regions (IGRs) (Jones-Rhoades *et al.* 2006). The expression of this gene begins with Pol II transcription to yield a primary miRNA transcript (pri-miRNA) that is capable of forming the characteristic imperfect "foldback" hairpin structure (Xie *et al.* 2005). The pri-miRNA transcript is cleaved by DCL1 in the nucleus with the help of the dsRBP HYPONASTIC LEAVES1 (HYL1) to produce the shorter precursor miRNA (pre-miRNA) dsRNA molecule. The first DCL1-catalyzed cleavage step in the miRNA biogenesis pathway is made just below the miRNA duplex region of the dsRNA stem loop (Lu and Fedoroff 2000). The miRNA duplex is then released from the pre-miRNA stem loop structure by the second cleavage step of the miRNA pathway which is again directed by the combined action of DCL1 and HYL1 (Vazquez *et al.* 2004). The two-nucleotide 3′ overhangs of the liberated miRNA duplex are methylated by the sRNA-specific methyltransferase HEN1. The duplex of siRNAs are also methylated by HEN1 and is assumed to protect all sRNA species from polyuridylation and degradation (Yu *et al.* 2005 and Yu *et al.* 2006). The miRNA duplex is then transported to the cytoplasm with several classes of miRNA relying on the action of the *Drosophila* Exportin-5 ortholog HASTY (HST) for this nuclear exportation step (Park *et al.* 2005). In the cytoplasm the mature single-stranded miRNA is loaded onto AGO1 the catalytic center of RISC to guide the slicer activity of AGO1 to repress the expression of complementary mRNA transcripts (Jones-Rhoades *et al.* 2006) which are predominantly mediated by transcript cleavage (Xie and Qi 2008).

In addition tasiRNAs also known to play significant role in RNAi pathway which arises from defined genetic loci (TAS loci) through miRNA-dependent biogenesis pathway (Vaucheret 2006) (Fig 1). The expression of tasiRNAs is initiated by Pol II transcription to yield TAS transcripts that contain miRNA target site(s). The miRNA-directed cleavage of a TAS transcript is thought to generate critical features that are recognized by RdRP6 (Allen *et al.* 2005). The cleaved TAS transcript is converted into dsRNA by RdRP6 a process that also requires the SUPRESSOR OF GENE SILENING 3 (SGS3) (Peragine *et* al. 2007). Successive cleavage of the resulting dsRNA by DCL4 produces a phased array of 21-nt siRNAs (Xie *et al.* 2005). Some of these siRNAs (the ta-siRNAs) are incorporated into RISCs to direct the cleavage of mRNA targets (Peragine *et al.* 2007).

The second cytoplasmic pathway protects cells against the invasion of foreign nucleic acids such as viruses and transgenes through the action of siRNAs that direct the cleavage of homologous RNAs (Lecellier and Voinnet 2004). Viruses engineered to contain an endogenous sequence can trigger silencing of the endogene which is termed 'virus-induced gene silencing' (VIGS) (Baulcombe 1999; Ruiz *et al.* 1998 and Godge *et al.* 2008) (Fig 2). This system involves a branched pathway that also converges on the production of dsRNA. Depending on the nature of the transgene and the virus dsRNA is either formed directly by intra- or intermolecular base pairing (hpRNA constructs IR transgenes) or indirectly through antisense transgenes or replication intermediates

(RNA viruses) or through the action of RdRP6 and other proteins (sense transgenes and DNA viruses) (Smith *et al.* 2000). The subsequent processing of the dsRNA into two classes of siRNAs (Hamilton *et al.* 2002) depends on different DCL activities (Tang *et al.* 2003). DCL3 produces the long siRNAs (24-26 nt) that involved in methylation and systemic silencing while DCL4 produces the short siRNAs (21 nt) (Dunoyer *et al.* 2005) that can either guide RISC-mediated RNA degradation or spread to adjacent cells where they trigger RdRP6-dependent systemic silencing (Himber *et al.* 2003 and Schwach *et al.* 2005).

The third nuclear pathway ensures genome integrity and defense against endogenously repeated DNAs transposons and retroelements. It is related to transcriptional gene silencing (TGS) because it involves DNA methylation and/or chromatin modification by endogenous heterochromatic siRNAs (Zilberman *et al.* 2003 and Chan *et al.* 2004). The key player of this RNAi pathway is heterochromatic siRNAs which are typically 24-nt small RNAs that are associated with genomic repetitive sequences such as transposons retroelements rDNAs and centromeric repeats (Hamilton *et al.* 2002 and Kasschau *et al.* 2007). Current model for heterochromatic siRNA biogenesis states that Pol IVa produces single stranded RNA transcripts from certain genomic loci including transposable elements and other repetitive regions (Lu *et al.* 2005). The Pol IVa transcripts move to the nucleolus where they are converted into dsRNAs by RdRP2 and subsequently processed by DCL3 to yield heterochromatic siRNAs in the nucleolar RNA processing center that colocalizes with Cajal bodies (Li *et al.* 2006). The heterochromatic siRNAs direct DNA and histone methylation in an AGO4-dependent pathway (Zilberman *et al.* 2004). The heterochromatic siRNA-directed DNA methylation also requires the Pol IVb (Herr *et al.* 2005) the SNF2-like ATPase DRD1 (Kanno *et al.* 2005) and the *de novo* cytosine methyltransferase DRM2 (Chan *et al.* 2004). Since AGO4 has been shown to physically interact with the largest subunit of Pol IVb and DCL3 AGO4 and Pol IVb colocalize in the nucleolar RNA processing center Pol IVb may be part the AGO4-containing RISC in the chromatin RNA silencing pathway (Li *et al.* 2006). It has been proposed that the AGO4/NRPD1b/siRNA complex directs target recognition for DRD1- and DRM2-dependent DNA methylation at specific loci (Pontes *et al.* 2006). AGO6 another member of the *Arabidopsis* AGO family has been shown to have partially redundant functions with AGO4 in heterochromatic siRNA-directed silencing (Zheng *et al.* 2007).

5. WORKING MODEL OF RNAi IN PLANTS

RNA interference was discovered in plants as a mechanism whereby invading nucleic acids such as transgenes and viruses are silenced through the action of small RNA molecules (Brodersen and Voinnet 2006). The most important biochemical features of RNAi pathways in plants include the formation of double stranded RNA; processing of dsRNA to small (s) 20–26nt dsRNAs with staggered ends; and inhibitory action of a selected sRNA strand within effector complexes acting on partially or fully complementary RNA or DNA. The present model for RNAi in plants (Fig. 3) includes three phases: initiation by producing small RNAs maintenance by assembly of effector complexes signal amplification and transitive silencing along with dissemination by non-cell autonomous silencing. These phases are briefly described as:

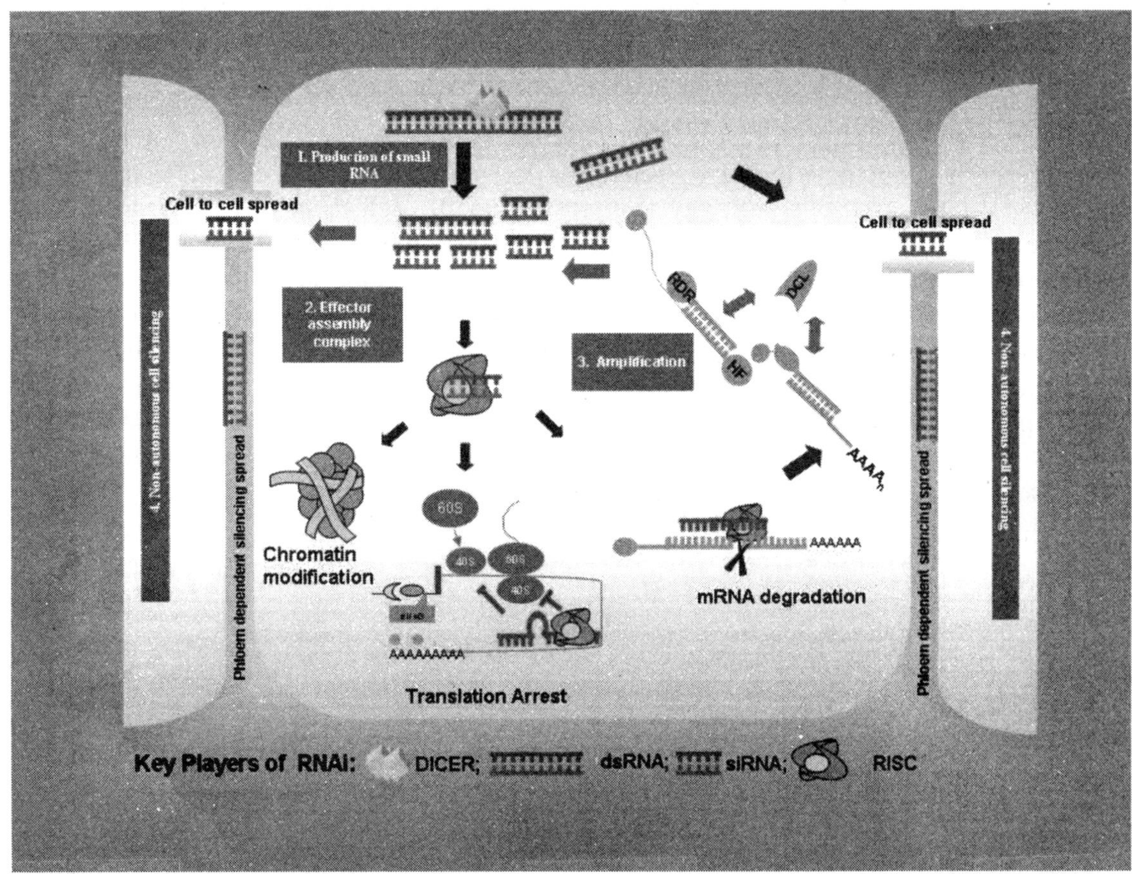

Fig. 3 Simplified working model of RNAi in plants. Most important steps include 1) Dicing leading to the production of small duplex RNAs (21-30 nt); 2) Assembly of effector complexes guiding chromatin remodeling, translational arrest and mRNA cleavages or degradation; 3) Amplification to generate new sources of dsRNA for dicing, leading to further silencing; 4) Non-cell autonomous cell silencing beyond the sites of silencing initiation via putative silencing signals.

5.1 PRODUCTION OF SMALL RNAs

RNA silencing begins with recognition of an RNA trigger by a type III endonuclease Dicer (Bernstein *et al.* 2001) leading to the production of small duplex RNAs 21 to 30 nts in length with the characteristic 2′nt overhang at the 3′ ends. Subsequently the small RNAs are loaded into an effector complex to guide specific RNA silencing (Tomari and Zamore 2005). Two types of RNA molecules have the potential to serve as a trigger of RNA silencing. dsRNAs are precursors of siRNAs whereas ssRNAs with step-loop structures serve as precursors of miRNAs (pre-miRNAs). *Arabidopsis* encodes DCL1-4 all of which recognize dsRNA triggers although the primary role of DCL1 is to process pre-miRNAs. Two classes of siRNAs of sizes 21 to 22 and 24 nt respectively are produced in

Arabidopsis plants and it is likely that the 21, 22 and 24 nt siRNAs are produced by DCL4, DCL2 and DCL3 respectively (Dunoyer *et al.* 2005; Gasciolli *et al.* 2005; Qi *et al.* 2005 and Xie *et al.* 2005). In *Arabidopsis* miRNAs and siRNAs are methylated at the 3′ end which is essential for protection of the small RNAs from degradation mediated by 3′-uridylation activity (Li *et al.* 2005 and Yu *et al.* 2005).

5.2 ASSEMBLY OF EFFECTOR COMPLEXES

To act as silencing triggers double-stranded siRNAs must be channeled through an ordered RISC assembly pathway that results in the selection of one strand and the destruction of the other (Pham and Sontheimer 2005). RISC which directs methylation of chromatin siRISC and miRISC which guide cleavage and translational arrest respectively of target mRNAs play a significant role in assembly of effector complexes (Fig. 4). All of these complexes contain one (guide) strand of the duplex small RNAs as

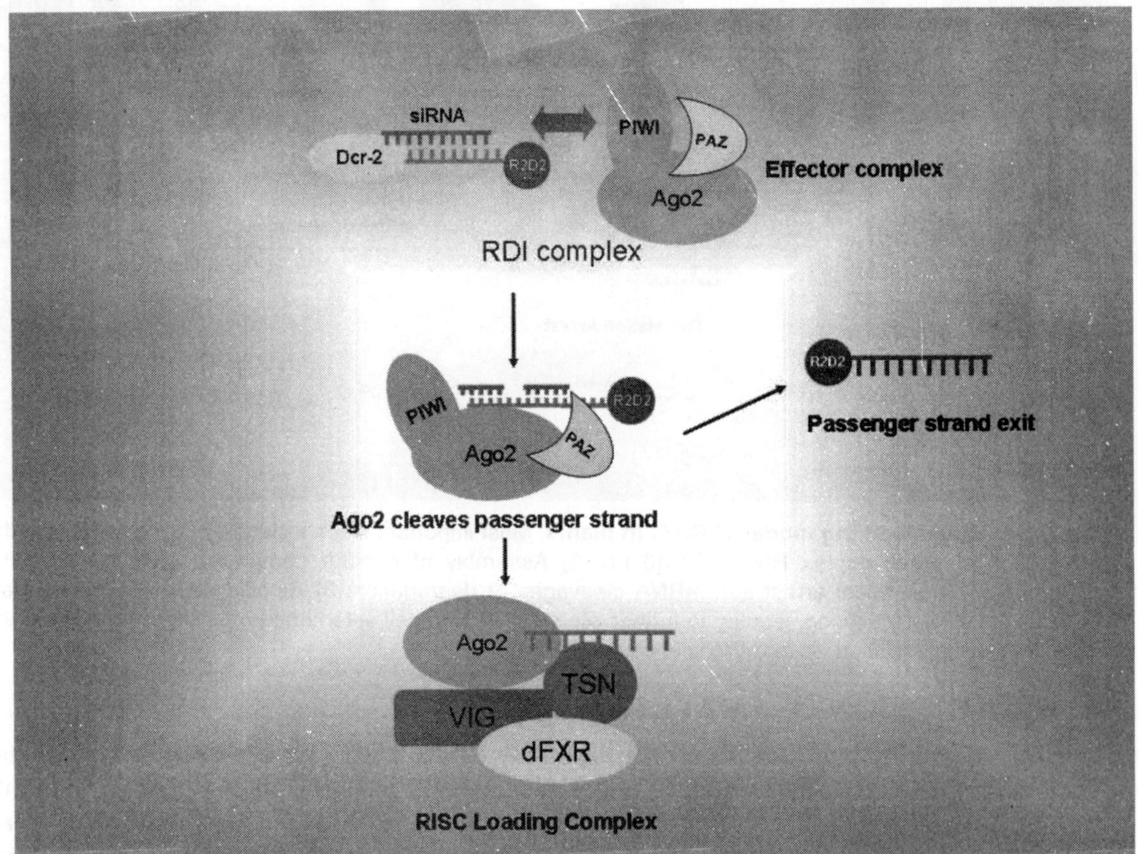

Fig. 4: Formation of RISC assembly complexes. The asymmetric siRNA molecule bound by Dcr2 and R2D2 which sense the stability at both ends of siRNA duplex. This initiation complex is known as RDI complex (Dcr2 –R2D2 initiation complex).Dcr2 is eventually exchanged with Ago2, which by virtue of its PIWI domain cleaves the passenger strand which results in the formation of an active RISC loading complex.

the specificity determinants and a member of the Argonaute protein (AGO) family (Li and Ding 2006). The PIWI domain of AGOs has structural similarities to RNaseH and the ability to cleave the target RNA base-paired with the guide strand siRNA called slicing has been demonstrated for *Arabidopsis* AGO1 (Baumberger and Baulcombe 2005 and Liu *et al.* 2005). It is likely that assembly of these effector complexes may follow a pathway in which siRISC assembly begins with binding of siRNA duplexes by the heterodimer of Dicer-2 and R2D2, a dsRNA binding protein with tandem dsRNA binding motifs in the RISC loading complex (RLC) (Tomari *et al.* 2004) (Fig. 4). Thus Dicer-2 is also required in RISC assembly downstream of siRNA production. Next RLC delivers the siRNA duplex into AGO2 which subsequently cleaves the passenger strand siRNA triggering its dissociation from the complex and activation of RISC that contains only the siRNA guide strand (Matranga 2005 and Rand *et al.* 2005).

5.3 AMPLIFICATION AND TRANSITIVE SILENCING

Fungi nematodes and plants encode eukaryotic RNA-dependent RNA polymerases (RdRP) that generate new sources of dsRNA for dicing leading to further silencing amplification. In both plants and *Caenorhabditis elegans,* RdRP amplification results in the spread of silencing along the target gene beyond the region initially targeted for silencing referred to as transitive RNAi (Sijen *et al.* 2001; Vaistij *et al.* 2002). RdRP genes are essential for RNA silencing in the germ line (*ego-1*) and somatic tissues (*rrf-1*) in *C. elegans* have been identified (Smardon *et al.* 2000 and Sijen *et al.* 2001). *Arabidopsis* encodes six RdRPs designated RdRP1–6 which together with individual DCLs control specific small RNA biogenesis pathways.

5.4 NON-CELL AUTONOMOUS SILENCING

The effects of RNA silencing can spread beyond the sites of silencing initiation *via* a putative specific silencing signal (Voinnet 2005). Two distinct steps have been observed in the spread of RNA silencing in plants. Accumulated evidences suggest a role for 21-nt siRNAs in the short-distance spread and 24-nt siRNAs in the phloem-dependent long-distance transport (Palauqui *et al.* 1997; Hamilton *et al.* 2002 and Himber *et al.* 2003). Although RdRP amplification is not required for the cell-to-cell spread extensive short-distance spread beyond 10 to 15 cells in plants requires the *RdRP6/DCL4* pathway and its product the 21-nt siRNAs (Himber *et al.* 2003; Dunoyer *et al.* 2005; Schwach *et al.* 2005). In contrast a predicted role for the 24-nt siRNAs (or their longer precursor dsRNA) produced by the *RdRP2/DCL3/AGO4* pathway in the long distance silencing spread remains to be rigorously examined. However both classes of siRNAs are found in the phloem indicating their potential to mediate silencing spread in plants (Yoo *et al.* 2004). Similarly it is also not clear if DNA methylation associated with the maintenance or persistent silencing of transgenes play a specific role in non-cell autonomous silencing.

6. METHODS TO INDUCE RNAi IN PLANTS

In RNAi research field one of the biggest challenges is the delivery of the active molecules that will trigger the RNAi pathway in plants. In this system a number of methods for delivery of dsRNA or siRNA into different cells and tissue include transformation with dsRNA-forming vectors for selected gene(s) by an *Agrobacterium* mediated transformations (Chuang and Meyerowtiz 2000 and Waterhouse *et al.* 2001); delivery cognate dsRNA of *uidA* GUS (β-glucuronidase) and TaGLP2a:GFP (green fluorescent protein) reporter genes into single epidermal cells of maize barley and wheat by particle bombardment (Schweizer *et al.* 2000) introducing a *Tobacco rattle virus* (TRV)-based vector in tomato plants by infiltration (Liu *et al.* 2002a); delivery of dsRNA into tobacco suspension cells by cationic oligopeptide polyarginine-siRNA complex; infecting plants with viral vectors that produce dsRNA (Dalmay *et al.* 2000a) and delivery of siRNA into cultured plant cells of rice cotton and slash pine for gene silencing by nanosense pulsed laser-induced stress wave (LISW) (Tang *et al.* 2006) are being used. Among these the most reliable and commonly used approaches for delivery of dsRNA to plants cells are agroinfiltration micro-bombardment and VIGS. These are discussed in the following sections.

6.1 AGROINFILTRATION

Agroinfiltration is a powerful method to study processes connected with RNAi. The injection of *Agrobacterium* carrying similar DNA constructs into the intracellular spaces of leaves for triggering RNA silencing is known as agroinoculation or agroinfiltration (Hily and Liu 2007). In most cases agroinfiltration is used to initiate systemic silencing or to monitor the effect of suppressor genes. In plants cytoplasmic RNAi can be induced efficiently by agroinfiltration similar to a strategy for transient expression of T-DNA vectors after delivery by *Agrobacterium tumefaciens*. The transiently expressed DNA encodes either an ss- or dsRNA which is typically a hairpin (hp) RNA. The infiltration of hairpin constructs are especially effective because their dsRNA can be processed directly to siRNAs while the constructs expressing ssRNA can also be useful to induce silencing (Johansen and Carrington 2001; Voinnet *et al.* 2001; Mlotshwa *et al.* 2002 and Tenllado *et al.* 2003) and for dissecting the mechanism of gene silencing especially concerned with its suppressors systemic silencing signal and also for simple protein purification (Johansen and Carrington 2001; Voinnet *et al.* 2001; Mlotshwa *et al.* 2002 and Tenllado *et al.* 2003). Besides they provide a rapid versatile and convenient way for achieving a very high level of gene expression in a distinct and defined zone.

6.2. MICRO-BOMBARDMENT

In this method a linear or circular template is transferred into the nucleus by micro-bombardment. Synthetic siRNAs are delivered into plants by biolistic pressure to cause silencing of GFP expression. Bombarding cells with particles coated with dsRNA or siRNA or DNA that encode hairpin constructs as well as sense or antisense RNA activate the RNAi pathway. The silencing effect of RNAi is occasionally detected as early as a day after bombardment and it continues up to 3 to 4 days post bombardment. Systemic spread of the silencing occurred 2 weeks later to manifest in the vascular tissues of the non-bombarded leaves of *Nicotiana benthamiana* that were closest to the bombarded ones. After one month or so the loss of GFP expression was seen in non-vascular tissues as well. RNA blot hybridization with systemic leaves indicated that the biolistically delivered siRNAs induced due to *de novo* formation of siRNAs which accumulated to cause systemic silencing (Klahre *et al.* 2002).

6.3. VIRUS INDUCED GENE SILENCING (VIGS)

Modified viruses as RNA silencing triggers are used as a mean for inducing RNA in plants. Different RNA and DNA viruses have been modified to serve as vectors for gene expression (Timmermans *et al.* 1994 and Pogue *et al.* 2002). Some viruses such as *Tobacco mosaic virus* (TMV) *Potato virus X* (PVX) and TRV can be used for both protein expression and gene silencing (Kumagai *et al.* 1995; Angell and Baulcombe 1999; MacFarlane and Popovich 2000 and Mallory *et al.* 2002). All RNA virus-derived expression vectors will not be useful as silencing vectors because many have potent anti-silencing proteins such as TEV (*Tobacco etch virus*) that directly interfere with host silencing machinery (Kumagai *et al.* 1995; Palmer and Rybicki 2001). Similarly DNA viruses have not been used extensively as expression vectors due to their size constraints for movement (Kjemtrup *et al.* 1998). However a non-mobile Maize streak Virus (MSV)-derived vector has been successfully used for long-term production of protein in maize cell cultures (Kumagai *et al.* 1995). Using viral vectors to silence endogenous plant genes requires cloning homologous gene fragments into the virus without compromising viral replication and movement. This was first demonstrated in RNA viruses by inserting sequences into TMV (Dallwitz and Zurcher 1996) and then for DNA viruses by replacing the coat protein gene with a homologous sequence (Kjemtrup *et al.* 1998). These reports used visible markers for gene silencing phytoene desaturase (*PDS)* and chalcone synthase (*CHS)* providing a measure of the tissue specificity of silencing as these have been involved in carotenoid metabolic pathway. The *PDS* gene acts on the antenna complex of the thylakoid membranes and protects the chlorophyll from photooxidation. By silencing this gene a drastic decrease in leaf carotene content resulted into the appearance of photobleaching symptom (Liu *et al.* 2002 and Turnage *et al.* 2002). Similarly over expression of *CHS* gene causing an albino phenotype instead of producing the anticipated deep orange color (Cogoni *et al.* 1994). As a result their action as a phenotypic marker helps in easy understanding of the mechanism of gene silencing. Table 1 shows some general characteristics for currently available virus-derived gene silencing vectors. Most viruses are plus-strand RNA viruses or satellites whereas *Tomato golden mosaic virus* (TGMV) and *Cabbage leaf curl virus* (CaLCuV) are DNA viruses. Though RNA viruses replicate in the cytoplasm, DNA viruses replicate in plant nuclei using the host DNA replication machinery. Both types of viruses induce diffusible homology-dependent systemic silencing of endogenous genes. However the extent of silencing spread and the severity of viral symptoms can vary significantly in different host plants and host/virus combinations. With the variety of viruses and the diversity of infection patterns transmission vectors and plant defenses it is not surprising that viruses differ with respect to silencing (Teycheney and Tepfer 2001). Because the continuing development of virus-based silencing vectors can extend VIGS to economically important plants it is useful to consider some of the characteristics of successful VIGS vectors.

TABLE 1 : RNAi effects on targeted region in some fungal plant pathogens.

Pathogen	*Targeted region*	*Result*	*References*
Magnaporthae oryzae	*eGFP*	Sequence specific degradation of mRNA	Kadotani *et al.* 2003
Cladosporium falvum	*cgl 1* and *cgl 2*	Blocking disease infection spread	Segers *et al.* 1999
Venturia inaequalis	*Multiple inverted repeats*	-	Fitzgerald *et al.* 2004
Fusarium graminearum	-	-	Nakayashiki 2005
Blumeria graminis	*Mlo*	Immunity	Schweizer *et al.* 2000

7. RNAi APPLICATION IN CROP PLANTS:

7.1. IMPROVEMENT OF PLANT NUTRITIONAL VALUE

Genetic engineering technologies have advantages over classical breeding not only by enlarging the scope of genes and the types of mutations to be manipulated but also by the ability to control the spatial and temporal expression patterns of the genes of interest. Control of spatial and temporal expression patterns is important because for many crops the tissue that is consumed as food (mostly seeds) is distinct from tissue that controls plant growth and productivity (mainly roots and shoots). However in many cases genes controlling specific traits do not operate in a tissue-specific manner but function in all or most plant organs. Therefore a mutation of a given gene that is good for the improvement of seed quality is often deleterious for the growth of other plant organs.

A well-studied example illustrating such a problem is increasing the level of lysine in plants. Lysine synthesis is strongly regulated by a feedback inhibition loop in which lysine inhibits the activity of dihydrodipicolinate synthase (DHPS) the first enzyme on the pathway specifically committed to lysine biosynthesis. Genetic mutations in the tobacco DHPS gene rendering its encoded DHPS lysine-insensitive cause lysine overproduction in all plant organs (Frankard *et al.* 1992 and Negrutiu *et al.* 1984). However although high lysine levels in seeds are beneficial increases in the level of this amino acid in vegetative tissues are undesirable because high levels of lysine cause abnormal vegetative growth and flower development that in turn reduces seed yield (Frankard *et al.* 1992 and Negrutiu *et al.* 1984). Targeting the expression of bacterial genes encoding lysine-insensitive DHPS to seeds of several transgenic model and crop plants using seed-specific promoters eliminated its undesirable effects in vegetative tissues resulting in plants with good growth characteristics and high lysine levels in their seeds (Falco *et al.* 1995; Karchi *et al.* 1994 and Mazur *et a*l. 1999). Tissue-specific manipulation in transgenic plants holds importance not only for gene overexpression but also for gene suppression approaches. This is because traits of interest in plant organs that are consumed as foods may also be negatively regulated by different genes controlling a variety of factors such as:

(*i*) enzymes that either degrade or convert the desired metabolites into undesirable ones; and (*ii*) enzymes or cellular processes or regulatory factors that interfere with the production of the desired metabolite.

Such genes could be suppressed but their constitutive suppression might have adverse effects on plant growth and reproduction.

To date the utilization of the RNAi technology to enhance specifically the nutritional value of plant organs that are consumed as foods is progressively increasing with several published successful efforts suggesting that this approach has tremendous potential. The first is also associated with improving lysine level in seeds. Because lysine accumulation in plants is negatively affected by its catabolism (degradation), constitutive knockout of lysine catabolism using a gene insertion knockout approach accelerates lysine accumulation in seeds when combined with the seed-specific expression of a feedback-insensitive DHPS (Zhu and Galili 2003). However seeds of plants that have accumulated elevated lysine levels germinate poorly because the excess lysine levels produced in the seeds are not efficiently degraded during seed germination (Zhu and Galili 2003). Reduction of lysine catabolism specifically during seed development by an RNAi approach indeed improves seed germination (Zhu and Galili 2004).

Seed-specific RNAi approaches have also successfully been used to generate dominant high lysine corn by suppressing the expression of 22-kD maize zein storage proteins a group of abundant proteins in maize seed but poor in lysine content (Segal *et al*. 2003). Reduction of lysine-poor zeins by traditional breeding was only successful in screening recessive lysine-rich mutants called opaque mutants particularly opaque-2. Opaque-2 encodes a maize b-zip type transcriptional factor that controls the expression of a subset of storage proteins including the 22-kD zein storage protein. Albeit rich in lysine this opaque-2 mutant is not very useful in agriculture because of its other adverse effects on seed quality and yield. Down-regulation of the maize lysine-poor 22-kD zein gene *via* RNAi does not alter the general functions of the Opaque-2 transcription factor but generates quality and normal maize seeds with high levels of lysine-rich proteins and free lysine. Maize is mostly used in western countries for livestock feed but in many developing countries it is also a staple crop for humans. It is thus expected that high lysine maize developed using RNAi approaches will be of significant economical and humanitarian significance world wide and that a similar RNAi-mediated approach will be used to enhance the lysine content in seeds of other staple crops.

Another nutritionally important crop is cotton. Cotton is mainly used for fiber production and is an important crop not only in developed countries but also in many developing countries where malnutrition and starvation are widespread (Sunilkumar *et al*. 2006). In these developing areas the cottonseeds that remain after fiber extraction could be extensively used as sources of protein and calories but they are largely underutilized because they contain a toxic gossypol terpenoid. Gossypol is also produced in vegetative cotton tissues where it protects cotton plants from insects and other pathogens (Sunilkumar *et al*. 2006). Hence the suppression of its production should occur specifically in cottonseeds. Indeed transgenic cotton plants expressing an

RNAi construct of the δ-cadinene synthase gene of gossypol synthesis fused to a seed-specific promoter caused seed-specific reduction of this metabolite while its content in non-seed tissues was comparable to the control plants (Sunilkumar *et al.* 2006). These cotton plants are thus expected to have similar insect and pathogen resistance to that of wild type cotton but to produce seeds with higher nutritional value.

While insufficient food calories is one of the major causes of malnutrition in developing countries an excess of digested calories leads to obesity and other diseases in developed countries (Maskarinec *et al.* 2006). Foods rich in inefficiently digested carbohydrates such as fiber are therefore considered to be health promoting (Williams 1995). The major nutritional source of plant-derived carbohydrates is starch which is composed of amylopectin and amylose polysaccharides synthesized by two competitive pathways. Yet in cooked plant foods undergoing cooling before eating amylose molecules tend to efficiently form digestion-resistant complexes that are part of healthy dietary fiber (Crowe *et al.* 2000). Aiming to increase the relative content of amylose in wheat grains an RNAi construct designed to silence the genes encoding the two starch-branching isozymes of amylopectin synthesis were expressed under a seed-specific promoter in wheat (Regina *et al.* 2006). This resulted in increased grain amylose content to over 70% of the total starch content. When fed to rats in a diet as a wholemeal the high amylose grains had positive effects on indices of gastrointestinal health elucidating the potential of RNAi technology to improve human health particularly in developing countries (Regina *et al.* 2006).

A significant part of human diets both in developed and developing countries is composed of vegetables. Among the vegetables tomato fruits are relatively rich in a number of vitamins as well other health promoting metabolites such as flavonoids and carotenoids including the strong antioxidant carotenoid or lycopene which provides the tomato fruit with its typical red color. Carotenoids are synthesized by the same biosynthetic pathway that synthesizes chlorophyll and it has been shown that genes controlling the light-mediated regulation of the photosynthetic machinery also influence tomato fruit quality by altering the levels of carotenoids and flavonoids (Adams-Phillips *et al.* 2004). The tomato high pigment (hp-2) phenotype which accumulates elevated levels of carotenoids and flavonoids is due to mutations in the regulatory gene DE-ETHIOLATED1 (DET1) which represses several light-dependent signaling pathways (Levin *et al.* 2003 and Mustilli *et al.* 1999). Despite their positive effects on fruit quality hp-2 mutants generally possess abnormal growth and various vegetative phenotypes and so are transgenic plants in which the DET1 gene is constitutively silenced (Davuluri *et al.* 2004; Levin *et al.* 2003 and Mustilli *et al.* 1999). Such negative phenotypes are not unexpected taking into account the central regulatory function of the DET1 gene in light signaling and photosynthesis. Yet RNAi-mediated suppression of DET1 expression under fruit-specific promoters has recently shown to improve carotenoid and flavonoid levels in tomato fruits with minimal effects on plant growth and other fruit quality parameters (Davuluri *et al.* 2005). This exemplifies again how coupling the highly efficient RNAi gene suppression machinery with tissue specific promoters provides a highly valuable trait that is impossible to obtain by conventional breeding.

RNAi technology has also been successful in genetic modification of the fatty acid composition of oil. A hairpin RNA- (hpRNA) mediated RNAi method was used in cotton to down-regulate two key fatty acid desaturase genes encoding stearoyl-acyl-carrier protein Δ 9-desaturase and oleoyl-phosphatidylcholine ω 6-desaturase (Liu *et al.* 2002). Knockdown of these two genes in cotton led to the increase of nutritionally improved high-oleic (HO) and high-stearic (HS) cottonseed oils that are essential fatty acids for human heart health. It is almost certain that more cases of improved crops via dsRNA-triggered RNAi technology will be reported in the coming years. With more target genes of metabolic pathways discovered in the practice of nutritional genomics (Grusak 2002) in plants RNAi and miRNA technologies will become essential tools to manipulate plant macro- and micro-nutrients for human health. The readers are referred to Larkin and Harrigan (2007) for examples of the application of metabolic profiling to studying a diverse range of transgenically modified plants. It is evident that metabolic profiling platforms will have value in assessing changes in pathways associated with the production of important nutrients and plant chemicals including but not restricted to fatty acids amino acids carbohydrates carotenoids flavonoids and terpenses such as gossypol.

7.2. RNAi IN PLANT DISEASE MANAGEMENT

Despite substantial advances in plant disease management strategies our global food supply is still threatened by a multitude of pathogens and pests. This changed scenario warrants us to respond more efficiently and effectively to this problem. The situation demands judicious blending of conventional, unconventional and frontier technologies. In this sense RNAi technology has emerged as one of the most potential and promising strategies for enhancing the building of resistance in plants to combat various fungal bacterial viral and nematode diseases causing huge losses in important agricultural crops. The nature of this biological phenomenon has been evaluated in a number of host-pathogen systems and effectively used to silence the action of pathogen. Many of the examples listed below illustrate the possibilities for commercial exploitation of this inherent biological mechanism to generate disease-resistant plants in the future by taking advantage of this approach.

7.2.1. Management of Plant Pathogenic Fungi

RNA-mediated gene silencing (RNA silencing) is used as a reverse tool for gene targeting in fungi. Homology-based gene silencing induced by transgenes (co-suppression) antisense or dsRNA has been demonstrated in many plant pathogenic fungi including *Cladosporium fulvum* (Hamada and Spanu 1998) *Magnaporthae oryzae* (Kadotani *et al.* 2003) *Venturia inaequalis* (Fitzgerald *et al.* 2004) *Neurospora crassa* (Goldoni *et al.* 2004) *Aspergillus nidulans* (Hammond and Keller 2005) and *Fusarium graminearum* (Nakayashiki 2005) (Table 1) whether it is suitable for large-scale mutagenesis in fungal pathogens remains to be tested. Hypermorphic mechanism of RNA interference implies that this technique can also be applicable to all those plant pathogenic fungi which are polyploid and polykaryotic in nature. And also offers a solution to the problem where frequent lack of multiple marker genes in fungi is experienced. Simultaneous silencing of several unrelated genes by introducing a single chimeric construct has been demonstrated in case of *Venturia inaequalis* (Fitzgerald *et al*. 2004).

HCf-1 a gene that codes for a hydrophobin of the tomato pathogen *C. fulvum* (Spanu 1997) was co-suppressed by ectopic integration of homologous transgenes. Transformation of *Cladosporium fulvum* with DNA containing a truncated copy of the hydrophobin gene *HCf-1* caused co-suppression of hydrophobin synthesis in 30% of the transformants. The co-suppressed isolates had a hydrophilic phenotype lower levels of *HCf-1* mRNA than wild type and contain multiple copies of the plasmid integrated as tandem repeats at ectopic sites in the genome (Hamada and Spanu 1998). The transcription rate of *HCf-1* in the co-suppressed isolates was higher than in the untransformed strains suggested that silencing acted at the post-transcriptional level (Hamada and Spanu 1998). This was due to ectopic integration of the transgene next to promoters which initiate transcription to form antisense RNA and that this in turn determines down-regulation of HCf-1. But gene silencing was not associated with DNA cytosine methylation (Hamada and Spanu 1998). Similarly the silencing of *cgl1* and *cgl2* genes using the cgl2 hairpin construct in *Cladosporium fulvum* has also been reported (Segers *et al*. 1999) though the effect was possibly restricted to highly homolougous genes (exons of *cgl 1* and *cgl 2* are 87% identical). However the less homologus *cgl 3* (53% overall identity to *cgl* 2) was not affected as the target specificity always depends upon the actual sequence alignment and more over short regions of high density that led to unwanted off-targets effects. Such a strategy could be exploited for protecting the consumable products of vegetables and fruits crops from the post harvest diseases caused by different plant pathogens in future.

Fitzgerald and collegues (2004) using hairpin vector technology have been able to trigger simultaneous high frequency silencing of a green fluorescent protein (*GFP*) transgene and an endogenous trihydroxynaphthalene reductase gene (*THN*) in *V. inaequalis*. *GFP* transgene acting as easily detectable visible marker while the trihydroxynaphthalene reductase gene (*THN*) playing role in melanin biosynthesis. High frequency gene silencing was achieved using hairpin constructs for the *GFP* or the *THN* genes transferred by *Agrobacterium* (71 and 61% respectively). *THN*-silenced transformants exhibited a distinctive light brown phenotype and maintained the ability to infect apple. Silencing of both genes with this construct occurred at a frequency of 51% of all the transformants. All 125 colonies silenced for the GFP gene were also silenced for THN (Fitzgerald *et al*. 2004). Similarly multiple gene silencing has been achieved in *Cryptococcus neoformans* using chimeric hairpin constructs (Liu *et al*. 2002) and in plants using partial sense constructs (Abbott *et al*. 2002).

The first effort towards the systematic silencing of *Magnaporthe grisea* a causal organism of rice blast was carried out in by Kadotani *et al*. (2003) by using the enhanced green florescent protein gene as a model. To assess the ability of RNA species to induce silencing in fungus plasmid construct expressing sense antisense and hairpin RNA were introduced into an *eGFP*-expressing transformants. The fluorescence of *eGFP* in the transformants was silenced much more efficiently by hairpin RNA of *eGFP* than by other RNA species. In the silenced transformants the accumulation of *eGFP* mRNA was drastically reduced. But not methylation of coding or promoter region was involved. The small interfering RNA molecules of 19-23 nucleotides were observed in both sense and antisense strands of *eGFP* gene (Kadotani *et al* 2003). Later on Nakayashiki and collegues (2005) developed a protocol for silencing the *mpg1* and polyketide synthase-like genes. *mpg1* gene is a hydrophobin gene which is essential for pathogenicity as it act as a cellular relay for adhesion and trigger for the development of appressorium (Talbot *et al.* 1996). Their work on this host-pathogen system revealed that they were successfully able to silence the above mentioned genes at varying degrees by pSilent-1-based vectors in 70–90% of the resulting transformants. Ten to fifteen per cent of the silenced transformants exhibited almost "null phenotype". This vector was also efficiently applicable to silence a *GFP* reporter in another ascomycete fungus *Colletotrichum lagenarium* (Nakayashiki 2005).

7.2.2. Management of Plant Pathogenic Bacteria

One of the striking examples of bacterial disease management where RNAi showed a remarkable type of gene regulation was documented by Escobar *et al.* (2001). They developed a crown gall disease management strategy that targets the process of tumourogensis (gall formation) by initiating RNAi of the *iaaM* and *ipt* oncogenes. Expression of these genes is a prerequisite for wild type tumor formation. Transgenic *Arabidopsis thaliana* and *Lycopersicon esculentum* transformed with RNAi constructs targeting *iaaM* and *ipt* gene(s) showed resistance to crown gall disease. Transgenic plants generated through this technology contained a modified version of these two bacterial gene(s) required to cause the disease and was the first report to manage a major bacterial disease through RNAi. The extra genes recognize and effectively shut down the expression of the corresponding bacterial gene during infection thus preventing the spread of infection. The incoming bacteria could not make the hormones needed to cause tumors and plants deficient in silencing were hyper-susceptible to *A tumefaciens* (Dunoyer *et al.* 2007). Successful infection relied on a potent anti-silencing state established in tumors whereby siRNA synthesis is specifically inhibited. The procedure can be exploited to develop broad-spectrum resistance in ornamental and horticultural plants which are susceptible to crown gall tumorigenesis. This approach can be advocated for the effective management of those pathogens which multiply very rapid and results in tumor formation such as *Albugo candida Synchytrium endobioticum Erwinia amylovora etc.* The natsiRNA (nat-siRNAATGB2) was strongly induced in *Arabidopsis* upon infection by *Pseudomonas syringae pv tomato* and down-regulates a PPRL gene that encodes a negative regulator of the RPS2 disease resistance pathway. As a result the induction of nat-siRNAATGB2 increases the RPS2-mediated race-specific resistance against *P. syringae* pv *tomato* in *Arabidopsis* (Katiyar-Agarwal *et al.* 2006). Recently the accumulation of a new class of sRNA 30 to 40 nucleotides in length termed long-siRNAs (lsiRNAs) was found associated with *P. syringae* infection. One of these lsiRNAs AtlsiRNA-1 contributes to plant bacterial resistance by silencing AtRAP a negative regulator of plant defense (Katiyar and Agarwal *et al.* 2007). A *Pseudomonas* bacterial flagellin derived peptide is found to induce the accumulation of miR393 in *Arabidopsis*.

miR393 negatively regulates mRNAs of F-box auxin receptors resulting in increased resistance to the bacterium (*P. syringae*) and the overexpression of miR393 was shown to reduce the plant's bacterial titer by 5-fold (Navarro *et al.* 2006).

7.2.3. Management of Plant Pathogenic Viruses

Antiviral RNAi technology has been used for viral disease management in human cell lines (Bitko and Barik 2001; Gitlin *et al.* 2002; Jacque *et al.* 2002 and Novina *et al.* 2002). Such silencing mechanisms (RNAi) can also be exploited to protect and manage viral infections in plants (Waterhouse *et al.* 2001 and Ullu *et al.* 2002). The effectiveness of the technology in generating virus resistant plants was first reported to PVY in potato harbouring vectors for simultaneous expression of both sense and antisense transcripts of the helper-component *proteinase* (*HC-Pro*) gene (Waterhouse *et al.* 1998). The *P1/HC-Pro* suppressors from the potyvirus inhabited silencing at a step down stream of dsRNA processing possibly by preventing the unwinding of duplex siRNAs or the incorporation into RISC or both (Chapman *et al.* 2004). The utilization of RANi technology has resulted in inducing immunity reaction against several other viruses in different plant-virus systems (Table 2).

In phyto-pathogenic DNA viruses like geminiviruses non-coding intergenic region of *Mungbean yellow mosaic India virus* (MYMIV) was expressed as hairpin construct under the control of the 35S promoter and used as biolistically to inoculate MYMIV-infected black gram plants and showed a complete recovery from infection which lasted until senescence (Pooggin *et al.* 2003). RNAi mediated silencing of geminiviruses using transient protoplast assay where protoplasts were co-transferred with a siRNA designed to replicase (Rep)-coding sequence of *African cassava mosaic virus* (ACMV) and the genomic DNA of ACMV resulted in 99% reduction in Rep transcripts and 66% reduction in viral DNA (Vanitharani *et al.* 2003). It was observed that siRNA was able to silence a closely related strain of ACMV but not a more distantly related virus.

About more than 40 viral suppressors have been identified in plant viruses (Ruiz and Voinnet 2007). Results from some of the well-studied virus suppressors indicated that suppressors interfere with systemic signaling for silencing (Mlotshwas *et al.* 2002). During last few years the *p69* encoded by *Turnip yellow mosaic virus* has been identified as silencing suppressors that prevented host RDR-dependent secondary dsRNA synthesis (Chen *et al.* 2004). *P14* protein encoded by aureus viruses suppressed both virus and transgene-induced silencing by sequestering both long dsRNA and siRNA without size specificity (Merai *et al.* 2005). Multiple suppressors have been reported in *Citrus tristeza virus* where *p20* and coat protein (*CP*) play important role in suppression of silencing signal and *p23* inhibited intracellular silencing (Lu *et al.* 2004). Multiple viral components viral RNAs and putative RNA replicase proteins were reported for a silencing or suppression of *Red clover necrotic mosaic virus* (Takeda et *al.* 2005). In this case the RNA silencing machinery deprived of DICER-like enzymes by the viral replication complexes appears to be the cause of the suppression. *Pns10* encoded by *Rice dwarf virus* suppressed local and systemic S-PTGS but not IR-PTGS suggesting that *Pns10* also targets an upstream step of dsRNA formation in the silencing pathway (Cao *et al.* 2005).

TABLE 2 : Effects of targeted region of RNAi in various plant-virus systems.

Host system	*Virus*	*Targeted region*	*References*
N. benthamiana, M. esculenta	*African cassava mosaic virus*	*pds, su, cyp79d2*	Fofana *et al.* 2004
Barley, wheat	*Barley stripe mosaic virus*	*pds*	Holzberg *et al.* 2002; Scofield *et al.* 2005
Soybean	*Bean pod mottle virus*	*pds*	Zhang and Ghabrial 2006
Barley, rice, maize	*Brome mosaic virus*	*pds, actin 1, rubisco activase*	Ding *et al.* 2006
Arabidopsis	*Cabbage leaf curl virus*	*gfp, CH42, pds*	Turnage *et al.* 2002
P. sativum	*Pea early browning virus*	*pspds, uni, kor*	Constantini *et al.* 2004
N. benthamiana	*Poplar mosaic virus*	*gfp*	Naylor *et al.* 2005
N. benthamiana, S. tuberosum	*Potato virus X*	*pds, gfp*	Ruiz *et al.* 1998; Faivre-Rampant *et al.* 2004
Nicotiana tabacum	*Satellite tobacco mosaic virus*	*Several genes*	Gossele *et al.* 2002
N. benthamiana, N. tabacum	*Tobacco mosaic virus*	*pds, psy*	Kumagai *et al.* 1995
N. benthamiana, Arabidopsis, tomato, *Solanum* species, Chilli pepper, opium poppy, *Aquilegia*	*Tobacco rattle virus*	*Rar1, EDS1, NPR1/NIM1, pds, rbcS, gfp*	Liu *et al.* 2002b; Ratcliff *et al.* 2001; Brigneti *et al.* 2004; Chung *et al.* 2004; Hileman *et al.* 2005; Gould and Kramer 2007
N. benthamiana	*Tomato bushy shunt virus*	*gfp*	Hou and Qiu 2003
N. benthamiana	*Tomato golden mosaic virus*	*su, luc*	Peele *et al.* 2001
N. benthamiana, Lycopersicon esculentum N. glutinosa, N. tabacum	*Tomato yellow leaf curl China virus-associated b DNA satellite*	*pcna, pds, su, gfp*	Tao and Zhou 2004

(Modified after Godge *et al.* 2008)

Niu and colleagues (2006) used a 273-bp (base pair) sequence of the *Arabidopsis* miR159 a pre-miRNA transcript expressing amiRNAs against the viral suppressor genes *P69* and *HC-Pro* to provide resistance against *Turnip yellow mosaic virus* and *Turnip mosaic virus* infection respectively. In addition a dimeric construct harboring two unique amiRNAs against both viral suppressors conferred resistance against these two viruses in inoculated *Arabidopsis* plants.

Similarly Qu *et al.* (2007) used a different amiRNA vector to target the 2 b viral suppressor of the *Cucumber mosaic virus* (CMV); a suppressor that interacted with and blocked the slicer activity of *AGO1* had also shown to confer resistance to CMV infection in transgenic tobacco. A strong correlation between virus resistance and the expression level of the 2 b-specific amiRNA was shown for individual plant lines.

It is evident from above-mentioned reports that the RNA components such as single strand template RNA dsRNA and/or siRNA of the silencing pathways are the preferred targets of most viral suppressors. However plant viruses are known to have evolved a counter-silencing mechanism by encoding proteins that can overcome such resistance (Li and Ding 2006; Díaz-Pendón and Ding 2008). These suppressors of gene silencing are often involved in viral pathogenicity mediate synergism among plant viruses and result in the induction of more severe disease. Simultaneous silencing of such diverse plant viruses can be achieved by designing hairpin structures that can target a distinct virus in a single construct (Díaz-Pendón and Ding 2008).

Contrarily the RNAi system may cause an increase in the severity of viral pathogenesis and/or encode proteins which can inactivate essential genes in the RNAi machinery (Elbashir *et al.* 2001 and Li *et al.* 2002) that helps them in their replication in the host genome (Hannon 2002).

7.2.4. Management of Plant Parasitic Nematodes

Several major plant parasitic nematodes such as the root-knot (*Meloidogyne* spp.) and cyst (*Heterodera* spp.) along with other minor nematodes cause significant damage to important crops like legumes vegetables and cereals in most parts of the world and continue to threaten these agricultural crops. So a natural eco-friendly defense strategy that delivers a cost-effective control of plant parasitic nematodes is needed which is difficult to achieve through conventional approaches. However the birth of RNAi technology from classical *C. elegans* studies has shown the ways and means to explore the possibilities of this mechanism for protecting plants from nematode damage. In this context two approaches have been advocated one of them relies on targeting plant genes that are involved with the infection process and the second approach targets essential genes within the nematode. RNAi can be induced in *C. elegans* by feeding it dsRNA so it was reasoned that expressing hpRNAs containing sequences of vital nematode genes in the host plant might deliver dsRNA to a feeding nematode to incapacitate or kill it.

After the demonstration of gene silencing using siRNA duplexes in the nematode (Fire *et al.* 1998) the use of RNAi has rapidly emerged as the technique of choice for plant nematologists to put their efforts especially for nematode management in agriculture. RNAi-mediated suppression of a gene plays an indispensable role in hampering the nematode development and may adversely affect the progression of pathogenesis in direct or indirect ways. There are accumulating evidences for the efficacy of RNAi in plant parasitic nematode management and a wide range of genes have been targeted for silencing in cyst and root-knot nematode species (Table 3).

RNAi in the context of phyto-parasitic nematodes was used as early as the beginning of this century when stimulation of oral ingestion by second-stage juveniles of cyst nematodes *H. glycines G. pallida* (Urwin *et al.* 2002) and root-knot nematode *M. incognita* (Bakhetia *et al.* 2005) was achieved by using octopamine. Later on resorcinol- and

TABLE 3 : RNAi effect on targeted region of plant parasitic nematodes.

Nematode	*Targeted region*	*RNAi affect*
M. incognita	Cysteine proteinase	Delayed development, Decrease in established nematodes population
	Dual oxidase	Decrease in established nematodes population and fecundity.
	Splicing factor, Integrase	Reduction in gall formation and Female nematode population
	Secreted peptide 16D 10	Reduction in gall formation and established nematode population
H. glycines	Cysteine proteinase	Increased male: female ratio.
	C-type lectin	Reduction in established nematodes population
	Major sperm protein	Reduction in mRNA level
	Aminopeptidase	Decrease in established nematodes population and increase in male: female ratio.
	_-1,4-endoglucanase	Decrease in established nematodes population
	Pectate lyase , Chorismate mutase	Increase in male: female ratio.
	Secreted peptide SYV46	Decrease in established nematode population
G. pallida	Cysteine proteinase	Increase in male: female ratio.
	FMR Famide-like peptides	Motility inhibited
G. rostochiensis	Chitin synthase	Delay in egg hatch
	_-1,4-endoglucanase	Decrease in established nematodes population
	Secreted amphid protein	Reduction in invasion ability to locate and invade plant roots
Heterodera schachtii	*Suc* transporter genes	Reduction of female nematode development

(Modified after Karakas 2008)

serotonin-inducing dsRNA uptake by second stage juvenile of *M. incognita* was found to be more effective than octopimine (Rosso *et al.* 2005). The genes targeted by RNAi to date are expressed in a range of different tissues and cell types. The ingested dsRNA can silence genes in the intestine (Urwin *et al.* 2002 and Shingles *et al.* 2007) female reproductive system (Lilley *et al.* 2005) sperm (Urwin *et al.* 2002 and Steeves *et al.* 2006) and both subventral and dorsal oesophageal glands (Chen *et al.* 2005 and Rosso *et al.* 2005; Huang *et al.* 2006; Bakhetia *et al.* 2007). Uptake of dsRNA from the gut is a proven route to systemic RNAi in *C. elegans*. The systemic nature of RNAi in plant parasitic nematodes following ingestion of dsRNA suggests that they share similar uptake and dispersal pathways.

However RNAi of a *chitin synthase* gene expressed in the eggs of *Meloidogyne artiella* was achieved by soaking intact eggs contained within their gelatinous matrix in a solution containing dsRNA (Fanelli *et al.* 2005). The enzyme plays a role in the synthesis of the chitinous layer in the eggshell. Depletion of its transcript by RNAi led to a reduction in stainable chitin in eggshells and a delay in hatching of juveniles from treated eggs. Similarly RNAi targeting for *cysteine proteinase* transcripts did not reduce parasitic population of established nematodes on plants but result into the alteration of their sexual fate in favour of males at 14 days after invasion (Urwin *et al.* 2002).

On the other hand *H. glycines* exposed to dsRNA corresponding to a protein with homology to C-type lectins did not affect sexual fate but 41% fewer nematodes were recovered from the plants (Urwin *et al.* 2002). But treatment with dsRNA corresponding to the major sperm protein (MSP) had no effect on nematode development or sexual fate 14 days after treatment. In addition to this reduction in transcript abundance for targeted mRNAs in the infective juvenile and for MSP transcripts when males reached sexual maturity and sperm are produced was observed (Urwin *et al.* 2002). In further extension of such types of experiments showed efficient FITC uptake by soaking *M. incognita* 90-95% of individuals swallowed the dye when the target was a dual oxidase (an enzyme comprised with a peroxidase domain EF-hands and NADPH oxidase domain and potentially involved in extracellular matrix development). The effect of RNAi was observed when root knot nematode (RKN) juveniles were fed on dual oxidase-derived dsRNA the reduction in the number and size of established females at 14 and 35 days post infection with an overall reduction of 70% in egg production was observed (Bakhetia *et al.* 2005). RNAi has also been induced for a *chitin synthase* gene that is expressed in the eggshells of *M. artiella* after soaking its developing eggs in a dsRNA (Fanelli *et al.* 2005).

Heterodera schachtii induces syncytial feeding structures in the roots of host plants and this requires the up-regulation of *Suc* transporter genes to facilitate increased nutrient flow to the developing structure. Targeting these genes and down-regulating them with RNA silencing resulted in a significant reduction of female nematode development (Hoffman *et al.* 2008). Indeed tobacco plants transformed with hpRNA constructs against two such root-knot nematode genes have shown such an effect: the target mRNAs in the plant parasitic nematodes were dramatically reduced and the plants showed effective resistance against the parasite (Fairbairn *et al.* 2007).

8. CONCLUSIONS AND FUTURE OUTLOOK OF RNAi

RNAi and miRNA technologies of gene silencing are newly developed genomics tools that have great advantages over antisense and co-suppression due to their higher silencing efficiency and shortened time period for screening for the targeted plants. These technologies are particularly useful in conjunction with the practice of gene or pathway discoveries through nutritional genomics trancriptomics proteomics and metabolomics in plants to improve human health. The RNA silencing has practical use because of the ability to reduce gene expression in a manner that is highly sequence specific as well as technologically facile and economical having potential in agriculture specifically for nutritional improvement of plants and the management of mascotous plant diseases. However the major obstacles hindering its immediate applications include selection of targeting sequences and in the delivery of siRNA. The key issues are

(1) how to select silencing targets for a particular disease and

(2) how to efficiently deliver siRNAs into specific cell types *in vivo*?

Tissue or organ-specific RNAi vectors have already been proven to be useful for targeted gene silencing in specific plant tissues and organs with minimal interference with the normal plant life cycle. New-generation RNAi vectors the miRNA vectors have been developed with high silencing accuracy and fewer side effects in plants. Genetic engineering of highly nutritional food crops requires both gene silencing and counter-silencing technologies. Besides RNAi technology can be considered an eco-friendly biosafe and ever green technology as it eliminates even certain risks associated with development of transgenic plants carrying first generation constructs (binary vectors and sense and antisense genes). As witnessed from earlier strategies for obtaining viral resistant plants the expression of protein product from the transgene of interest risked hetero-encapsidation through protein-protein interactions between target and non-target viral gene product resulted in the development of a non-aphid transmissible strain of Zucchini yellow mosaic virus to aphid-transmissible strain from a transgene expressing a plum pox capsid protein (Lecoq *et al.* 1993). Since RNAi triggers the formation of dsRNA molecules that target and facilitate the degradation of the gene of interest as well as the transgene itself to avoid problems arising from the synthesis of gene sequences as well as non-coding regions of gene thus limiting undesirable recombination events. Keeping in view the potentialities of RNAi technology and lesson from classical examples demonstrated that why and how this technology has emerged to combat plant pathogens in the near future as it has already added new dimensions in the chapter of plant disease management. Further development of vectors that can suppress the RNAi pathway but overexpress transgenes in a tissue-specific manner will revolutionize this field. Such vectors could be based on various viral RNA silencing suppressors and their derivatives. Future directions will focus on developing finely tuned RNAi-based gene silencing vectors that are able to operate in a temporally and spatially controlled manner. However a better and comprehensive understanding of RNAi should allow future plant researchers to work effectively and efficiently in order to improve crop plants nutritionally and manage various mascotous intruders of crop plants.

REFERENCES

Abbott JC Barakate A Pincon G Legrand M Lapierre C Mila I Schuch W Halpin C (2002) Simultaneous suppression of multiple genes by single transgenes. Down-regulation of three unrelated lignin biosynthetic genes in tobacco. *Plant Physiology* **128:** 844-853

Adams-Phillips L Barry C Giovannoni J (2004) Signal transduction systems regulating fruit ripening. *Trends in Plant Science* **9:** 331–338.

Akbergenov R Si-Ammour A Blevins T Amin I Kutter C Vanderschuren H Zhang P Cruissem W Meins FJ Hohn T Pooggin MM (2006) Molecular characterization of geminivirus derived small RNAs in different plant species. *Nucleic Acids Research* **34:** 462-471

Allen E Xie Z Gustafson AM Carrington JC (2005) microRNA-directed phasing during *Trans*acting siRNA biogenesis in plants. *Cell* **12:** 207-221

Allen E Xie Z Gustafson AM Sung GH Spatafora JW Carrington JC (2004) Evolution of microRNA genes by inverted duplication of target gene sequences in Arabidopsis thaliana. *Nature Genetics* **36:** 1282-1290

Almeida R Allshire RC (2005) RNA silencing and genome Regulation. *Trends in Cell Biology* **15:** 251-258

Anantharaman V Koonin EV Aravind L (2002) SPOUT: a class of methyltransferases that includes spoU and trmD RNA methylase suprefamilies and novel superfamilies of predicted procaryotic RNA methylases. *Journal of Molecular Microbiology and Biotechnology* **4:** 71-75

Angell SM Baulcombe DC (1999) Technical advance: potato virus X amplicon mediated silencing of nuclear genes. *The Plant Journal* **20:** 357-362

Bakhetia M Charlton W Atkinson HJ McPherson MJ (2005) RNA interference of dual oxidase in the plant nematode *Meloidogyne incognita. Molecular Plant-Microbe Interaction* **18:** 1099-1106

Bakhetia M Urwin PE Atkinson HJ (2007) qPCR analysis and RNAi define pharyngeal gland cell-expressed genes of *Heterodera glycines* required for initial interactions with the host. *Molecular Plant-Microbe Interaction* **20:** 306-312

Bartel B Bartel DP (2003) MicroRNAs: at the root of plant development? *Plant Physiology* **132:** 709-717

Bartel DP (2004) MicroRNAs: genomics biogenesis mechanism and function. *Cell* **116:** 281-297

Baulcombe DC (1999) Fast forward genetics based on virus-induced gene silencing. *Current Opinion in Plant Biology* **2:** 109-113

Baulcombe DC (2004) RNA silencing in plants. *Nature* **431:** 356-363

Baumberger N Baulcombe DC (2005) Arabidopsis ARGONAUTE1 is an RNA Slicer that selectively recruits microRNAs and short interfering RNAs. *Proceedings of the National Academy of Sciences USA* **102:** 11928-11933

Beclin C Boutet S Waterhouse P Vaucheret H (2002) A branched pathway for transgene-induced RNA silencing in plants. *Current Biology* **12:** 684-688

Bernstein E Caudy AA Hammond SM Hannon GJ (2001) Role for a bidentate ribonuclease in the initiation step of RNA interference. *Nature* **409:** 363-366

Bitko V Barik S (2001) Phenotypic silencing of cytoplasmic genes with sequence specific double stranded short interfering RNA and its applications in the reverse genetics of wild type negative strand RNA virus. *BMC Microbiology* **1:** 34-44

Bouche N Lauressergues D Gasciolli V Vaucheret H (2006) An antagonistic function for *Arabidopsis* DCL2 in development and a new function for DCL4 in generating viral siRNAs. *The EMBO Journal* **25:** 3347-3356

Boutet S Vazquez F Liu J Beclin C Fagard M Gratias A Morel J B Crete P Chen X Vaucheret H (2003) *Arabidopsis* HEN1: a genetic link between endogenous miRNA controlling development and siRNA controlling transgene silencing and virus resistance. *Current Biology* **13:** 843-848

Brigneti G Martin-Hernandez AM Jin H Chen J Baulcombe DC Baker B Jones JD (2004) Virus-induced gene silencing in *Solanum* species. *The Plant Journal* **39:** 264-272

Brodersen P Voinnet O (2006) The diversity of RNA silencing pathways in plants. *Trends in Genetics* **22;** 268-280

Cao X Zhou P Zhang X Zhu S Zhong X Xiao Q Ding B and Li Y (2005) Identification of an RNA silencing suppressors from a plant double stranded RNA virus. *Journal of Virology* **79:** 13018-13027

Catalanotto C Azzalin G Macino G Gogoni C (2000) Gene silencing in worms and fungi. *Nature* **404:** 245

Chan SW Zilberman D Xie Z Johansen LK Carrington JC Jacobsen SE (2004) RNA silencing gene control *de novo* DNA methylation. *Science* **303:** 1336

Chapman EJ Prokhnevsky A I Gopinath K Dolja VV Carrington JC (2004) Viral RNA silencing suppressors inhibit the micro-RNA pathway at an interphase step. *Genes and Development* **18:** 1179-1186

Chen J Li WX Xie D Peng JR Ding SW (2004) Viral virulence protein suppresses RNA silencing-mediated defense but up regulates the role of microRNA in host gene regulation. *Plant Cell* **16:** 1302-1313

Chen Q Rehman S Smant G Jones JT (2005) Functional analysis of pathogenicity proteins of the potato cyst nematode *Globodera rostochiensis* using RNAi. *Molecular Plant-Microbe Interactions* **18:** 621-625

Chen S Hofius D Sonnewald U Bornke F (2003) Temporal and spatial control of gene silencing in transgenic plants by inducible expression od double stranded RNA. *The Plant Journal* **36:** 731-740

Chuang CF Meyerowtiz EM (2000) Specific and heritable genetic interference by double-stranded RNA in *Arabidopsis thaliana*. *Proceedings of the National Academy of Sciences USA* **97:** 4985-4990

Chung E Seong E Kim YC Chung EJ Oh SK Lee S Park JM Joung YH Choi D (2004) A method of high frequency virus induced gene silencing in chili pepper *Capsicum annuum* L. cv. *Bukang*). *Molecular Cell* **17:** 377-380

Cogoni C Macino G (1999) Homology dependent gene silencing in plants and fungi: a number of variations on the same theme. *Current Opinion in Microbiology* **2:** 657-662

Cogoni C Romano N Macino G (1994) Suppression of gene expression by homologous transgenes. *Antonie Leeuwenhoek International Journal of General Molecular Microbiology* **65:** 205-209

Constantin GD Krath BN MacFarlane SA Nicolaisen M Johansen IE Lund OS (2004) Virus-induced gene silencing as a tool for functional genomics in a legume species. *Plant Journal* **40:** 622-631

Covey SN Al-Kaff NS (2000) Plant DNA viruses and gene silencing. *Plant Molecular Biology* **43:** 307-322

Dallwitz MJ Zurcher EJ (1996) Plant viruses online. In: Brunt AA Crabtree K Dallwitz MJ Gibbs AJ Watson L Zurcher EJ (Eds) *Descriptions and Lists from the VIDE Database* CAB International UK 1484 pp.

Dalmay T Hamilton AJ Mueller E Baulcombe DC (2000) Potato virus X amplicons in Arabidopsis mediate genetic and epigenetic gene silencing. *The Plant Cell* **12:** 369-380

Dalmay T Hamilton AJ Rudd S Angell S Baulcombe DC (2000) An RNA-dependent RNA polymerase gene in *Arabidopsis* is required for posttranscriptional gene silencing mediated by a transgene but not by a virus. *Cell* **101:** 543-553

Dalmay T Horsefield R Braunstein TH Baulcombe DC (2001) SDE3 encodes an RNA helicase required for post-transcriptional gene silencing in *Arabidopsis*. *The EMBO Journal* **20:** 2069-2078

Davies WP (2003). An historical perspective from the Green Revolution to the gene revolution. Nutrition Reviews **61:** S124–S134

Davuluri G R van Tuinen A Mustilli A C Manfredonia A Newman R Burgess D Brummell D A King S R Palys J Uhlig J (2004). Manipulation of DET1 expression in tomato results in photomorphogenic phenotypes caused by post-transcriptional gene silencing. *Plant Journal* **40:** 344–354.

Davuluri G R van Tuinen A Fraser P D Manfredonia A Newman R Burgess D Brummell D A King S R Palys J Uhlig J (2005) Fruit-specific RNAi-mediated suppression of DET1 enhances carotenoid and flavonoid content in tomatoes. *Nature Biotechnology* **23:** 890–895.

de Bakker MD Raponi M Arndt GM (2002) RNA-meditaed gene silencing in non-pathogenic and pathogenic fungi. *Current Opinion in Microbiology* **5:** 323-329

de Boer SH (2003) Perspective on genetic engineering of agricultural crops for resistance to disease. *Canadian Journal of Plant Pathology* **25:** 10-20

Deleris A Gallego-Bartolome J Bao J Kasschau KD Carrington JC Voinnet O (2006) Hierarchical action and inhibition of plant Dicer-like proteins in antiviral defense. *Science* **313:** 68-71

Díaz-Pendón JA Ding SW (2008) Direct and Indirect Roles of Viral Suppressors of RNA Silencing in Pathogenesis. *Annual review of Phytopathology* **46:** 303-326

Dinesh-Kumar SP Anandalakshmi R Marathe R Schiff M Liu Y (2003) Virus-induced gene silencing. *Methods in Molecular Biology* **236:** 287-294

Ding XS Schneider WL Chaluvadi SR Rouf Mian RM Nelson RS (2006) Characterization of a Brome mosaic virus strain and its use as a vector for gene silencing in monocotyledonous hosts. *Molecular Plant Microbe Interaction* **19:** 1229-1239

Doench JG Petersen CP Sharp PA (2003) siRNAs can function as miRNAs. *Genes and Development* **17:** 438-442

Dorokhov YL Frolova OY Skurat EV Ivanov PA Gasanova TV Sheveleva AA Ravin NV Makinen KM Klimyuk VI Skryabin KG Gleba YY Atabekov JG (2006) A novel function for a ubiquitous plant enzyme pectin methylesterase: the enhancer of RNA silencing. *FEBS Letters* **580:** 3872-3878

Dugas DV Bartel B (2004) microRNA regulation of gene expression in plants. *Current Opinion in Plant Biology* **7:** 512-520

Dunoyer P Himber C Ruiz-Ferrer V Alioua A Voinnet O (2007) Intra- and intercellular RNA interference in *Arabidopsis thaliana* requires components of the microRNA and heterochromatic silencing pathways. *Nature Genetics* **39:** 848-856

Dunoyer P Himber C Voinnet O (2005) DICER-LIKE 4 is required for RNA interference and produces the 21-nucleotide small interfering RNA component of the plant cell-to-cell silencing signal. *Nature Genetics* **37:** 1356-1360

Dunoyer P Lecellier CH Parizitto EA Himber C Voinnet O (2004) Probing the microRNA and small interfering RNA pathways with virus-encoded suppressors of RNA silencing. *The Plant Cell* **16:** 1235-1250

Dykxhoorn DM Lieberman J (2005) The silent revolution: RNA interference as basic biology research tool and therapeutic. *Annual Review of Medicine* **56:** 401-423

Elbashir SM Lendeckel W Tuschl T (2001) RNA interference is mediated by 21 and 22-nucleotide RNAs. *Genes and Development* **15:** 188-200

Escobar MA Civerolo EL Summerfelt KR Dandekar AM (2001) RNAi-mediated oncogene silencing confers ressitance to crown gall tumorigenesis. *Proceedings of the National Academy of Sciences USA* **98:** 13437-13442

Fagard M Boutet S Morel JB Bellini C Vaucheret H (2000) AGO1 QDE-2 and RDE-1 are related proteins required for post-transcriptional gene silencing in plants quelling in fungi and RNA interference in animals. *Proceedings of the National Academy of Sciences USA* **97:** 11650-11654

Fairbairn DJ Cavalloro AS Bernard M Mahalinga-Iyer J Graham MW Botella JR (2007) Host-delivered RNAi: an effective strategy to silence genes in plant parasite nematodes. *Planta* **226:** 1525–1533

Faivre-Rampant O Gilroy EM Hrubikova K Hein I Millam S Loake GJ Birch P Taylor M Lacomme C (2004) Potato virus X-induced gene silencing in leaves and tubers of potato. *Plant Physiology* **134** 1308-1316

Falco S C Guida T Locke M Mauvais J Sanders C Ward R T Webber P (1995) Transgenic canola and soybean seeds with increased lysine. *Biotechnology* (NY) **13:** 577–582.

Fanelli E Di Vito M Jones JT De Giorgi C (2005) Analysis of chitin synthase function in a plant parasitic nematode *Meloidogyne artiellia* using RNAi. *Gene* **349:** 87-95

Fire A Xu S Montgomery MK Kostas SA Driver SE Mello CC (1998) Potent and specific genetic interference by double-stranded RNA in *Caenorhabditis elegans. Nature* **391:** 806-811

Fitxgerald A Van Kha JA Plummer KM (2004) Simultaneous silencing of multiple genes in the apple scab fungus *Venturia inaequalis* by expression of RNA with chimeric inverted repeats. *Fungal Genetics and Biology* **41:** 963-971

Fofana IB Sangare A Collier R Taylor C Fauquet CM (2004) A geminivirus-induced gene silencing system for gene function validation in cassava. *Plant Molecular Biology* **56:** 613-624

Frankard V Ghislain M and Jacobs M (1992) Two feedback-insensitive enzymes of the aspartate pathway in Nicotiana sylvestris. *Plant Physiology* **99:** 1285–1293.

Frederick Meins F Si-Ammour AJ Blevins T (2005) RNA silencing systems and their relevance to plant development. *Annual Review of Cell and Developmental Biology* **21:** 297-318

Fulci V Macino G (2007) Quelling: posttranscriptional gene silencing guided by small RNAs in Neurospora crassa. *Current Opinion in Microbiology* **10:** 191-203

Galili G Galili S Lewinsohn E and Tadmor Y (2002) Genetic molecular and genomic approaches to improve the value of plant foods and feeds. *Critical Reviews in Plant Sciences* **21:** 167–204

Gasciolli V Mallory AC Bartel DP Vaucheret H (2005) Partially redundant functions of Arabidopsis DICER-like enzymes and a role for DCL4 in producing trans-acting siRNAs. *Current Biology* **15:** 1494-1500

Gasciolli V Mallory AC Bartel DP Vaucheret H (2005) Partially Redundant Functions of DICER-like enzymes and a role for DCL4 in producing- acting siRNAs. *Current Biology* **15:** 1494 -1500

Gitlin L Karelsky S Andino R (2002) Short interference confers intracellular antiviral immunity in human cells. *Nature* **4:** 418-430

Godge MR Purkayastha A Dasgupta I Kumar PP (2008) Virus-induced gene silencing for functional analysis of selected genes. *Plant Cell Reporter* **27:** 209-219

Goldoni M Azzalin G Macino G Cogoni C (2004) Efficient gene silencing by expression of double stranded RNA in *Neurospora crassa*. *Fungal Genetics and Biology* **41:** 1016-1024

Gossele VV Fache II Meulewaeter F Cornelissen M Metzlaff M (2002) SVISS-a novel transient gene silencing system for gene function discovery and validation in tobacco. *The Plant Journal* **32:** 859-866

Gould B Kramer EM (2007) Virus-induced gene silencing as a tool for functional analyses in the emerging model plant *Aquilegia* (columbine Ranunculaceae). *BMC Plant Methods* **12:** 6

Grusak M A (2002) Phytochemicals in plants: Genomics-assisted plant improvement for nutritional and health benefits. *Current Opinion in Biotechnology* **13:** 508–511

Guang S Bochner AF Pavelec DM Burkhart KB Harding S Lachowiec J Kennedy S (2008) An argonaute transports siRNAs from the cytoplasm to the nucleus. *Science* **321:** 537-541

Guo HS Ding SW (2002) A Viral protein inhibits the long range signaling activity of the silencing signal. *EUROPEAN MOLECULAR BIOLOGICAL ORGAN Journal* **21:** 398-407

Guo HS Fei JF Xie Q Chua NH (2003) A chemical-regulated inducible RNAi system in plants. *The Plant Journal* **34:** 383-392

Hall TTM (2005) Structure and function of Argonaute proteins. *Structure* **13:** 1403-1408

Hamada W Spanu PD (1998) Co-suppression of the hydrophobin gene *Hcf-1* is correlated with antisense RNA biosynthesis in *Cladosporium fulvum*. *Molecular and General Genetics* **259:** 630-638

Hamilton A Voinnet O Chappell L Baulcombe DC (2002) Two classes of short interfering RNA in RNA silencing. *The EMBO Journal* **21:** 4671-793

Hammond SM Bernstein E Beach D Hannon GJ (2000) An RNA-directed nuclease mediates post-transcriptional gene silencing in *Drosophila* cells. *Nature* **404:** 293–296

Hammond TM Keller NP (2005) RNA silencing in *Aspergillus nidulans* is independent of RNA-dependent RNA polymerase. *Genetics* **169:** 607-617

Han MH Goud S Song L Fedoroff N (2004) The *Arabidopsis* double-stranded RNA-binding protien HYL1 plays a role in microRNA-mediated gene regulation. *Proceedings of the National Academy of Sciences USA* **101:** 1093-1098

Hannon GJ (2002) RNA interference. *Nature* **418:** 244-251

Herr AJ (2005) Pathways through the small RNA world of plants. *FEBS Letters* **579:** 5879-5888

Herr AJ Jensen MB Dalmay T Baulcombe DC (2005) RNA polymerase IV directs silencing of endogenous DNA. *Science* **308:** 118–120.

Herr AJ Molnar A Jones A Baulcombe DC (2006) Defective RNA processing enhances RNA silencing and influences flowering of *Arabidopsis*. *Proceedings of the National Academy of Sciences USA* **103:** 14994-15001

Hileman LC Drea S Martino G Litt A Irish VF (2005) Virus induced gene silencing is an effective tool for assaying gene function in the basal eudicot species *Papaver somniferum* (opium poppy). *The Plant Journal* **44:** 334-341

Hilly JM Liu Z (2007) An Overview of small RNAs. In: Bassett CL (Ed) *Regulation of Gene Expression in Plants* Springer-Verlag Berlin pp 123-147

Himber C Dunoyer P Moissiard G Ritzenthaler C Voinnet O (2003) Transitivity dependent and -independent cell-to-cell movement of RNA silencing. *The EMBO Journal* **22:** 4523-4533

Hoffman Y Aflalo C Zarka A Gutman J James TY Boussiba S (2008) Isolation and characterization of a novel chytrid species (phylum Blastocladiomycota) parasitic on the green alga *Haematococcus*. *Mycological Research* **112:** 70-81

Hoisington D Khairallah M Reeves T Ribaut J M Skovmand B Taba S and Warburton M (1999) Plant genetic resources: What can they contribute toward increased crop productivity? *Proceedings of the National Academy of Sciences of the USA* **96:** 5937–5943.

Holzberg S Brosio P Gross C Pogue GP (2002) Barley stripe mosaic virus-induced gene silencing in a monocot plant. *Plant Journal* **30:** 315-327

Hou H Qiu W (2003) A novel co-delivery system consisting of a tomato bushy stunt virus and a defective interfering RNA for studying gene silencing. *Journal of Virology Methods* **111:** 37-42

Huang G Allen R Davis EL Baum TJ Hussey RS (2006) Engineering broad root-knot resistance in transgenic plants by RNAi silencing of a conserved and essential root-knot nematode parasitism gene. *Proceedings of the National Academy of Sciences USA* **103:** 4302-14306

Hutvagner G Zamore PD (2002) A microRNA in a multiple-turnover RNAi enzyme complex. *Science* **297:** 2056-2060

Irvine DV Zaratiegui M Tolia NH Goto DB Chitwood DH Vaughn MW Joshua-Tor L Martienssen RA (2006) Argonaute slicing is required for heterochromatic silencing and spreading. *Science* **313:** 1134-1137

Jacque JM Triques K Stevenson M (2002) Modulation of HIV-1 replication by RNA interference. *Nature* **418:** 435-438

James C (2006) http://www.gmocompass.org/eng/agri_biotechnology/gmo

Johansen LK Carrington JC (2001) Silencing on the spot. Induction and suppression of RNA silencing in the Agrobacterium-mediated transient expression system. *Plant Physiology* **126:** 930-938

Jones-Rhoades MW Bartel DP (2004) Computational identification of plant microRNAs and their targets including a stress-induced miRNA. *Molecular Cell* **14:** 787-799

Jones-Rhoades MW Bartel DP Bartel B (2006) MicroRNAS and their regulatory roles in plants. *Annual Review of Plant Biology* **57:** 19–53.

Kadotani N Nakayashiki H Tosa Y Mayama S (2003) RNA silencing in the pathogenic fungus *Magnaporthe oryzae*. *Molecular Plant-Microbe Interaction* **16:** 769-776

Kanno T Aufsatz W Jaligot E Mette MF Matzke M Matžke AJ (2005) A SNF2-like protein facilitates dynamic control of DNA methylation. *EMBO Reporter* **6:** 649–655.

Karakas M (2008) RNA interference in plant parasitic nematodes. *African Journal of Biotechnology* **7:** 2530-2534

Karchi H Shaul O Galili G (1994) Lysine synthesis and catabolism are coordinately regulated during tobacco seed development. *Proceedings of the National Academy of Sciences of the USA* **91:** 2577–2581.

Kasschau KD Carrington JC (1998) A counter defensive strategy of plant viruses: suppression of posttranscriptional gene silencing. *Cell* **95:** 461-470

Kasschau KD Fahlgren N Chapman EJ Sullivan CM Cumbie JS Givan SA Carrington JC (2007) Genome-wide profiling and analysis of Arabidopsis siRNAs *PLoS Biology* **5:** e57.

Katiyar-Agarwal S Gao S Vivian-Smith A Jin H (2007) A novel class of bacteria-induced small RNAs in Arabidopsis. *Genes and Development* **21:** 3123–3134

Katiyar-Agarwal S Morgan R Dahlbeck D Borsani O Villegas JA Zhu J Staskawicz BJ Jin H (2006) A pathogen-inducible endogenous siRNA in plant immunity. *Proceedings of the National Academy of Sciences USA* **103:** 47-52

Kjemtrup S Sampson KS Peele CG Nguyen LV Conkling MA (1998) Gene silencing from plant DNA carried by a geminivirus. *Plant Journal* **14:** 91-100

Klahre U Crete P Leuenberger SA Iglesias VA Meins F (2002) High molecular weight RNAs and small interfering RNAs induce systemic post transcriptional gene silencing in plants. *Proceedings of the National Academy of Sciences USA* **99:** 11981-11986

Kobayashi K Patricia Zambryski P (2007) RNA silencing and its cell-to-cell spread during Arabidopsis embryogenesis. *The Plant Journal* **50:** 597 - 604

Kumagai MH Donson J della-Cioppa G Harvey D Hanley K Grill LK (1995) Cytoplasmic inhibition of carotenoid biosynthesis with virus-derived RNA. *Proceedings of the National Academy of Sciences USA* **92:** 1679-1683

Kurihara Y Takashi Y Watanabe Y (2006) The interaction between DCL1 and HYL1 is important for efficient and precise processing of pri-miRNA in plant microRNA biogenesis. *RNA* **12:** 206-212

Larkin P J Harrigan G G (2007) Opportunities and surprises in crops modified by transgenic technology: Metabolic engineering of benzylisoquinoline alkaloid gossypol and lysine biosynthetic pathways. *Metabolomics* **3:** 371-382

Lau NC Lim LP Weinstein EG Bartel DP (2001) An abundant class of tiny RNAs with probable regulatory roles in *Caenorabditis elegans*. *Science* **294:** 858-862

Lecellier CH Voinnet O (2004) RNA silencing: no mercy for viruses? *Immunology Review* **198:** 285-303

Lecoq H Ravelonandro M Wips-Scheibel C Monsion M Raccah B Dunez J (1993) Aphid transmission of an non-aphid transmissible strain of *Zucchini yellow mosaic potyvirus* from transgenic plants expressing the capsid protein of plum pox potyvirus. *Molecular Plant-Microbe Interactions* **6:** 403-406

Lee M (1998) Genome projects and gene pools: New germplasm for plant breeding? Proceedings of the *National Academy of Sciences of the USA* **95:** 2001–2004.

Levin I Frankel P Gilboa N Tanny S Lalazar A (2003) The tomato dark green mutation is a novel allele of the tomato homolog of the DEETIOLATED1 gene. *Theoretical and Applied Genetics* **106:** 454–460.

Lewis BP Shih I Jones-Rhoades MW Bartel DP Burge CB (2003) Prediction of mammalian microRNA targets. *Cell* **115:** 787-798

Li C Pontes O El-Shami M Henderson I Bernatavichute Y Chan S Lagrange T Pikaard C Jacobsen S (2006) An ARGONAUTE 4-containing nuclear processing center colocalized with cajal bodies in *Arabidopsis thaliana*. *Cell* **126:** 93 – 106

Li F Ding SW (2006) Virus counter defense: diverse strategies for evading the RNA silencing immunity. *Annual Review of Microbiology* **60:** 503-531

Li J Yang Z Yu B Liu J Chen X (2005) Methylation protects miRNAs and siRNAs from a 3_-end uridylation activity in Arabidopsis. *Current Biology* **15:** 1501-1507

Lilley CJ Goodchild SA Atkinson HJ Urwin PE (2005) Cloning and characterization of a *Heterodera glycines* minopeptidase cDNA. *International Journal of Parasitology* **35:** 1577-1585

Liu H Cottrell TR Pierini LM Goldman WE Doering TL (2002) RNA interference in the pathogenic fungus *Cryptococcus neoformans*. *Genetics* **160:** 463–470

Liu Q Singh S P Green A G (2002). High-stearic and high-oleic cottonseed oils produced by hairpin RNA-mediated post-transcriptional gene silencing. *Plant Physiology* **129:** 1732–1743.

Liu J Carmell MA Rivas FV Marsden CG Thomson JM (2004) Argonaute2 is the catalytic engine of mammalian RNAi. *Science* **305:** 1437-1441

Liu Y Schiff M Marathe R Dinesh-Kumar SP (2002b) Tobacco *Rar1 EDS1* and *NPR1/NIM1* like genes are required for *N* mediated resistance to tobacco mosaic virus. *The Plant Journal* **30:** 415-429

Liu YL Schiff M Dinesh-Kumar SP (2002a) Virus induced gene silencing in tomato. *The Plant Journal* **31:** 777-786

Llave C Kasschau KD Rector MA Carrington JC (2002) Endogenous and silencing associated small RNAs in plants. *The Plant Cell* **14:** 1605-1619

Lu C Federoff N (2000) A mutation in the *Arabidopsis* HYL1 gene encoding a dsRNA binding protein affects responses to abscisic acid auxin and cytokinin. *The Plant Cell* **12:** 2351-2366

Lu C S.S. Tej S. Luo C.D. Haudenschild B.C. Meyers P.J. Green (2005) Elucidation of the small RNA component of the transcriptome *Science* **309:** 1567–1569.

Lu R Martin-Hernandez AM Peart JR Malcuit I Baulcombe DC (2003) Virus induced gene silencing in plants. *Methods* **30:** 296-303

MacFarlane SA Popovich AH (2000) Efficient expression of foreign proteins in roots from tobravirus vectors. *Virology* **267:** 29-35

Mallory AC Parks G Endres MW Baulcombe D Bcwman LH (2002) The amplicon-plus system for high-level expression of transgenes in plants. *Nature Biotechnogy* **20:** 622-625

Maskarinec G Takata Y Pagano I Carlin L Goodman M T Le Marchand L Nomura A M Wilkens L R Kolonel L N (2006) Trends and dietary determinants of overweight and obesity in a multiethnic population. *Obesity (Silver Spring)* **14:** 717–726

Matzke M Aufsatz W Kanno T Daxinger L Papp I Mette MF Matzke JM (2004) Genetic analysis of RNA-mediated transcriptional gene silencing. *Biochimica et Biophysica Acta* **1677:** 129-141

Mazur B Krebbers E Tingey S (1999) Gene discovery and product development for grain quality traits. *Science* **285:** 372–375

McManus MT Sharp PA (2002) Gene Silencing in mammals by small interfering RNAs. *Nature Review Genetics* **3:** 737-747

Merai Z Kerenyi Z Molnar A Barta E Valcozi A Bistray G Havelda Z Burgyan J Silhavy D (2005) Aureusvirus P14 is an efficient RNA silencing suppressor that binds double stranded RNAs without size specificity. *Journal of Virology* **79:** 7217-7226

Milhavet O Gary DS Mattson MP (2003) RNA Interference in Biology and Medicine. *Pharmacological Reviews* **55:** 629–648

Mlotshwas S Voinnet O Mette MF Matzke M Vaucheret H Ding SW Pruss G Vance GB (2002) RNA silencing and its mobile silencing signal. *The Plant Cell* **14:** S289-301

Mocellin S Provenzano M (2004) RNA interference: learning gene knock-down from cell physiology. *Journal of Translational Medicine* **2:** 1-6

Morel JB Godon C Mourrain P Beclin C Boutet C Feuerbach F Proux F Vaucheret H (2002) Fertile hypomorphic ARGONAUTE (ago1) mutants impaired in post-transcriptional gene silencing and virus resistance. *Plant Cell* **14:** 629-639

Moss EG Tang L (2003) Conservation of the heterochronic regulator Lin-28 its developmental expression and microRNA complementary sites. *Developmental Biology* **258:** 432-442

Mourrain P Beclin C Elmayan T Feuerbach F Godon C Morel JB Jouette D Lacombe AM Nikic S Picault N Remoue K Sanial M Vo TA Vaucheret H (2000) *Arabidopsis SGS2* and *SGS3* genes are required for posttranscriptional gene silencing and natural virus resistance. *Cell* **101:** 533–542

Mustilli A C Fenzi F Ciliento R Alfano F Bowler C (1999) Phenotype of the tomato high pigment-2 mutant is caused by a mutation in the tomato homolog of DEETIOLATED1. *Plant Cell* **11:** 145–157.

Nakayashiki H (2005) RNA silencing in fungi: mechanisms and applications. *FEBS Letters* **579:** 5950-5970

Nakayashiki H Hanada S Nguyen BQ Kadotani N Tosa Y Mayama S (2005) RNA silencing as a tool for exploring gene function in ascomycete fungi. *Fungi Genetics and Biology* **42:** 275-283

Napoli C Lemieux C Jorgensen R (1990) Introduction of a chimeric chalcone synthase gene into petunia results in reversible co-suppression of homologous genes in *trans*. *The Plant Cell* **2:** 279-289

Navarro L Dunoyer P Jay F Arnold B Dharmasini N Estelle M Vionnet O Jones JD (2006) A plant miRNA contributes to antibacterial resistance by repressing auxin signalling. *Science* **312:** 436–439

Naylor M Reeves J Cooper JI Edwards ML Wang H (2005) Construction and properties of a gene-silencing vector based on Poplar mosaic virus (genus *Carlavirus*). *Journal of Virology Methods* **124:** 27-36

Negrutiu I Cattoir-Reynearts A Verbruggen I Jacobs M. (1984) Lysine overproducer mutants with an altered dihydrodipicolinate synthase from protoplast culture of *Nicotiana sylvestris* (Spegazzini and Comes). *Theoretical and Applied Genetics* **68:** 11–20

Niu QW Lin SS Reyes JL Chen KC Wu HW Yeh SD Chua NH (2006) Expression of artificial microRNAs in transgenic *Arabidopsis thaliana* confers virus resistance. *Nature Biotechnology* **24:** 1420-1428

Novina CD Murray MF Dykxhoorn DM Beresford PJ Riess J Lee SK Collman RG Lieberman J Shanker P Sharp PA (2002) siRNA-directed inhibition of HIV-1 infection. *Nature Mediterranean* **8:** 681-686

Nykänen A Haley B Zamore PD (2001) ATP requirements and small interfering RNA structure in the RNA interference pathway. *Cell* **107:** 309-321

Omarov RT Ciomperlik JJ Scholthof HB (2007) RNAi-associated ssRNA-specific ribonucleases in *Tombusvirus* P19 mutant-infected plants and evidence for a discrete srRNA-containing effector complex. *Proceedings of the National Academy of Sciences USA* **104:** 1714-1719

Palmer KE Rybicki EP (2001) Investigation of the potential of maize streak virus to act as an infectious gene vector in maize plants. *Archives of Virology* **146:** 1089-1104

Park M Y Wu G Gonzalez-Sulser A Vaucheret H Poethig RS (2005) Nuclear processing and export of microRNAs in *Arabidopsis. Proceeding of National Academy of Sciences. USA.* **102:** 3691-3696.

Park W Li J Song R Messing J Chen X (2002) Carpel factory a Dicer homolog and HEN1 a novel protein act in miRNA metabolism in *Arabidopsis thaliana. Current Biology* **12:** 1484-1495

Pasquinelli AE Reinhart BJ Slack F Martindale MQ Kuroda MI Maller B Hayward DC Ball EE Degnan B Müller P Spring J Srinivasan A Fishman M Finnerty J Corbo J Levine M Leahy P Davidson E Ruvkun G (2000) Conservation of the sequence and temporal expression of *let-7* heterochronic regulatory RNA. *Nature* **408:** 86-89

Peele C Jordan CV Muangsan N Turnage M Egelkrout E Eagle P Hanley-Bowdoin L Robertson D (2001) Silencing of a meristematic gene using geminivirus-derived vectors. *Plant Journal* **27:** 357-366

Peragine A Yoshikawa M Wu G Albrecht HL Poethig RS (2004) SGS3 and SGS2/SDE1/RDR6 are required for juvenile development and the production of trans-acting siRNAs in *Arabidopsis. Genes and Development* **18:** 2368-2379

Pham JW Sontheimer EJ (2005) Molecular requirements for RNA-induced silencing complex assembly in the *Drosophila* RNA interference pathway. *Journal of Biological Chemistry* **280:** 39278-39283

Pickford AS Cogoni C (2003) RNA-mediated gene silencing. *Cellular and Molecular Life Science* **60:** 871-882

Pogue GP Lindbo JA Garger SJ Fitzmaurice WP (2002) Making an ally from an enemy: plant virology and the new agriculture. *Annual Review of Phytopathology* **40:** 45-74

Pontes O Li CF Nunes PC Haag J Ream T Vitins A Jacobsen SE. Pikaard CS (2006) The Arabidopsis chromatin-modifying nuclear siRNA pathway involves a nucleolar RNA processing center *Cell* **126:** 79-92.

Pooggin M Shivaprasad PV Veluthambi K Hohn T (2003) RNAi targetting of DNA viruses. *Nature Biotechnology* **21:** 131-32

Qi Y Denli AM Hannon GJ (2005) Biochemical specialization within Arabidopsis RNA silencing pathways. *Molecular Cell* **19:** 421-428

Qu J Ye J Fang RX (2007) Artificial microRNA-mediated virus resistance in plants. *Journal of Virology* **81:** 6690-6699

Rand TA Petersen S Du F Wang X (2005) Argonaute2 cleaves the anti-guide strand of siRNA during RISC activation. *Cell* **123:** 621-629

Ratcliff F Martin-Hernandez AM Baulcombe DC (2001) Tobacco rattle virus as a vector for analysis of gene functions by silencing. *Plant Journal* **25:** 237-245

Ratcliff FG MacFarlane SA Baulcombe DC (1999) Gene silencing without DNA. RNA-mediated cross-protection between viruses. *Plant Cell* **11:** 1207-1216

Regina A Bird A Topping D Bowden S Freeman J Barsby T Kosar-Hashemi B Li Z Rahman S Morell M (2006) High-amylose wheat generated by RNA interference improves indices of large-bowel health in rats. *Proceedings of the National Academy of Sciences of the USA* **103:** 3546–3551

Rhilla R Singh RL Singh US Singh R Duveiller E Singh HB (2001) Recent advances in the management of plant diseases using chemical. *Indian Journal of Plant Pathology* **19:** 1-23

Rhoades MW Reinhart BJ Lim LP Burge CB Bartel B Bartel DP (2002) Prediction of plant microRNA targets. *Cell* **110:** 513-520

Ronemus M Vaughn MW Martienssen RA (2006) MicroRNA-targeted and small interfering RNAmediated mRNA degradation is regulated by argonaute dicer and RNA-dependent RNA polymerase in *Arabidopsis*. *Plant Cell* **18:** 1559-1574

Rosso MN Dubrana MP Cimbolini N Jaubert S Abad P (2005) Application of RNA interference to root-knot nematode genes encoding esophageal gland proteins. *Molecular Plant-Microbe Interactions* **18:** 615-620

Ruiz MT Voinnet O Baulcombe DC (1998) Initiation and maintenance of virus-induced gene silencing. *Plant Cell* **10:** 937-946

Ruiz-Ferrer V Voinnet O (2007) Viral suppression of RNA silencing: 2b wins the golden fleece by defeating argonaute. *Bioassays* **29:** 319-323

Schauer SE Jacobsen SE Meinke DW Ray A (2002) DICER-LIKE1: blind men and elephants in *Arabidopsis* development. *Trends in Plant Science* **7:** 487- 491

Schwach F Vaistij FE Jones L Baulcombe DC (2005) An RNA-dependent RNA polymerase prevents meristem invasion by potato virus X and is required for the activity but not the production of a systemic silencing signal. *Plant Physiology* **138:** 1842-1852

Schwarz DS Hutvagner G Du T Xu Z Aronin N and Zamore PD (2003) Asymmetry in the assembly of the RNAi enzyme complex. *Cell* **115:** 199–208.

Schweizer P Pokorny J Schulze-Lefert P Dudler R (2000) Double stranded RNA interference with gene functions at the single cell in cereals. *The Plant Journal* **24:** 895-903

Scofield SR Huang L Brandt AS Gill BS (2005) Development of a virus-induced gene silencing system for hexaploid wheat and its use in functional analysis of the Lr21-mediated leaf rust resistance pathway. *Plant Physiology* **138:** 2165-2173

Segal G Song R Messing J (2003) A new opaque variant of maize by a single dominant RNA-interference-inducing transgene. *Genetics* **165:** 387–397

Segers GC Hamada W Oliver RP Spanu PD (1999) Isolation and characteristaion of five different hydrophobin-encoding cDNA from the fungal tomato pathogen *Cladosporium fulvum*. *Molecular and General Genetics* **261:** 644-652

Shahinul Islam SM Miyazaki T Tanno F Itoh K (2005) Dissection of gene function by RNA silencing. *Plant Biotechnology* **22:** 443–446

Shingles J Lilley CJ Atkinson HJ Urwin PE (2007) *Meloidogyne incognita*: molecular and biochemical characterization of a cathepsin L cysteine proteinase and the effect on parasitism following RNAi. *Experimental Parasitology* **115:** 114-120

Sigova A Rhind N and Zamore PD (2004) A single Argonaute protein mediates both transcriptional and posttranscriptional silencing in *Schizosaccharomyces pombe*. *Genes and Development*. **18:** 2359-2367.

Sledz AC Williams BRG (2005) RNA interference in biology and disease. *Blood* **106:** 781-794

Smith N Singh S Wang MB Stoutjesdijk P Green A Waterhouse PM (2000) Total silencing by intron-spliced hairpin RNAs. *Nature* **407:** 319-320

Song JJ Joshua-Tor L (2006) Argonaute and RNA – getting into the groove. *Current Opinion in Structural Biology* **16:** 5-11

Spanu P (1997) HCf-1 a hydrophobin from the tomato pathogen *Cladosporium fulvum*. *Gene* **93:** 89-96

Steeves RM Todd TC Essig JS Trick HN (2006) Transgenic soybeans expressing siRNAs specific to a major sperm protein gene suppress *Heterodera glycines* reproduction. *Functional Plant Biology* **33:** 991-999

Storici F (2008) RNA- mediated DNA modification and RNA- template DNA repair. *Current Opinion in Molecular Therapeutics* **10:** 224-230

Sunilkumar G Campbell L M Puckhaber L Stipanovic R D Rathor K S (2006) Engineering cottonseed for use in human nutrition by tissue-specific reduction of toxic gossypol. *Proceedings of the National Academy of Sciences of the* USA **103:** 18054–18059

Sunkar R Zhu JK (2004) Novel and stress-regulated microRNAs and other small RNAs from *Arabidopsis*. *Plant Cell* **16:** 2001-2019

Takeda A Tsukuda M Mizumoto H Okamoto K Kaido M Mise K Okuno T (2005) A pant RNA virus suppressor RNA silencing through RNA replication. *The EMBO Journal* **24:** 3147-3157

Talbot NJ Kershaw MJ Wakley GE de Vries OMH Wessels JGH Hamer JE (1996) MPG1 encodes a fungal hydrophobin involved in surface interactions during infection-related development of *Magnaporthe grisea*. *Plant Cell* **8** :985-999

Tang G Reinhart BJ Bartel D Zamore PD (2003) A biochemical framework for RNA silencing in plants. *Genes and Development* **17:** 49-63

Tang W Weidner DA Hu BY Newton RJ Hu X (2006) Efficient delivery of small interfering RNA to plant cells by a nanosecond pulsed laser-induced wave for post transcriptional gene silencing. *Plant Science* **171:** 375-81

Tao X Zhou X (2004) A modified viral satellite DNA that suppresses gene expression in plants. *The Plant Journal* **38:** 850-860

Tenllado F Martinez-Garcia B Vargas M Diaz-Ruiz J R (2003) Crude extracts of bacterially expressed dsRNA can be used to protect plants against virus infection. *BMC Biotechnology* **3:** 3-14

Teycheney PY Tepfer M (2001) Virus specific spatial differences in the interference with silencing of the *chs-A* gene in non-transgenic petunia. *Journal of General Virology* **82:** 1239-1243

Timmermans M Das O Messing J (1994) Geminivirus and their uses as extrachromosomeal replicons. *Annual Review of Plant Physiology* **45:** 79-112

Tomari Y Du T Haley B Schwarz DS Bennett R (2004) RISC assembly defects in the Drosophila RNAi mutant armitage. *Cell* **116:** 831-841

Tomari Y Zamore PD (2005) Perspective: machines for RNAi. *Genes and Development* **19**: 517-529

Turnage MA Muangsan N Peele CG Robertson D (2002) Geminivirus-based vectors for gene silencing in *Arabidopsis. The Plant Journal* **30:** 107-117

Ullu E Djikeng A Shi H Tschudi C (2002) RNA interference: advances and questions. *Philosophical Transactions of the Royal Society of London British Biological Science* **29:** 65-70

Urwin PE Lilley CJ Atkinson HJ (2002) Ingestion of double-stranded RNA by pre-parasitic juvenile cyst nematodes leads to RNA interference. *Molecular Plant-Microbe Interactions* **15:** 747-752

Vaistij FE Jones L Baulcombe DC (2002) Spreading of RNA targeting and DNA methylation in RNA silencing requires transcription of the target gene and a putative RNA dependent RNA polymerase. *Plant Cell* **14:** 857-867

Vanitharani R Chellappan P Fauquet CM (2003) Short interfering RNA-mediated interference of gene expression and viral DNA accumulation in cultured plant cells. *Proceedings of the National Academy of Sciences USA* **100:** 9632-9636

Vaucheret H (2006) Post-transcriptional small RNA pathways in plants: mechanisms and regulations. *Genes and Development* **20:** 759–771.

Vaucheret H Vazquez F Crete P Bartel DP (2004) The action of ARGONAUTE1 in the miRNA pathway and its regulation by the miRNA pathway are crucial f or plant development. *Genes and Development* **18:** 187-1197

Vazquez F Gasciolli V Crete P Vaucheret H (2004) The nuclear dsRNA binding protein HYL1 is required for microRNA accumulation and plant development but not posttranscriptional transgene silencing. *Current Biology* **14:** 346-351

Vazquez F Vaucheret H Rajagopalan R Lepers C Gasciolli V Mallory AC Hilbert JL Bartel DP Crete P (2004) Endogenous transacting siRNAs regulate the accumulation of Arabidopsis mRNAs. *Molecular Cell* **16:** 69-79

Verdel A Jia S Gerber S Sugiyama T Gygi S Grewal SI and Moazed D (2004) RNAi-mediated targeting of heterochromatin by the RITS complex. *Science* **303:** 672–676.

Voinnet O (2001) RNA silencing as a plant immune system against viruses. *Trends in Genetics* **17:** 449-459

Voinnet O (2004) Shaping small RNAs in plants by gene duplication. *Nature Genetics* **36:** 1245-1246

Voinnet O (2005) Non-cell autonomous RNA silencing. *FEBS Letters* **579:** 5858-5871

Voinnet O Lederer C Baulcomb DC (2000) A Viral movement protein prevents spread of the gene silencing signal in *Nicotiana benthamiana*. *Cell* **103:** 157-167

Volpe TA Kidner C Hall IM Teng G Grewal SI Martienssen RA (2002) Regulation of heterochromatin silencing and histone H3 lysine-9 methylation by RNAi. *Science* **29:** 1833-1837

Wadhwani, C Malik C P Kaur B (2008) RNA interference: Small and Silent In Biotechnology-Perspectives and Prospects (Eds. Malik, C P, Wadhwani, C and kaur, B). MD Publications, New Delhi Pp187-254

Waterhouse PM Graham MW Wang MB (1998) Virus resistance and gene silencing in plants can be induced by simultaneous expression of sense and antisense RNA. *Proceedings of the National Academy of Sciences USA* **95:** 13959-13964

Waterhouse PM Halliwell CA (2003) Exploring plant genomes by RNA induced gene silencing. *Nature Review Genetics* **4:** 29-38

Waterhouse P M Wang M B Lough T (2001) Gene silencing as an adaptive defense against viruses. *Nature* **411:** 834-842

Wesley SV Helliwell CA Smith NA Wang MB Rouse DT Liu Q Gooding PS Singh SP Abbott D Stoutjesdijk PA Robinson SP Gleave AP Green AG Waterhouse P (2001) Construct design for efficient effective and high-throughput gene silencing in plants. *Plant Journal* **27:** 581-590

Wilda M Fuchs U Wossmann W Borkhardt A (2002) Killing of leukemic cells with a BCR/ABL fusion gene by RNA interference (RNAi). *Oncogene* **21:** 5716-5724

Williams C L (1995) Importance of dietary fiber in childhood. *Journal of the American Dietetic Association* **95:** 1140–1146

Wu F Yu L Cao W Mao Y Liu Z He Y (2007) The N-terminal double-stranded RNA binding domains of *Arabidopsis* HYPONASTIC LEAVES1 are sufficient for pre-microRNA processing. *Plant Cell* **19:** 914-925

Xie Z Allen E Fahlgren N Calamar A Givan SA Carrington JC (2005) Expression of Arabidopsis MIRNA genes. *Plant Physiology* **138:** 2145-2154.

Xie Z Allen E Wilken A Carrington JC (2005) DICER-LIKE 4 functions in *trans*-acting small interfering RNA biogenesis and vegetative phase change in *Arabidopsis thaliana*. *Proceedings of the National Academy of Sciences USA* **102:** 12984-12989

Xie Z Fan B Chen C Chen Z (2001) An important role of an inducible RNA-dependent RNA polymerase in plant antiviral defense. *Proceedings of the National Academy of Sciences USA* **98:** 6516-6521

Xie Z Johansen LK Gustafson AM Kasschau KD Lellis AD Zilberman D Jacobsen SE Carrington JC (2004) Genetic and functional diversification of small RNA pathways in plants. *PLoS Biol*ogy **2:** E104

Xie Z Qi X (2008) Diverse small RNA-directed silencing pathways in plants. Biochimica et Biophysica Acta (in Press)

Yang Z Ebright YW Yu B Chen X (2006) HEN1 recognizes 21-24 nt small RNA duplexes and deposits a methyl group onto the 2_ OH of the 3′ terminal nucleotide. *Nucleic Acids Research* **34:** 667-675

Yoo BC Kragler F Varkonyi-Gasic E Haywood V Archer-Evans S (2004) A systemic small RNA signaling system in plants. *Plant Cell* **16:** 1979-2000

Yu B Yang Z Li J Minakhina S Yang M (2005) Methylation as a crucial step in plant microRNA biogenesis. *Science* **307:** 932-935

Yu D Fan B MacFarlene SA Chen Z (2003) Analysis of the Involvement of an inducible *Arabidopsis* RNA-dependent RNA polymerase in antiviral defense. *Molecular Plant-Microbe Interactions* **16:** 206-216

Zheng X Zhu J Kapoor A Zhu JK (2007) Role of Arabidopsis AGO6 in siRNA accumulation DNA methylation and transcriptional gene silencing. *EMBO Journal* **26:** 1691-1701.

Zhu X Galili G (2003) Increased lysine synthesis coupled with a knockout of its catabolism synergistically boosts lysine content and also transregulates the metabolism of other amino acids in Arabidopsis seeds. *Plant Cell* **15:** 845–853.

Zhu X Galili G (2004) Lysine metabolism is concurrently regulated by synthesis and catabolism in both reproductive and vegetative tissues. *Plant Physiology* **135:** 129–136

Zilberman D Cao X Jacobsen SE (2003) *ARGONAUTE4* control of locus-specific siRNA accumulation and DNA and histone methylation. *Science* **299:** 716-719

Zilberman D Cao X Johansen LX Xie Z Carrington JC Jacobsen SE (2004) Role of Arabidopsis ARGONAUTE4 in RNA-directed DNA methylation triggered by inverted repeats. *Current Biology* **14:** 1214-1220.

4

BACTERIAL BIOSORPTION OF HEAVY METALS

SEDKY H. A. HASSAN, YASSER MAHMOUD AWAD, MOHAMMAD HUMAYUN KABIR, SANG EUN OH AND JIN HO JOO

Recently, water as an essential resource for all living organisms, is shown to confront high risks caused by heavy metals released from the industrial wastes runoff to water streams and river. So, realistic science approach has been directed towards biosorption of heavy metals from waste water and aqueous solution. Biosorption using biosorbents (*i.e.*, living or dead bacterial biomass, or other low cost biomaterials) has consider as a worldwide driven research in the field of water environment clean up technologies. Biosorbents have the abilities to bind metal pollutants and therefore, can remove the pollutants from water. This article report biosorption recent information and biosorbents efficiencies in heavy metals removal from contaminated compared to costly conventional methods water. In addition, brief reviewed of metal biosorption and bioaccumulation, biosorption mechanisms, factors affecting the biosorption and effect of pretreatment in addition to the future of biosorption will be discussed in the present review.

INTRODUCTION

Due to increasing concern for environmental safety, cooperative efforts are committed to clean up contaminated water not only for food safety which is recognized as of global interest but also for human health. Globally, it has been recognized that heavy metals pollution may be one of the most serious challenges confronting water as an essential natural resource for all living organisms.

Environmental pollution caused by metals has accelerated dramatically since the beginning of the industrial revolution (Nriagu, 1990). In South Korea, mine residues and waste have been abandoned on hillsides and discharged by rain directly to stream water and adjacent agricultural soils (Chen *et al.* 2000; Jung and Thornton, 1997). The heavy metals run off causes detrimental effects on surrounding soil and water bodies, and is menace to human health with outbreaks of diseases as itai-itai due to Cd accumulation in rice (Ok *et al.* 2007). Conventional methods as ion exchange, chemical precipitation, reverse osmosis and evaporative recovery are very expensive and/or often inefficient for the removal of heavy metal from the aqueous solution. Therefore, it is necessary to develop novel technologies for metal removal (Chong and Volesky, 1995).

Currently, using microbial biomass as biosorbents for, heavy metal removal from industrial waste streams has been considered as a potential alternative technology (Costa and Leite, 1991). Biosorption can separate dissolved metals out of diluted solution in high performance and rapid intrinsic kinetics. Therefore, it is an ideal treatment of high volume complex wastewater with low level (Edyvean *et al.*, 1997). Biosorption is also based on the use of sorbent (solid phase) which derived from different kinds of biomaterials or biomass. Moreover, it is obvious that natural ion exchange of biosorbents increased the adsorption capacity of metal on those biomaterials (Volesky, 2003). However, many researchers have worked on low cost adsorbents with high adsorption capacity as chitosan, zeolites, waste slurry and lignin. In addition, these inexpensive adsorbents show high removal capacities for each metal compared to activated carbon (Babel and Kurniawan 2003).

CONVENTIONAL TECHNOLOGIES FOR HEAVY METAL REMEDIATION

In the last few decades, it has become clear that conventional treatments of polluted sites is too expensive. So we should rely on other low-cost technologies with high efficiency to remediate heavy metals from contaminated water. However, in this review, different chemical and physical treatments will be discussed as conventional methods to remove heavy metals from industrial wastewater.

Conventional methods include chemical precipitation, chemical oxidation or reduction, ion exchange, filtration, electrochemical treatment, reverse osmosis, membrane technologies and evaporation recovery (Atkinson *et al.* 1998; Crini, 2006).

Regarding chemical precipitation is commonly used for heavy metal removal from water via addition of anion for metals participating in the form of suspended solid particles, and then these particles are removed by settling, filtration or centrifugation. However, precipitation lacks specificity and is ineffective in heavy metal ions at low concentration (Gray, 1999). In this concern ferrous sulphate, ferric chloride, aluminum sulphate and polyaluminium chloride are also used as coagulants in the pretreatment of currency printing ink wastewater (Nandy *et al.* 2003). Also, Electro-coagulation is also effective in removing dissolved metals from wastewater using an electrical current by neutralizing ions and then, metals become destabilized and precipitate in a stable form (Gray, 1999).

Ion exchange is an adsorption process that has the capacity to achieve ppb levels of heavy metals in aqueous solution through surface ions exchange of an ion exchange resin. However, ion exchange method is nonselective and is highly sensitive to solution pH and also, it is expensive compared to the other methods (Gray, 1999). Furthermore, membrane technology is used to separate heavy metals from wastewater without generating any pollution load. In this part, commercial nanofiltration membrane is used for cadmium and nickel removal from aqueous solutions (Murthy and Chaudhari, 2009). Moreover, nanofiltration gained its popularity in metal separation/rejection by reverse osmosis (RO) and ultrafiltration (UF) (Lamba *et al* 2002; Qin *et al* 2004). However, the cost of operation, maintenance in system application and necessary pretreatment is high.

Reverse osmosis and electro-dialysis involves the use of semi-permeable membranes for the recovery of metal ions from diluted wastewater. Whilst, in electro-dialysis, the selective membranes as alternation of cation and anion membranes are fitted between the electrodes in electrolytic cells, and therefore, the associated ion migrates under continuous electrical current, allowing the recovery of metals. On the other hand, fouling of membrane by slightly soluble components in solution and in wastewater which had high suspended solids are obviously disadvantages of this method (Huang and Koseoglu 1993).

Concerning adsorption process, metal ions are transferred and physically or chemically attached to the surface of a solid adsorbent (Volesky, 2003). Activated carbon and aluminosilicates are commonly used as adsorbents for metal removal. Also, a combination of activated alumina adsorption and micro filtration was used to treat contaminated water samples from Deloro water treatment plant containing 100 ppm arsenic (Motrazavi *et al.* 1993). Adsorption is an effective method for removal of low-level metal ions. However the regeneration might be expensive and could influence the performance of the adsorbent.

Unquestionable, activated carbon has been widely used in wastewater treatment but remains an expensive method whilst the greater cost as higher as its quality (Babel and Kurniawan 2003). On the other hand, these treatment methods may be ineffective or extremely expensive; mostly when the metals in solution ranged 1–100 mg/l (Nourbakhsh *et al* 1994).

In recent years, research is focused on developing eco-friendly and effective methods using biological treatments as biosorption to remove heavy metals from waste water in comparison to conventional treatment which not environmentally recommended and usually dependent on the concentration of the waste (Vijayaraghavan and Yun, 2008).

METAL UPTAKE BY MICROORGANISMS

Removal of heavy metal ions from polluted wastewaters by bacterial biomass has been studied extensively (Hassan *et al.* 2009 ; Kao W-C *et al.* 2008; Tuzen *et al.* 2008 ; Akhtar *et al.* 2007; Kadukova and Vircikova, 2005; Chaisuksant, 2003; Leung *et al.* 2001; Veglio *et al.* 1997; Akthar and Mohan, 1995; Fourest *et al.* 1994; Volesky, 2003, 1994, 1990). Microorganisms have the ability to remove heavy metals by two process; (1) biosorption "an energy-independent binding of metals to cell wall", and (2) bioaccumulation "an energy-dependent process of metal uptake into the cells" (Karna *et al.* 1999: Gadd, 1992).

Both living and non-living microbial biomass of bacteria, fungi and algae have been used in removing toxic metal ions (Kumar *et al.* 2008; Karna *et al.* 1999; Kumar *et al.* 1998; Volesky, 1990, 1994).

Metal removal by microorganisms is a complex process that depends on the chemistry of the metal ions, cell wall composition of microorganisms, cell physiology, and physicochemical factors such as pH, temperature, time, ionic strength and metal concentration (Hassan *et al* 2009; Gabr *et al* 2008; Kao W-C et *al.*2008; Tuzen *et al* 2008 ; Godlewska-Zylkiewicz 2006; Mapoleto *et al* 2005 ; Chaisuksant 2003; Goyal *et al* 2003 ; Volesky 2003; Klimmek *et al* 2001; Leung *et al* 2001). Metabolism-independent process of metal binding sites to the cell walls of bacterial biomass and external surfaces is the main mechanism present in the case of non-living biomass. Metabolism-independent or biosorption essentially involves adsorption process such as ionic and physicochemical adsorption. A variety of functional groups located on the bacterial cell walls are known to be included in metal biosorption. These involved carboxyl, amine, hydroxyl, phosphate and sulfhydryl groups. The removal of heavy metals by living and non-living biosorbents have been studied (Hassan *et al* 2009; Gabr *et al* 2008; Kadukova and Vircikova 2005; Mapoleto *et al* 2005; Chaisuksant 2003; Pardo *et al* 2003; Goyal *et al* 2003; Volesky 2003; Veglio *et al* 1997; Fourest *et al* 1994).

Although, metal biosorption by bacterial biomass occurs through complexation, coordination, physical adsorption, chelation, ion exchange, inorganic precipitation and/or a combination of these process (Akhtar *et al* 2007; Chen *et al* 2005; Kratochvil and Volesky 1998; Tsezos *et al.* 1995). Such process includes the active participation of several anionic ligands present on the biomass, like phosphoryl, carboxyl, carbonyl, sulfydryl and hydroxyl groups, was used to immobilize metal ions (Volesky, 2003). Also, the ability of any biosorbent depends on biomass characteristics, physicochemical properties of the heavy metals and the environmental conditions i.e., the solution pH, temperature, and interaction with other ions etc (Volesky 2003).

BIOACCUMULATION (METAL UPTAKE BY LIVING CELLS)

Bioaccumulation is an intracellular metal accumulation process by the living cells. This process involves metal binding on intracellular compounds, metal binding proteins, intracellular precipitation, methylation and other mechanisms. The uptake process usually involves adsorption of heavy metal ions at the bacterial cell wall or cell membrane through interactions with various functional groups and/or transport into the cell with subsequent transformation.

The term bioaccumulation describes an active process by living cells in which the removal of heavy metals depends on the metabolic activity of a living organism. It has been proved that, in some cases, that growing cells are able to remove metals continuously through internal detoxification mechanisms (Malik, 2004).

The activation energy for bioaccumulation, which is about 63 KJ/mol, corresponds to a biochemical process. This process involves transport of the metal across the cell membrane and its subsequent transformation (Davis *et al.* 2003; Veglio and Beolchini

1997; Blackwell *et al.* 1995). Most mechanisms of metal transport appear to rely on the electrochemical proton gradient across the cell membrane, which has a chemical and electrical potential, both of which are responsible for the transport of ionized solutes across membranes. The metal ion uptake is a function of cell age, composition of growth medium, contact time, pH of metal solution, initial metal ion concentration and concentration of cells in an aqueous solution. In addition, it seems that bioaccumulation is also dependent on the culture conditions and growth medium condition. Kinetics of bioaccumulation of metal ions is usually biphasic in nature: consisting of an initial rapid phase which can contribute up to 90% of the uptake and last for 10 to 30 minutes. The second phase is slower and has been found to last up to four hours (Kadukova and Vircikova, 2005). It can be inhibited by metabolic inhibitors such as low temperatures and lack of energy sources (Genter, 1996; Cho *et al.* 1994).

Biosorption is a quick process independent of presence of specific nutrients, while bioaccumulation is a slow and nutrient dependent. Living systems can be unreliable due to many problems of maintaining active biomass populations under high variable wastewater conditions. The extent of bioaccumulation is lower than that of the biosorption. Besides, bioaccumulation represents only 0.5–2% of dry cell weight (Raraz, 1995; Wood and Wang, 1985). From terminological point of view, bioaccumulation term is usually used for metal uptake by living cells, while biosorption is used for passive sequestering by dead cells (Volesky, 2003).

BIOSORPTION (METAL UPTAKE BY NON LIVING CELLS)

Biosorption can be defined as the ability of biological materials to accumulate heavy metals from wastewater through metabolically mediated or physicochemical pathways of uptake or a property of certain types of inactive, non-living microbial biomass to bind and concentrate heavy metals from even very dilute aqueous solution (Volesky, 2003; Fourest and Roux, 1992).

Biological materials (bacteria, fungi, algae and plant biomass) are known for their potential to adsorb heavy metals (Kratochvil and Volesky, 1998; Chang and Hong, 1994). Biosorption or bioadsorption is in fact a passive uptake of heavy metals by biomass.

Mechanisms of cell surface sorption are independent of cell metabolism; they are based upon physicochemical interactions between heavy metal ions and functional groups of the cell wall of microorganisms. The needed activation energy for biosorption is about 10–20 KJ/mol, which is in agreement with the physical nature of the process (Vijayaraghavan and Yun, 2008; Kadukova and Vircikova, 2005).

Biosorption is usually a rapid process responsible for metal removal by non-living biomass. It does not depend on temperature, metabolic energy, or the presence of metabolic inhibitors. Norton *et al.* (2004) reported that metal uptake is not significantly affected by biomass viability. Therefore, any metabolic activity process is currently considered to be a negligible part of biosorption.

Biosorbents in the waste materials discarded either from fermentation industries or from rubber plantaions or marine biomass as a source of raw biosorptive materials..

Therefore, the biomass can be either living or non-living but non living is preferred as it is easy to handle, and the processes is not growth dependent as well as it has no physiological constraints.

Consequently, biomass exhibits this property, acting just as chemical substance, as an ion exchanger of biological origin. It is particularly the cell wall structure of certain algae, fungi and bacteria, which was found responsible for this phenomenon (Volesky, 2003, 1990). Also, the microorganism's cell wall consists mainly of polysaccharides, lipids and proteins, which have many binding possibilities for metals. Such compounds possess numerous functional groups, including carboxylate, hydroxide, amine, imidazole, sulfate and sulfhydryl, with various charge distributions and geometries. So, these functional groups can selectively bind certain metal ions. In this part, binding is attributed to ion exchange, adsorption, complexing, microprecipitation and crystallization processes which occurrs on the cell wall (Sheng *et al.* 2004; Malik, 2004; Davis *et al.* 2003; Veglio and Beolchini, 1997; Blackwell *et al.* 1995). The cell wall constituents play an important role in metal sequestering due to the complexity of the used biomaterials. Also it is quite possible that some of these mechanisms act simultaneously, to a degree dependent on the biosorbent and the aqueous environment. Since this process is independent of the metabolism, binding of metals is very quick, up to 1 min (Delgado *et al.* 1998; Bustard *et al.* 1997) and mostly reversible (Veglio and Beolchini, 1997). Advantages and disadvantages of biosorption by non-living biomass are as follows :

Advantages

There are a number of advantages associated with the use of dead cells (Ahluwalia and Goyal, 2007; Tsezos *et al.* 1995). The dead microbial biomass is less sensitive to toxic heavy metals, and biosorption may not be inhibited by their presence. In contrast, inhibition of heavy metals removal due to the toxicity of these materials only occurs in the case of metabolically mediated accumulation by living cells. The advantages of biosorption of heavy metals using dead cells of bacterial biomass include the following features :

- Growth-independent. Nutrients are not required for the growth of bacterial cells. Therefore, there are no problems of feeding or metabolic products.
- Dead cells or non-living biomass is not subjected to toxicity limitation of cells.
- We can use the biomass from fermentation industries waste.
- There is no metabolic activity or physiological activity since the bacteria or fungal are dead cells.
- Low operational cost.
- Because non-living biomass behaves as an ion exchanger; the process is very rapid and takes place within few minutes to few hours. Metal loading on biomass is often highly efficient.

- Easily operating and controlling. We can control operating conditions such as pH, temperature, time and metal concentrations.
- Sterilization conditions are not required for this process.
- Pre treatment of bacterial biomass for improvement biosorptive capacity.
- Possibility of metal recovery.
- Regeneration and reuse of biosorbent.
- Biomass can be stored for a period of time.
- High efficiency of removal from dilute metal solutions.

Disadvantages

- The biological process improvement potential (e.g. through genetic engineering of cells) is limited because cells are not metabolizing. Because the production of the adsorptive agent occurs during pre-growth, there is no biological control over characteristic of biosorbent. This will be particularly true if waste biomass from a fermentation unit, is being utilized.
- There is no potential for biological altering the metal valency state, *i.e.*, less soluble forms are even for degradation of organometallic complexes.
- The major disadvantage of biosorption technology is that it has had limited practical applications to date and large scale production of the biomass has been established (Feng and Aldrich, 2004). This may be remedied in the future through progressive research and development as well as advanced pilot-scale projects.

The adsorption is easy to understand when it refers to a single metal situation; however, the assessment of sorption becomes complicated in a multi-ion situation which is generally encountered in effluent. Biosorption efficiency depends upon many factors including the capacity, affinity and specificity of the biosorbent and their physical and chemical conditions in effluents.

BIOSORPTION ISOTHERMS

Several studies on metal accumulation by microorganisms have been used the adsorption isotherms as a way to describe metal accumulation with metal concentration increase. The relationship between the amount of metal ion which is adsorbed on the cell (q_{eq}) and the metal ion concentration remaining in solution (C_{eq}), is described by an isotherm equation. Also, the isotherm represents the equilibrium distribution of metal ions between the aqueous and solid phase. Thereby, the equilibrium distribution is important in the determination of the maximum biosorption capacity. Several isotherm models are available to describe this equilibrium distribution. In addition, biosorption isotherm curve can be applied for plotting equilibrium concentration of heavy metals versus the adsorbed amount by bacterial biomass. The two most commonlly used models for describing this type of system are those given by Langmuir and Freundlich.

Langmuir isotherm

Langmuir adsorption isotherm is based on the following assumption (Volesky, 2003; Fourest and Roux, 1992):

(1) Metal ions are chemically adsorbed at a fixed number of well-defined sites,
(2) Each site can only hold one ion,
(3) All sites are energetically equivalent,
(4) There is no interaction between the ions,
(5) The maximum adsorption occurs when a saturated monolayer of solute molecules is present on the adsorbent surface,
(6) The energy of adsorption is constant,
(7) There is no migration of adsorbate molecules in the surface plane (localized adsorption).

Langmuir equation, which is valid for monolayer sorption on the surface containing definite number of identical sites and has been used extensively for dilute solutions, has the following form.

$$q_c = \frac{q_{\max} bC_c}{1 + bC_e} \qquad \text{...(1)}$$

While, the linear form of Langmuir equation is

$$C_{eq}/q = 1/q_{max}b + C_{eq}/q_{max} \qquad \text{...(2)}$$

Where q_{max} is the langmuir constant (mg/g) reflect the maximum adsorption capacity of the metal ion per unit weight of biomass to form a complete monolayer on the surface bound at high C_{eq}. The value of Langmuir constant b(L/mg), represents a ratio of adsorption rate constant to desorption rate constant, which also gives an indication of the affinity of the metal for binding sites on the biosorbent, q_{max} and b can be determined from the linear form of Langmuir equation Eqe (2) by plotting Ceq/qeq vs Ceq (Hassan *et al.* 2009; Norton *et al.* 2004; Volesky , 2003; Aksu and Dönmez 2001; Puranik and Paknikar 1999; Veglio *et al.* 1997; Chong and Volesky, 1995; Volesky and Holan, 1995).

Langmuir model was chosen to estimate the maximum adsorption capacity corresponding to complete monolayer coverage on the biomass surface. On the other hand, Freundlich model was chosen to estimate the adsorption intensity of the biosorbent towards the biomass.

FREUNDLICH ADSORPTION ISOTHERM

Freundlich expression is an empirical equation based on the sorption on a heterogeneous surface. Furthermore, Freundlich isotherm describes more complicated adsorption model where interactions occur between adsorbed molecules. It models monolayer adsorption on a heterogeneous surface and takes the following mathematical form:

$$q_{cq} = KC_{edl}^{1/n} \qquad \ldots(3)$$

The linearized form of this model is;

$$\ln q_{eq} = \ln K_f + 1/n \ln C_{eq} \qquad \ldots(4)$$

Freundlich equation generates an exponential shaped theoretical equilibrium curve and it is represented through eq. 3, where K and n are the Freundlich constants characterizing of the system. Equation 3, can be linearized in logarithmic form (eq. 4) and Freundlich constants can be determined.

Freundlich isotherm is also more widely used but provided no information on the monolayer adsorption capacity. In contrast to Langmuir model, Freundlich equation measures the adsorption capacity K_f, and the intensity of adsorption (n). (Hassan *et al.* 2009; Goyal *et al.* 2003; Volesky, 2003; Chong and Volesky, 1995; Mattuschka and Strauble, 1993; Aksu and Dönmez 2001; Puranik and Paknikar, 1999).

BRUNAUER–EMMETT AND TELLER (BET EQUATION)

BET isotherm is used occasionally and is represented multi-layer adsorption at heterogeneous sites with varying binding affinities at the adsorbent surface. Also, it assumed that a Langmuir isotherm applies to each layer (Kapoor and Viraraghavan, 1995; Morley and Gadd, 1995). Langmuir model assumed that the adsorption take place only on occupied adsorption sites (monolayer). On the other hand, BET model assumes that a number of adsorbate molecules layers are formed at the surface of sorbent and Langmuir equation can be applied to each layer. The equilibrium condition will therefore, involve several types of surfaces in the sense of molecules layers number on each surface. Also, multiple layering adsorption is similar to Langmuir model, the rate of adsorption on any layer is equal to the rate of desorption from that layer. The BET isotherm equation is:

$$Q_e = \frac{BC_cQ^0}{(C_s - C_e)\,[(1 + (B-1)\,C_e)/C_s]}$$

$$q_e = \frac{C}{(C_s - C)q_c} = \frac{1}{BQ^0} + \frac{B\ -1}{BQ^0}\frac{C}{C_s} \text{ (liner form)}$$

In which Q^0 is the metal uptake per unit weight of adsorbent in forming a complete monolayer on the surface, B is a constant relating to the energy of interaction with the surface. C_e is the measured metal concentration in solution at equilibrium and C_S is the saturation concentration of the solute.

BET and Freundlich models can be interpreted best at low concentration adsorbent/adsorbate interactions (Morley and Gadd, 1995). Kiran and Kaushik (2008) showed applicability example of BET model for Cr(VI) biosorption using *Lyngbya putealis* exopolysaccharides and they claimed that multilayer adsorption occurred in this system. In addition, Preetha and Viruthagiri (2005) applied BET model to describe the biosorption of Zn^{2+} by *Rhizopus arrhizus*.

Sips isotherm

Sips isotherm is an empirical model to study the adsorption process (Al-Asheh *et al.* 2000: Sips, 1948). It is the combination of both Langmuir and Freundlich isotherms, thus it is also known as Langmuir–Freundlich isotherm. Sips isotherm equation is given as below:

$$q_e = \frac{q_m (bC_e)^{1/n}}{1 + (bC_e)^{1/n}}$$

Where, *qm* is the total binding sites (mg/g), *b* is the median association constant (L/mg) and $1/n$ is the heterogeneity factor. *Ce* is the equilibrium concentration (mg/L), while *qe* is the adsorbed amount at equilibrium (mg/g). When 1/n<1, indicates the adsorbent has heterogeneous binding sites, while $1/n$ closer or even to 1 indicates homogeneous adsorbent (Al-Asheh *et al.* 2000).

Sips model assumes that the surface is homogeneous, but the sorption is a cooperative process due to adsorbate–adsorbate interactions.

Kumar and Porkodi (2006) concluded that the Sips isotherm provided best fitting of experimental data, followed by the Langmuir and the Redlich–Peterson isotherm equations. Oo *et al.* (2009) reported that the $1/n$ values for the adsorption of copper (II) (0.5737) and lead (II) (0.3252) confirmed with Freundlich isotherm which assumes heterogeneous adsorption. Sips isotherm provides a reasonably accurate prediction of heavy metal biosorption experimental results with high value of coefficient correlation (*R*2) (Apiratikul and Pavasant, 2008; Basha *et al.* 2008; Senthilkumar *et al.* 2007; Vijayaraghavan *et al.* 2006).

Redlich–Peterson isotherm

Redlich–Peterson isotherm contains three parameters and is an improvement over Langmuir and Freundlich isotherms. However, Redlich–Paterson is another empirical equation, which capable to represent adsorption equilibria over a wide range of concentration. This equation has the following form;

$$q_e = \frac{K_{RP} C_e}{1 + a_{RP} C_e^{\beta}}$$

In which, K_{RP} and d_{RP} are the Redlich–Peterson constants, and _ is basically in the range of zero to one. If _ is equal to 1, it reduces to Langmuir isotherm equation. while, in case where the value of *d* is much higher than one, Redlich–Peterson isotherm equation can be approximated by a Freundlich-type equation. Consequently, this equation is commonly used in the prediction of heavy metals bioadsorption equilibria data (Padmavathy, 2008; Preetha and Viruthagiri, 2007; Dursun, 2006; Ho 2006; Ho and Ofomaja, 2006; Vijayaraghavan *et al.* 2006).

NON-IDEAL COMPETITIVE ADSORPTION (NICA)

NICA is a multi-component competitive binding model that accounts for both ion-specific non-ideality and chemical heterogeneity of the biosorbent. Also, it has been used for, e.g. metal adsorption by bacteria (Plette *et al.* 1996) in addition to Scatchard plots have been used to show biosorption equilibria.

Often, either of these models is applicable to describe metal sorption by microorganisms (Table 1), but since the models were originally developed for other systems, they should be interpreted with a great care for such complex biological systems as microorganisms (Kapoor and Viraraghavan, 1995). For more information about the adsorption isotherms referred to the review of Gimbert *et al.* (2008).

TABLE 1 : Use of models to describe metal accumulation by microorganism

Organism	*Metal(s)*	*Model*	*References*
Pseudomonas cepacia	Cd, Zn	Langmuir	Savvaidis *et al.* 1992
Bacillus subtilis,	Cd, Zn, Cu	Langmuir	Cotoras *et al.*, 1992
Rhizopus arrhizus	Ag, Ni, Cu, Cd, Zn	Langmuir	Fourest *et al.* 1994
Rhizopus arrhizus	Cu	Langmuir	Brady and Tobin, 1995
Mucor hiemalis	Cr III	Langmuir	Pillichshammer *et al.* 1995
Aspergillus carbonarius	Cu	Langmuir	Al-Asheh and Duvnjak, 1995
B.megaterium, B. circulans,	Cr VI	Langmuir	Srinath *et al.* 2002
Chlorella kessleri	Cu	Langmuir	Kadukova and Vircıkova, 2005
Spirulina sp	Cd, Cu, Cr111	Langmuir	Chojnacka *et al.* 2005
Pseudomonas aeruginosa	Pb, Cd, Cu	Langmuir	Chang *et al.*1997
Citobacter strain MCMB-181	Cd, Zn, Pb	Langmuir	Puranik and Paknikar, 1999
Pseudomnas fluroescens	Ni	Langmuir	Lopez *et al.* 2001
Pseudomnas aeruginosa	Ni, Cu	Langmuir	Sar *et al.*1999
Sphaerotilus natans	Cu,Cd	Langmuir	Esposito *et al.*2001
Pseudomonas putida	Pb, Cd, Cu, Zn	Langmuir	Pardo *et al.*2003
Corynobacterium glutamicum	Pb	Langmuir	Choi and Yun, 2004
Arthrobacter sp	Ni, Pb, Cu, Mn	Langmuir	Veglio *et al.*1997
Arthrobacter sp	Cu, Cd,	Langmuir	Pagnanelli *et al.*2002
Zoogloea ramigera	Cu	Freundlich	Aksu *et al.*1992
Streptococcus equisimilis	Cr VI	Freundlich	Goyal *et al.*2003
Escherichia coli,		Freundlich	Mullen *et al.*1989
Thiothrix strain A1	Cu, Ni, Zn	Freundlich	Shuttleworth and Unz, 1993
Chlorella vulgaris	Cu	Freundlich	Aksu *et al.*1992

Organism	Metal(s)	Model	References
Bacillus sp	Pb, Cu	Freundlich, Langmuir	Tunali *et al.*2006
Pseudomonas putida	Cu, Zn	Freundlich, Langmuir	Chen *et al.*2005
Streptomyces coelicolor	Ni, Cu	Freundlich, Langmuir	Öztürk *et al.*2004
Enterobacter sp	Cu,Cd	Freundlich, Langmuir	Lu *et al.*2003
Candida sp	Cu	Freundlich, Langmuir	Aksu and Dommez, 2001
Ochrobacterium anthrobi	Cd, Cu, Cr111	Freundlich, Langmuir	Ozdemir *et al.*2003
Zoogloea ramigera	Pb, Ni, Cu,Fe111	Freundlich, Langmuir	Sag and Kutsal, 1995
Pseudomnas strain	U	Freundlich, Langmuir	Sar and Souza, 2001
Rhizopus arrhizus	Cd, Zn, Cu	BET	Morley and Gadd, 1995
Rhizopus arrhizus	Cu	BET	Gadd, 1988
Rhodobacter sphaeroides	Cd, Pb	NICA	Seki *et al.*1998
Algae Gelidium	Cu(II), Pb(II), and Cd(II)	Langmuir	Vilar *et al.*2007
Aspergillus niger	Cu(II) and Pb(II)	Langmuir, Freundlich and Redlich–Peterson	Dursun ,2006
	Cr(VI)	Langmuir and Freundlich	Mungasavalli *et al.*2007
Aspergillus sydoni			Kumar *et al.*2008
Bacillus cereus	Pb(II) and Cu(II)	Langmuir and Freundlich	Jian-hua *et al.*2007
Escherichia coli	Ni(II) and Cd(II)	Freundlich	Zhou *et al.*2007
Caulerpa lentillifera	Cu(II), Cd(II), and Pb(II)	Dubinin-Radushkevich, Sips	Apiratikul *et al.*2008
Caulerpa lentillifera	Cu(II), Cd(II), Pb(II), and Zn(II)	Langmuir and Freundlich	Pavasant *et al.*2006
Eichhornia crassipes	Cr(VI)	Freundlich	Choi and Yun, 2006
Penicillium janthinellum	Cr(VI)	Langmuir and Freundlich	Kumar *et al.*2008

Organism	*Metal(s)*	*Model*	*References*
Pseudomonas aeruginosa	Cr(III) and Cr(VI)	Langmuir and Freundlich	Kang *et al.*2007
Pseudomonas sp	.Cd(II), and Cr(VI)	Langmuir and Freundlich	Ziagova *et al.*2007
Rhizopus arrhizus	Cr(VI)	Langmuir, Freundlich and Redlich–Peterson	Preetha and Viruthagiri, 2007
Staphylococcusxylosus	Cd(II), and Cr(VI)	Langmuir and Freundlic	Ziagova *et al.*2007

FACTORS AFFECTING BIOSORPTION

pH in aqueous solution

For biosorption of heavy metal ions, pH is one of the most important environmental factors. Solution pH value strongly influences not only the site dissociation of the biomass surface, but also the solution chemistry of the heavy metals as hydrolysis, complexation by organic and/or inorganic ligands, redox reactions, precipitation, the speciation and the biosorption availability of the heavy metals (Esposito *et al.* 2001; Wang *et al.* 1997).

The biosorptive capacity of metal cations increases with increasing pH of the sorption system, but not in a linear relationship. On the other hand, extreme pH value can cause precipitation of metal complexes, so it should be avoided during experiments. As pH increases, functional sites on the microbial surface become deprotonated and may bind cationic metals instead. This results in the common trend of increasing accumulation with increasing pH. However, in many cases, the accumulation starts to decrease again at higher pH. Therefore, at low pH values the cell surface becomes more positively charged, then reducing the attraction between metal ions and functional groups on the cell wall. In contrast, higher pH results in facilitation of the metal biosorption, since the cell surface is more negatively charged. The metals will transform into hydroxide complexes at high pH values, however, it could not be considered as a biosorption behavior of the cells. Mapoleto and Torto (2004) proved that the biosorption capacity of Cd^{2+}, Cr^{3+}, Cr^{6+}, Cu^{2+}, Pb^{2+} and Zn^{2+} is dependent on pH. For all studied metal ions, the optimum pH values in all cases were found to be greater than 5. The optimum pH for Cd^{2+} and Pb^{2+} biosorption is 5.8. Also, Özer and Özer (2003) reported that optimal pH value for Pb^{2+} and Ni (II) ion uptake is 5.0 by *S. cerevisiae*.

Cr(VI) and some other metals such as arsenic are known to exist as anions meaning that the metal carry negative charge. These anions exhibit different pH features from metal cations. The biosorption was favourable at low pH for these anions. In this concern, cell wall ligands of bacterial biomass are protonated at low pH values, and compete significantly with metal binding. With pH increase, more ligands such as amino and carboxyl groups, would be exposed leading to attraction between these negative charges and the metals and hence increases in biosorption on to the cell surface. As the pH increased further, the overall surface charge on cells could become negative and therefore, biosorption of anions decreased again (Khambhaty *et al.* 2009; Wang and Chen, 2006; Park *et al.* 2005; Tewari *et al.* 2005).

Time

The biosorption process of heavy metal by bacterial biomass usually completes rapidly. The biosorption of metals as cadmium, cobalt, copper, zinc, lead, nickel, chromium and uranium by nonliving cells of bacterial biomass, is a rapid process and often reaches equilibrium within several minutes (Özer and Özer, 2003; Kapoor and Viraraghavan, 1997).

However, it is necessary to optimize the contact time, considering the efficiency of desorption and regeneration of the biomass. Gabr *et al.* (2008) optimized the sorption time for Pb(II) and Ni (II) by heavy metal resistant strain *Pseudomonas aeruginosa* ASU 6a. Besides, the result showed that a 30 min sorption period was the best option to ensure the metal removal from solution and good recovery from biosorbent. Similar results were observed by Hassan *et al.* (2009) who concluded that the optimum time for biosorption of Cu and Cd by *P. stutzeri* was 30 min.

IONIC STRENGTH (IS)

Since Na, K, Ca, Mg are present in many industrial wastewaters, it is useful to estimate the extent and interferes of light metals with biosorption of heavy metal by bacterial biomass (Goyal *et al.* 2003; Volesky 2003; Esposito *et al.* 2001; Pagnanelli *et al.* 2000). Most of light metals bind weakly to the negatively charged biomass particles, through electrostatic attraction in which a significant ionic strength effect may be observed in the presence of high concentration of light metals. Calcium, which may form specific, coordinate bonds, might be having stronger effect on heavy metal biosorption (Goyal *et al.* 2003; Volesky, 2003; Esposito *et al.* 2001; Pagnanelli *et al.* 2000). However, the effect of ionic strength can be explained via the competition of light metals with the heavy metals for electrostatic binding to the biomass. Kaewsarn (2002) reported that the effect of Na^+ on Cu(II) uptake was negligible at a concentration of 10 mM, and K^+, Mg^{2+} and Ca^{2+} reduced the removal efficiency by 4%, 11% and 13%, respectively. In addition, Chang and Hong, (1994) noted that the mercury uptake by *P. aeruginosa* decreased with the increment of ionic strength. Also, Cho *et al.* (1994) showed that no significant decrease in the binding of Cd^{2+} and Zn^{2+} occurred up to the ionic strength of 10^{-3} M. In contrast, Chaisuksant, (2003) illustrated that the presence of light metal ions (Na^+, K^+, Mg^{1+} and Ca^{2+}) in solution had an insignificant effect on cadmium and copper sorption capacity by pretreated biomass of marine alga *Gracilaria fisheri*. In general, the increaseof ionic strength decreased biosorption efficiency because of the increment of electrostatic charge (Volesky, 2003).

SURFACE AREA AND ITS EFFECT ON BIOSORPTION

The passive part of microbial metal accumulation can be influenced by surface properties such as charge and orientation of the metal binding functional groups on the cell surface. Microorganisms have a high surface area-to-volume ratio as their small size, and therefore this high surface area provide a large contact area that can interact with metals in the surrounding environment. Microbial metal accumulation has received much attention during the last years due to the potential use of microorganisms for cleaning metal-polluted water. However, less attention has been paid to the role of microorganisms for metal mobility in soil even though the same processes might occur there. Metal accumulation by microorganisms is influenced by the surface characteristics of microorganisms, but also themselves can lead to changes in the surface properties such as charge changes (Ivanov *et al.* 1996; Chen and Ting, 1995; Collins and Stotzky, 1992).

PRETREATMENT OF BIOSORBENT

Untreated or living cells of bacterial biomass may not be affected on heavy metal removal (Huang and Huang, 1996). Pre-treatment of bacterial biomass by some physical and chemical reagents may result in the improvement of biosorption capability (Galun *et al.* 1987). These types of pretreatments modify cell surface of bacterial biomass which is essential for biosorption either by removing or masking the groups as well as exposing more metal binding sites (Gupta *et al.* 2000). Furthermore, many physical pretreatment methods have been used such as heat treatment, autoclaving, freeze drying and boiling (Gabr *et al.* 2008; Göksungur *et al.* 2005; Seki *et al.* 2005; Mapoleto and Torto, 2004; Robles *et al.* 2000; Pethkar and Panikar, 1998). Chemical methods include the treatment with various organic and inorganic reagents, such as acid and alkali, methanol, formaldehyde, etc. Those methods are found to improve metal biosorption to some extent. Because of the important role of cell wall in the metal biosorption by non-viable cells, metal biosorption may be enhanced by heat or chemical sterilization or by crushing. Thus degraded cells would offer a larger available surface area and expose the intracellular components and more surface binding sites due to the destruction of the cell membranes (Errasquín and Vázquez, 2003; Robles *et al.* 2000). Autoclaved bacterial biomass caused a decrease in biosorption efficiency. The bioadsorption capacity of autoclaved *Mucor rouxii* decreased compared to the live fungus, and it can be attributed to the loss of intracellular uptake (Yan and Viraraghavan, 2000). Whistler and Daniel (1985) reported that the heat treatment could cause a loss of amino-functional groups on the fungal surface through the non-enzymatic browning reaction. In contrast, Galun *et al.* (1987) reported that *Penicillium* biomass pretreatment at 100°C for 5 minutes increased the biadsorption of lead, cadmium, nickel and zinc ias well as the increase was attributed to the exposure of latent binding sites after pre-treatment

The heat inactivation of the fungal biomass of *Lentinus edodes* was resulted in an increase in the biosorption capacity for mercury(II), cadmium(II) and zinc(II) metal ions compared to live cells (Bayramoglu and Arica 2008). Similar observations were reported for other biomasses, including fungi and yeast (Vilar *et al.* 2006; Bayramoglu *et al.* 2005; Brady and Tobin, 1995).

The findings obtained by Tunali and Akar (2006) indicated that pretreatment methods by heat or detergent or alkali altered the adsorption capacity of *B. cinerea* biomass which ranged from 1.75 ± 0.4590 to 12.98 ± 0.9623 mg g^{-1} biomass at 100 mg l^{-1} initial concentration of Zn(II) compared to the native biomass. The highest Zn(II) loading value (12.98 ± 0.9623 mg g^{-1}) was obtained by heat inactivated biomass, while pre-treatment detergent and sodium hydroxide treatments which increased the uptake of Zn(II) from 2.18 ± 0.6245 to 6.38 ± 0.6520 and 4.14 ± 1.0678 mg g^{-1}, respectively. The increase of biosorption rate was attributed to the activation and availability of more binding sites after pre-treatment. Preheating of biomass could cause a loss of amino-functional groups from the surface of bacterial or fungal biomass. Therefore, these functional groups are among the groups in the composition of polysaccharides which contribute to the binding of heavy metals (Yan and Viraraghavan, 2000).

The alkaline pre-treatment of bacterial biomass, including NaOH, KOH, NH_4OH, alkaline detergents or other alkaline reagents such as tolune, ruptures and destroy the ce11 walls of the microbes and exposes additional functional groups for metal ion binding, thus enhancing the biosorption capacity of biomass (Nadeem *et al.* 2008 ; Ochie, 2008 ; Tong 2007; Tan and Cheng, 2003;Ginisty *et al.* 1998).

Alkaline treatment destroys autolytic enzymes that cause purifications of biomass and removes both lipids and proteins that masked reactive sites. Sar *et al.* (1999) observed that metals (Cu^{2+} and Ni^{2+}) uptake capacity of lyophilized *Pseudomonas auruginosa* cells was enhanced when pretreated with NaOH, NH_4OH or toluene. Furthermore, Liu *et al.* (2004) reported that the effect of alkali or physical pretreatment increased the capacity for Zn(II) biosorption by the indigenous *T. thiooxidans* biomass compared to untreated biomass.

The acidic pre-treatment also improved the biosorption efficiency of bacterial biomass. Therefore, this may be due to the availability of more binding sites and elimination of impurities and ions that block heavy metals binding sites (Abbas *et al.* 2008; Gupta and Rastogi, 2008; Vijayaraghavan and Yun, 2008; Araujo *et al.* 2007). In addition Huang and Huang (1996) reported that acid pretreatment can strongly enhance the adsorption capacity of *Aspergillus oryzae* mycelia. It was assumed that the acid treatment may dissolve polysaccharide compounds in the outer layer of the fungal cell wall, thus producing additional binding sites. However, in the investigation by Kapoor and Viraraghavan (1995), acid pre-treatment of *A. niger* significantly reduced the biosorption of lead, cadmium, nickel and zinc.

Detergent treatments can increase the removal of metals from the solution, because the disruption of cell components increased cell permeability and exposed a wider variety of potential binding sites (Gadd, 1993).

Increased metal capacities of biosorbents due to the successful introduction of additional functional groups into the biomass cell wall were also reported by Seki *et al.*

(2005) and Marshall *et al.* (1999). Consequently, an improvement in the binding capacity of the alga *Lyngbya taylorii* (two- to six-fold) for Cd^{2+}, Ni^{2+}, Pb^{2+} and Zn^{2+} was observed upon phosphorylation (esterification of the hydroxyl groups of the polysaccharides) of the biomass (Klimmek *et al.* 2001). On the other hand, the blockage (via methylation) of negatively charged carboxylic groups resulted in an increase of the sorption amounts of anionic metallic species Cr (VI) and As (V), Pt (IV) (Seki *et al.* 2005; Parsons *et al.* 2003; Jianlong, 2002). Although, these treatments could change the properties of biological solids structure and therefore, their ability for biosorption. In this regard, freeze-drying does not seem to modify the structure of the wall cells (Pardo *et al.* 2003).

MECHANISM OF BIOSORPTION AND BIOACCUMULATION

The main biosorption of heavy metals mechanism is referred to non directed physico-chemical interactions that occur between dissolved metal ions in aqueous phase and cellular components of living or dead organisms which act as a solid phase. Involved interactions may be physical adsorption, ion exchange, complexation, coordination, microprecipitation, chelations crystallization or diffusion (Vijayaraghavan and Yun, 2008; Volesky, 2003; Yan and Viraraghavan, 2000; Volesky and Holan, 1995; Fourest and Roux, 1992; Gadd, 1988). Besides, the biosorption mechanisms of heavy metals showed that heavy metals are adsorbed by physicochemical interactions between metal ions and the bacterial surface. The different mechanisms of biosorption by bacterial biomass are shown in Figure 1.

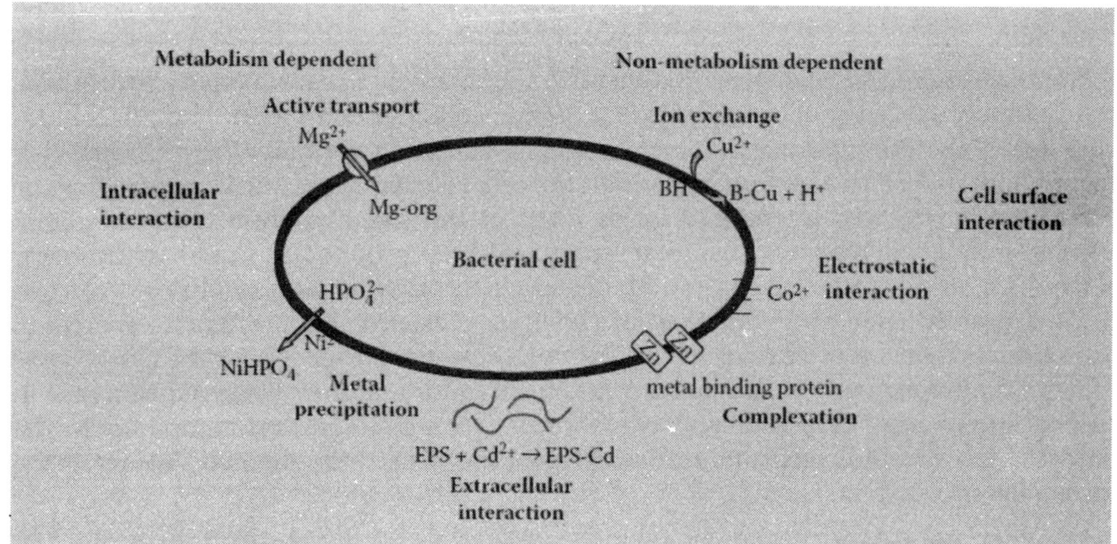

Fig. 1. Schematic diagram showing the different mechanisms of bacterial biosorption (Adapted from Prasad *et al. 2006*).

Binding of metals to cell surface components and intracellular accumulation were identified to proceed through two mechanisms in metals biosorption by living microorganisms (Gadd, 1993). The intracellular accumulation or metabolic processes may result in the accumulation of relatively large amounts of metal but these processes are slow and are mostly dependent on nutrients and environmental conditions (Malik, 2004). Also, cell surface and cell wall binding are a passive process and take place with both live and dead cells. Biosorption is very rapid, often less than a few minutes (Pardo *et al.* 2003) and these kinds of metal removal occur by ion exchange or physical adsorption processes involving specific chemical sites on cell wall which contained negative charge (Volesky and Holan, 1995). While, Akhtar *et al.* (2007) suggested that there are two phases of adsorption phenomenon were involved in biosorption of uranium in live *Trichoderma harzianum*: the first one, was related to physic sorption, and the second was due to both structural change and surface transformation. Besides, copper biosorption by live yeast *Saccaromyces cerevisae* was found to be biphasic and consisted of an initial, and rapid surface binding, then followed by a second slow intracellular uptake (Huang and Huang, 1996).

Gabr *et al.* (2008) reported some of the functional groups which contain negative charge like carboxylate, phosphate, sulfahydryle and amino groups which are present in cell wall of *P. aeruginosa* ASU 6a, and these function groups play an important role in biosorption of lead and nickel from aqueous solution. In addition, the cell wall of *P. aeruginosa* S22 contains proteins, lipids, which reflect the range of distinct potential complexation sites, such as carboxylate, phosphate, sulfahydryle and amino groups. Also amino acids as histadine, and cysteine, are reported to have strong affinity for metal ions (Komy *et al.* 2006). Aspartic acid and glutamic acid were also reported by Huange and Huang (1996) as amino acid residues might be responsible for metal adsorption.

Shumate *et al.* (1978) explored that metal ion uptake could react to form a precipitate, and discretely settled within the cells or as affine colloidal entrapped by extra cellular polymers. They also reported that metal ion may be transported into the cells and may react to form a product that remains within the cells. Furthermore, metal cations may be complexed by negatively charged sugar units at the end of polysaccharides chain. Besides, they cited that extra cellular polyphosphate groups associated with sugar metabolism, and complex metal ion by chelating through negatively charged oxygen atoms. Muraleedharan and Venkobachar (1990) used electron spin resonance spectra to study the mechanisms of biosorption. Moreover, the obtained spectra for the biosorbents before and after exposed to the metal solutions. In addition, they observed decrease in the concentration of free radicals with samples to metal solution compared to the control. This also indicated the participation of the negatively charged groups in the biosorption processes.

Another important mechanism which was involved in biosorption, is the ion exchange, where the cell wall of the biomaterial is generally regarded as a complex ion exchanger similar to commercials resins. It was reported that the cell walls contain a mixture of monovalent, divalent cations and that the displacement of one cation by another (e.g. H^+) is dependant on the strength o the individual ligand complex (Al-Asheh and Duvnjak, 1997).

BACTERIAL BIOSORPTION OF HEAVY METALS

Removal of toxic metal ions may contribute to the detoxification of polluted environments. Therefore, the investigation of metal uptake by bacteria is fundamental for the application of biosorption, since it gives information about metal ions removal efficiency in the process (Vegliό, and Beolchini, 1997). As a necessary factor for the design of equipment, the maximum adsorption capacity of bacteria is usually used by the parameter q_{max}. Table 2 shows the maximum biosorption capacity of heavy metals by bacteria reported in literature. However, the direct comparison of experimental data is not possible, due to differences in experimental conditions employed (such as; pH, equilibrium time, temperature, biomass concentration, heavy metal concentrations and the bacterial biomass). Even so, Table 2 provides some information to evaluate the possibile usage of bacterial biomass for the removal of heavy metal ions. Also, the biosorption capacity is not only dependent on the type of heavy metal ions, but also on the type of bacteria, due to the variations in cellular constituents.

TABLE 2 : Important results from the literature for metal biosorption by various bacterial biomass

Metal	*Biosorbent*	*pH*	*q_{max} (mg/g)*	*References*
Copper	*Bacillus sp. (ATS-1)*	5	16.3	Tunali *et al.* 2006
	Bacillus subtilis IAM 1026	5	20.8	Nakajima *et al.* 2001
	Enterobacter sp. J1	5	32.5	Lu *et al.* 2006
	Micrococcus luteus IAM 1056	5	33.5	Nakajima *et al.* 2001
	Pseudomonas aeruginosa PU21	5	23.1	Chang *et al.* 1997
	Pseudomonas cepacia	7	65.3	Savvaidis *et al.* 2003
	Pseudomonas putida	6	6.6	Pardo *et al.* 2003
	Pseudomonas putida	5.5	96.9	Uslu and Tanyol, 2006
	Pseudomonas putida CZ	4.5	15.8	Chen *et al.* 2005
	Pseudomonas stutzeri IAM 12097	5	22.9	Nakajima *et al.* 2001
	Sphaerotilus natans	6	60	Beolchini *et al.* 2006
	Streptomyces coelicolor	5	66.7	Öztürk *et al.* 2004
	Thiobacillus ferrooxidans	6	198.5	Ruiz-Manriquez *et al.* 1997
	Paenibacillus jamilae	6	7.8	Morillo *et al.* 2008
	Pseudomonas pseudoalcaligenes	5	48.4	Leung *et al.* 2001
	Pseudomonas aeruginosa AT18	5	113.64	Pérez Silva *et al.* 2009
	Pseudomonas stutzeri	5	33.16	Hassan *et al.* 2009
Cadmium	*Aeromonas caviae*	7	155.3	Loukidou *et al.* 2004
Cadmium	*Bacillus circulans*	7	26.5	Yilmaz and Ensari, 2005
	Enterobacter sp. J1	6	46.5	Lu *et al.* 2006

Metal	*Biosorbent*	*pH*	*q_{max} (mg/g)*	*References*
	Pseudomonas aeruginosa PU21	6	42.4	Chang *et al.* 1997
	Pseudomonas putida	6	8	Pardo *et al.* 2003
	Pseudomonas sp.	7	278	Ziagova *et al.* 2007
	Staphylococcus xylosus	6	250	Ziagova *et al.* 2007
	Streptomyces pimprina	5	30.4	Puranik *et al.* 1995
	Paenibacillus jamilae	6	20.4	Morillo *et al.* 2008
	Pseudomonas aeruginosa	8.5	6.18	Tuzen *et al.* 2008
	Streptomyces rimosus	8	64.9	Selatnia *et al.* 2004
Lead	*Bacillus sp. (ATS-1)*	3	92.3	Tunali *et al.* 2006
	Corynebacterium glutamicum	5	567.7	Choi and Yun, 2004
	Enterobacter sp. J1	5	50.9	Lu *et al.* 2006
	Pseudomonas aeruginosa	5	79.5	Chang *et al.* 1997
	Pseudomonas aeruginosa	5	0.7	Lin and Lai, 2006
	Pseudomonas putida	5.5	270.4	Uslu and Tanyol, 2006
	Pseudomonas putida	6.5	56.5	Pardo *et al.* 2003
	Streptomyces rimosus	6.5	135	Selatnia *et al.* 2004
	Streptoverticillium cinnamoneum	4	57.7	Puranik and Paknikar, 1997
	Paenibacillus jamilae	Not available	303.03	Morillo *et al.* 2008
	Pseudomonas aeruginosa	8.5	6.07	Tuzen *et al.* 2008
	Pseudomonas pseudoalcaligenes	5	271.7	Leung *et al.* 2001
	Pseudomonas aeruginosa ASU 6a	6	123	Gabr *et al.* 2008
Nickel	*Streptomyces coelicolor*	8	11.1	O¨ztu¨rk *et al.* 2004
	Streptomyces rimosus	6.5	36	Selatnia *et al.* 2004
	P. aeruginosa ASU 6a	7	113.6	Gabr *et al.* 2008
	Pseudomonas aeruginosa	8.5	5.25	Tuzen *et al.* 2008
	Escherichia coli	7	34.3	Kao *et al.* 2008
	Paenibacillus jamilae	6	17.6	Morillo *et al.* 2008
Zinc	*Paenibacillus jamilae*	6	12.31	Morillo *et al.* 2008
	Escherichiu coli	7	659	Kao *et al.* 2008
	Pseudomonas aeruginosa AT18	7.-7.2	87.72	Pérez Silva *et al.* 2009
	Pseudomonas putida	7	6.9	Pardo *et al.* 2003
	Pseudomonas putida	5	17.7	Chen *et al.* 2005
	Streptomyces rimosus	3	30	Mameri *et al.* 1999
	Streptomyces rimosus	3	80	Mameri *et al.* 1999
	Streptoverticillium cinnamoneum	5.0-6.0	21.3	Puranik and Paknikar, 1997

FUTURE PROSPECTS

Bacterial biosorption of heavy metals, is an inexpensive technology to remediate or stabilize those pollutants from contaminated aqueous solution, and has received a great concern not only as effective adsorbents from low-cost biomaterials but also as bacteria had the abilities to adsorb different types of pollutants. In this concern, bacterial biomass (living/non-living cells) can be applied to the mixed contaminated water for the removal of several heavy metals within short time.

Biosorption is confronting a great challenge due to nontechnical pitfalls involved in the commercialization of technology (Volesky and Naja, 2005). Besides, biosorption has two future approaches for metal removal; first one is to use living cells as hybrid technology for pollutant remediation (Tsezos, 2001). However, the second is to develop effective commercial biosorbents (Volesky, 2007). Consequently, the application of biosorption using living cells has obtained attention compared to dead cells in addition to growing microbes as a useful alternative for heavy metal contaminants from industrial waste (Malik, 2004). The optimization of biosorption characteristics and physical conditions is required as additional trend for developing biomaterial immobilization as reuse, recycling and metal-microbes interactions (Wang and Chen, 2006).

During the last few years, various adsorbents have received a great attention as they have metal-binding capacities and are able to clean up heavy metals from contaminated water at low cost and local availability. Low cost biosorbents is utilized in detoxifying effluents of metal plating, mining and ore operations, beside numerous researches illustrated that biosorption is a beneficial alternative to conventional treatments for heavy metals removal from aqueous solution (Vijayaraghavan and Yun, 2008; Wang and Chen, 2006).

It is clear that biosorption of heavy metals is self-driven and don't disturb the sites in cleaning process. Also, biosorption could contribute to conventional methods as a second or pretreatments that enhanced the efficiency of heavy metal removal from industrial wastes and wastewater. Further researches are required to focus on low cost biosorbents as it is difficult to create a definite protocol that could be applied to any contaminated water and wastewater. So, each biosorbent must be customized and adapted to site-specific conditions. Modeling, regeneration and immobilization of biosorbents deserve indepth investigations.

REFERENCES

Abbas M Nadeem R Zafar MN Arshad M (2008) Biosorption of chromium (III) and chromium (VI) by untreated and pretreated *Cassia fistula* biomass from aqueous solutions. *Water Air Soil Pollut* **191**139-148

Ahluwalia SS and Goyal D (2007) Microbial and plant derived biomass for removal of heavy metals from wastewater *Bioresour Technol* **98:** 2243-2257

Akhtar K., Akhtar M.W. and Khalid A.M., (2007) Removal and recovery of uranium from aqueous solutions by *Trichoderma harzianum*. *Wat. Res.* **41:** 1366-1378

Aksu Z and Dönmez G (2001) Comparison of copper (II) biosorptive properties of live and treated *Candida* sp. *J Environ Sci Health* **36**: 367–381

Aksu Z, Sag Y and Kutsal T, (1992) The biosorption of Copper(II) by *C. vulgaris* and *Z. ramigera. Environ Technol* **13**: 579–586

Akthar M, Sastry K and Mohan P, (1996) Mechanism of metal ion biosorption by fungal biomass. *BioMetals.* **9**: 21–28

Akthar MN, Mohan PM (1995) Bioremediation of toxic metal ions from polluted lake waters and industrial effluents by fungal biosorbent. *Curr Sci* **69**: 1028-1030

Al-Asheh S and Duvnjak Z (1997) Sorption of cadmium and other heavy metals by pine bark. *J Hazard Mater* **56**: 35-51

Al-Asheh S and Duvnjak Z, (1995) Adsorption of copper and chromium by *Aspergillus carbonarius Biotechnol Prog* **11**: 638–642

Al-Asheh S, Banat F, Al-Omari R and Duvnjak Z (2000) Prediction of binary sorption isotherms for the sorption of heavy metals by pine bark using single isotherm data. *Chemosphere* **41**: 659–665

Apiratikul R and Pavasant P (2008) Batch and column studies of biosorption of heavy metals by *Caulerpa lentillifera Bioresour Technol* **99**: 2766–2777

Araujo GCL Lemos SG Ferreira AG Freitas H Nogueira ARA (2007) Effect of pre-treatment and supporting media on Ni(II), Cu(II), Al(III) and Fe(III) sorption by plant root material. *Chemosphere* **68**: 537-545

Atkinson BW Bux F and Kasan HC (1998) Considerations for application of biosorption technology to remediate metal-contaminated industrial effluents. *Wat SA* **24**:129–35

Babel S and Kurniawan T A (2003) Low-cost adsorbents for heavy metals uptake from contaminated water: a review *J Hazard Mater* **97**:219–243

Basha S Murthy ZVP and Jha B (2008) Sorption of Hg(II) from aqueous solutions onto *Carica papaya*: application of isotherms. *Ind Eng Chem Res* **47**: 980–986

Bayramoglu G Arica MY (2008) Removal of heavy mercury(II), cadmium(II) and zinc(II) metal ions by live and heat inactivated Lentinus edodes pellets. *Chem Eng J* **143**: 133-40

Bayramoglu G Celik G Yalcin E Yilmaz M and Arica MY (2005) Modification of surface properties of *Lentinus sajor-caju*mycelia by physical and chemical methods: Evaluation of their Cr^{+6} removal efficiencies from aqueous medium. *J Hazard Mater* **119**: 219–229

Beolchini F Pagnanelli F Toro L Vegliò F (2006) Ionic strength effect on copper biosorption by Sphaerotilus natans: equilibrium study and dynamic modelling in membrane reactor. Wat Res **40**:144–152

Blackwell KJ Singleton I and Tobin JM (1995) Metal cation uptake by yeast: a review. *Appl Microbiol Biotechnol* **43**: 579–584

Brady JM and Tobin JM (1995) Binding of hard and soft metal ions to *Rhizopus arrhizus* biomass, *Enzym Microb Technol* **17**: 791–796

Bustard M Donnellan N Rollan A McHale AP (1997) Studies on the biosorption of uranium by a thermotolerant, ethanol-producing strain of *Kluyveromyces marxianus*. *Bioprocess Eng* **17**: 45–50

Chaisuksant Y (2003) Biosorption of cadmium (II) and copper (II) by pretreated biomass of marine alga *Gracilaria fisheri Environ Technol* **24:**1501– 1508

Chang JS and Hong J (1994) Biosorption of mercury by the inactivated cells of *Pseudomonas aeruginosa* PU 21 (Rip 64). *Biotechnol Bioeng* **44**: 999–1006

Chang JS Law R and Chang CC (1997) Biosorption of lead, copper and cadmium by biomass of *Pseudomonas aeruginosa* PU21 *Wat Res* **31**: 1651-1658

Chen P and Ting Y (1995) Effect of heavy metal uptake on the electrokinetic properties of *Saccharomyces cerevisiae*. *Biotechnol Lett* **17**: 107–112

Chen XC Wang YP Lin Q Shi JY Wu WX and Chen YX (2005) Biosorption of copper(II) and zinc(II) from aqueous solution by *Pseudomonas putida* CZ1. *Colloids Surf B Biointerfaces* **46**: 101-107

Chen H M Zheng C R Tu C and Shen Z G (2000) Chemical methods and phytoremediation of soil contaminated with heavy metals. *Chemosphere* **41:** 229-234

Cho DY Lee ST Park SW and Chung AS (1994) Studies on the biosorption of heavy metals onto *Chlorella vulgaris*. *J. Environ. Sci Health Part A*; **29**: 389– 409

Choi SB and Yun YS (2006) Biosorption of cadmium by various types of dried sludge: an equilibrium study and investigation of mechanisms. *J Hazard Mater* **138:** 378–383

Choi SB and Yun YS, (2004) Lead biosorption by waste biomass of *Corynebacterium glutamicum* generated from lysine fermentation process *Biotechnol Lett* **26**: 331-336

Chojnacka K Chojnacki A and Gorecka H (2005) Biosorption of Cr^{3+}, Cd^{2+} and Cu^{2+} ions by blue green algae *Spirulina* sp.: kinticks, equilibrium and the mechanism of the process. *Chemosphere* **59**: 75-84

Chong KH and Volesky B (1995) Description of 2-metal biosorption equilibria by Langmuir-type models. *Biotechnol Bioeng* **47:** 451-460

Collins Y and Stotzky G (1992) Heavy metals alter the electrokinetic properties of bacteria, yeasts, and clay minerals. *Appl Environ Microbiol* **58**: 1592–1600

Costa ACA and Leite SGC (1991) Metal biosorption by sodium alginate immobilized *Chlorerella homosphaera Biotechnol Letter* **13:** 559-562

Cotoras D Viedma P Cifuentes L and Mestre A (1992) Sorption of metal ions by whole cells of *Bacillus* and *Micrococcus Environ Technol Lett* **13**: 551–559

Crini G (2006) Non-conventional low-cost adsorbents for dye removal: a review *Biores Technol* **97:** 1061–85

Davis T A Volesky B and Mucci A (2003) A review of the biochemistry of heavy metal biosorption by brown algae. *Wat Res* **37:** 4311–4330

Delgado A Anselmo AM and Novais JM (1998) Heavy metal biosorption by dried powdered mycelium of *Fusarium flocciferum*. *Wat Environ Res* **70**: 370–376

Dursun AY (2006) comparative study on determination of the equilibrium, kinetic and thermodynamic parameters of biosorption of copper (II) and lead(II) ions onto pretreated *Aspergillus niger Biochem Eng J* **28:** 187–195

Edyvean RGJ Williams CJ Wilson MW and Aderhold D (1997) Biosorption using unusual biomass. In: Wase, D.A.J. and Foster, C., Editors. *Biosorbents for Metals Ions,* Taylor & Francis, London, pp 165–181

Errasquín E L Vázquez C (2003) Tolerance and uptake of heavy metals by *Trichoderma atroviride* isolated from sludge. *Chemosphere* **50:** 137-143

Esposito A Pagnanelli F Lodi A Solisio C and Veglió F (2001) Biosorption of heavy metals by *Sphaerotilus natans*: an equilibrium study at different pH and biomass concentrations. *Hydrometallurgy* **60:** 129-141

Feng D and Aldrich C (2004) Adsorption of heavy metals by biomaterials derived from the marine algae *Ecklonia maxima Hydrometallurgy* **73**:1-10

Fourest E and Roux JC (1992) Heavy metal biosorption by fungal mycelial by-products: mechanisms and influence of pH. *Appl Microbiol Biotechnol* **37:** 399–403

Fourest E Canal C and Roux J C (1994) Improvement of heavy metals biosorption by mycelial dead biomass (*Rhizopus arrhizus, Mucor meihi* and *Penicillium chrysogenum*) pH control and cationic activation. *FEMS Microbiol Rev* **14**: 325–332

Gabr RM Hassan SHA Shoreit AAM (2008) Biosorption of lead and nickel by living and non-living cells of *Pseudomonas aeruginosa* ASU 6a. *Int. Biodeterior. Biodegradation* **62:** 195-203

Gadd G (1992) Metals and microorganisms: a problem of definition. *FEMS Microbiol Lett* **100**: 197–204

Gadd GM (1993) Interactions of fungi with toxic metals. *New Phytol* **124:** 25-60

Gadd G M (1988) Accumulation of metal by microorganisms and algae. In: Rehm H, editor. Biotechnology: a complete treatise. vol. 6B, special microbial processes vol. 4. Verlagsgesellschaft, Weinheim: VCH; p 401– 430

Galun M Galun E Siegel BZ Keller P Lehr H and Siegel SM (1987) Removal of metal ions from aqueous solutions by *Penicillium* biomass: kinetic and uptake parameters. Water Air Soil Pollut **33:** 359–371

Genter RB (1996) Ecotoxicology of inorganic chemical stress to algae. Algal ecology. Praha7 Academic Press. (pp 403–468).

Gimbert F Morin-Crini N Renault F Badot P-M Crini G (2008) Adsorption isotherm models for dye removal by cationized starch-based material in a single component system: Error analysis. *J Hazard Mater* **157:** 34-46

Ginisty P Besnainou B Sahut C and Guezennec J (1998) Biosorption of cobalt by *Pseudomonas halodentrificans*: influence of cell wall treatment by alkali and alkaline-earth metals and ion-exchange mechanisms. *Biotechnol Lett* **20:**1035-1039

GodlewskaZylkiewicz B (2006) Microorganisms in inorganic chemical analysis. *Anal Bioanal Chem* **38:**114-123

Göksungur Y Üren S and Güvenc U (2005) Biosorption of cadmium and lead ions by ethanol treated waste baker's yeast biomass. *Bioresour Technol* **96:** 103–109

Goyal N Jain SC and Banerjee UC (2003) Comparative studies on the microbial adsorption of heavy metals. *Adv Environ Res* **7:** 311–319

Gray NF (1999) Water Technology, John Wiley and Sons, New York pp. 473–474.

Gupta VK Rastogi A (2008) Equilibrium and kinetic modelling of cadmium(II) biosorption by nonliving algal biomass *Oedogonium sp* from aqueous phase *J Hazard Mater* **153:** 759-766

Gupta R Ahuja P Khan S Saxena RK Mohapatra M (2000) Microbial biosorbents: meetings challenges of heavy metals pollution in aqueous solution. *Current Sciece* **78:** 967–973

Hassan SHA Kim SJ Jung A-Y Joo JH Oh SE and Yang JE (2009) Biosorptive capacity of Cd(II) and Cu(II) by lyophilized cells of *Pseudomonas stutzeri J Gen Appl Microbiol* **55:** 27-34

Ho Y (2006) Isotherms for the sorption of lead onto peat: comparison of linear and non-linear methods. *Polish Journal of Environmental Studies* **15:** 81–86

Ho Y and Ofomaja AE (2006) Biosorption thermodynamics of cadmium on coconut copra meal as biosorbent. *Biochem Eng J* **30:** 117–123

Huang C and Huang C P (1996) Application of *Aspergillus oryze* and *Rhizopus oryzae* for Cu(II) removal. *Wat Res* **30:** 1985-1990

Huang YC and Koseoglu SS (1993) Separation of heavy metals from **industrial waste** streams by membrane separation technology. *Waste Manag* **13:** 481-501

Ivanov A Khassanova L Collery P Khassanova Z Choisy C and Etienne JC (1996) Microorganisms as a tool of studying copper metal ions induced changes in electrophysical cell. *Properties Cell Mol Biol* **42,** 825–831

Jian-hua P Rui-xia L and Hong-xiao T (2007) Surface reaction of bacillus cereus biomass and its biosorption for lead and copper ions, *J Environ Sci* **19:** 403–408

Jianlong W (2002) Biosorption of copper (II) by chemically modified biomass of *Saccharomyces cerevisiae*. *Process Biochem* **37:** 847–850

Jung M C and Thornton I (1997) Environmental contamination and seasonal variation of metals in soils, plants and water in the paddy fields around a Pb-Zn mine in Korea. *Applied Geochm* **16:** 1377-1386

Kadukova J and Vircikova E (2005) Comparison of differences between copper bioaccumulation and biosorption. *Environ Int* **31:** 227-232

Kaewsarn P (2002) Biosorption of copper (II) from aqueous solutions by pretreated biomass of marine algae *Padina* sp. *Chemosphere* **47:** 1081–1085

Kang S Lee J and Kima K (2007) Biosorption of Cr(III) and Cr(VI) onto the cell surface of pseudomonas aeruginosa. *Biochem Eng J* **36:** 54–58

Kao W-C Huang C-C and Chang J-S (2008) Biosorption of nickel, chromium and zinc by MerP-expressing recombinant *Escherichia coli J Hazard Mater* **158:** 100-106

Kapoor A and Viraraghavan T (1995) Fungal biosorption an alternative treatment option for heavy metal bearing wastewaters: a review. *Bioresour Technol* **53:**195–206

Kapoor A and Viraraghavan T (1997) Heavy metal biosorption sites in *Aspergillus niger. Bioresour Technol* **61:** 221-227

Karna RR Uma L Subramanian G and Mohan PM (1999) Biosorption of toxic metal ions by alkali-extracted biomass of a marine cyanobacterium, *Phormidium valderianum* BDU 30501. *World J Microbiol Biotechnol* **15:** 729-732

Khambhaty Y Mody K Basha S Jha B (2009) Biosorption of Cr(VI) onto marine *Aspergillus niger:* experimental studies and pseudo- second order kinetics. *World J Microbiol Biotechnol* DOI 101007/s11274-009-0028-0

Kiran B and Kaushik A (2008) Cyanobacterial biosorption of Cr(VI): Application of two parameter and Bohart Adams models for batch and column studies. *Chem Eng J* **144:** 391-399

Klimmek S Stan HJ Wilke A Bunke G and Buchholz R (2001) Comparative analysis of the biosorption of cadmium, lead, nickel, and zinc by algae. *Environ Sci Technol* **35**: 4283–4288

Komy ZR Gabar RM Shoriet AAM Mohammed RM (2006) Characterization of acidic sites of *Pseudomonas* biomass capable of binding protons and cadmium and removal of cadmium via biosorption. *World J Microbiol Biotechnol* **22:**975–982

Kratochvil D and Volesky B (1998) Biosorption of Cu from ferruginous wastewater by algal biomass. *Wat Res* **32:** 2760-2768

Kumar KV and Porkodi K (2006) Relation between some two- and three-parameter isotherm models for the sorption of methylene blue onto lemon peel. *J Hazard Mater* **138:** 633-635

Kumar R Bishnoi NR Garima and Bishnoi K (2008) Biosorption of chromium(VI) from aqueous solution and electroplatingwastewater using fungal biomass. *Chem Eng J* **135:** 202–208

Kumar V Wati L Nigam P Yadav BS Singh D and Marchant R (1998) Decolorization and biodegradation of anaerobically digested surgarcane molasses spent wash effluent from biomethanation plants by white-rot fungi. *Process Biochem* **33:** 83-88

Kumar V Wati L Nigam P Yadav BS Singh D and Marchant R (1998) Decolorization and biodegradation of anaerobically digested surgarcane molasses spent wash effluent from biomethanation plants by white-rot fungi. *Process Biochem* **33:** 83-88

Lamba N Murthy ZVP Kumar R (2002) Membrane processing of an aqueous waste stream fromcatalyst manufacturing plant, *Sep Sci Technol* **37:** 191–202

Leung WC Chua H and Lo WH (2001) Biosorption of heavy metals by bacteria isolated from activated sludge. *Appl Biochem Biotechnol* **91:**171-84

Lin CC and Lai YT (2006) Adsorption and recovery of lead (II) from aqueous solutions by immobilized *Pseudomonas aeruginosa* PU21 beads. *J Hazard Mater* **137:** 99–105

Liu H-L Chen B-Y Lan Y-W and Cheng Y-C (2004) Biosorption of Zn(II) and Cu(II) by the indigenous *Thiobacillus thiooxidans Chem Eng J* **97:** 195-201

Lopez A Larao N Priergo J and Marques A (2001) Effect of pH on the biosorption of nickel and other heavy metals by *Pseudomonas fluorescens* 4F39. *J Ind Microbiol Biotechnol* **24:** 146– 151

Loukidou MX Zouboulis A I Karapantsios T D and Matis K A (2004) Equilibrium and kinetic modeling of chromium (VI) biosorption by *Aeromonas caviae Colloids and Surfaces A: Physicochem Eng Aspects* **242:** 93–104

Lu W-B Shi J-J Wang C-H Chang J-S (2006) Biosorption of lead, copper and cadmium by an indigenous isolate *Enterobacter* sp. J1 possessing high heavy-metal resistance. *J Hazard Mater* **134:**80–86

Lu Z H Dameron C T and Solioz M (2003) The *Enterococcus hirae* paradigm of copper homeostasis: copper chaperone turnover, interactions, and transactions. *Biometals* **16:** 137-143

Malik A (2004) Metal bioremediation through growing cells. *Environ Int* **30**: 261–278

Mameri N Boudries N Addour L Belhocine D Lounici H Grib H Pauss A (1999) Batch zinc biosorption by a bacterial nonliving *Streptomyces rimosus* biomass. *Wat Res* **33:**1347–1354

Mapoleto M and Torto N (2004) Trace enrichment of metal ions in aquatic environments by *Saccharomyces cerevisiae Talanta* **64**: 39–47

Mapoleto M Torto N and Prior B (2005) Evaluation of yeast strains as possible agents for trace enrichment of metal ions in aquatic environments. *Talanta* **65**: 930–937

Marshall WE Wartelle LH Boler DE Johns MM and Toles CA (1999) Enhanced metal adsorption by soybean hulls modified with citric acid. *Bioresour Technol* **69**: 263–268

Mattuschka B and Strauble G (1993) Biosorption of metals by a waste biomass. *J Chem Tech and biotechnology* **58**, 57-63

Morillo JA García-Ribera R Quesada T Aguilera M Ramos-Cormenzana A and Monteoliva-Sánchez M (2008) Biosorption of heavy metals by the EPS produced by *Paenibacillus jamilae World J Microbiol Biotechnol* **24**:2699–2704

Morley GF and Gadd GM (1995) Sorption of toxic metals by fungi and clay minerals. *Mycol Res* **99:** 1429–1438

Motrazavi S Volchek K and Whittaker H (1993) The removal of arsenic from water by adsorption and microfiltraion, proceeding of the 10th technical seminar on chemical spills, St John, New Brunswick, pp9-29

Mullen M D Wolf D C Ferris F G Beveridge T J Flemming C A and Bailey G W (1989) Bacterial sorption of heavy metals. *Appl Environ Microbiol* **55**: 3143-3149

Mungasavalli DP Viraraghavan T and Jin YC (2007) Biosorption of chromium fromaqueous solutions by pretreated *Aspergillus niger*: batch and column studies. *Colloids Surf A Physicochem Eng Aspects***301:** 214–223

Muraleedharan TR and Venkobachar C (1990) Mechanism of biosorption of copper (I) by *Ganoderma lucidum Biotechnol Bioeng* **35:** 320–325

Murthy ZVP Chaudhari L B (2009) Separation of binary heavy metals from aqueous solutions by nanofiltration and characterization of the membrane using Spiegler–Kedem model. *Chem Eng J* **150** : 181–187

Nadeem R Ansari TM Khalid AM (2008) Fourier Transform Infrared Spectroscopic characterization and optimization of Pb(II) biosorption by fish (*Labeo rohita*) scales. *J Hazard Mater***156:** 64-73

Nakajima A Yasuda M Yokoyama H Ohya-Nishiguchi H Kamada H (2001) Copper biosorption by chemically treated *Micrococcus luteus* cells*World J Microbiol Biotechnol* **17**:343–347

Nandy T Shastry S Pathe PP and Kaul S N (2003) Pre-treatment of currency printing ink wastewater through coagulation-flocculation process. *Water Air Soil Pollut* **148:** 15–30

Norton L Baskaran K and McKenzie T (2004) Biosorption of zinc from aqueous solutions using biosolids. *Adv Environ Res* **8:** 629-635

Nriagu J O (1990) Humans influence on the global cycling of trace metals. In: Heavy metals in the environment. Frame, J G, editor 1 CEP consultants Ltd Edinburg

Ochie A V (2008) Recent Progress on Biosorption of Heavy Metals from Liquids Using Low Cost Biosorbents: Characterization, Biosorption Parameters and Mechanism Studies. *CLEAN - Soil Air Water* **36:** 937 – 962

Ok Y S Lee H and Kim J G (2007) Enhanced uptake of Cadmium by native plant (*Artemisia princeps* var. Orientalis) using Ethylenediaminetetrascetic acid. *J Biol Sci* **7:** 681-684

Oo CW Kassim MJ and Pizzi A (2009) Characterization and performance of Rhizophora apiculata mangrove polyflavonoid tannins in the adsorption of copper (II) and lead (II). *Ind Crops Prod* doi:101016/jindcrop200903002

Özar A and Özer D (2003) Comparative study of the biosorption of Pb (II), Ni (II) and Cr (II) ions onto *S. cerviciae*: determination of biosorption heats. *J Hazard Mater* **100:** 219-229

Ozdemir G Ozturk T Ceyhan N Isler R and Cosar T (2003) Heavy metal biosorption by biomass of *Ochrobacterium anthropi* producing exopolysaccharide in activated sludge. *Bioresour Technol* **90:** 71-74

Öztürk A Artan T and Ayar A (2004) Biosorption of nickel (II) and copper (II) ions from aqueous solution by *Streptomyces coelicolor* A3 (2). *Colloids Surf B Biointerfaces* **34:** 105-111

Padmavathy V (2008) Biosorption of nickel(II) ions by baker's yeast: kinetic, thermodynamic and desorption studies. *Bioresour Technol* **99:** 3100–3109

Pagnanelli F Esposito A Veglio F (2002) Multi-metallic modelling for biosorption of binary systems. *Wat Res* **36:** 4095-4105

Pagnanelli F Petrangeli Papini M Trifoni M Toro L and Veglio F (2000) Biosorption of metal ions on *Arthrobacter* sp.: Biomass characterization and biosorption modeling. *Environ Sci Technol* **34**: 2773–2778

Pardo R Herguedas M Barrado E and Vega M (2003) Biosorption of cadmium, copper, lead and zinc by inactive biomass of *Pseudomonas putida Anal Bioanal Chem* **376:** 26-32

Park D Yun YS Jo JH Park JM (2005) Mechanism of hexavalent chromium removal by dead fungal biomass of *Aspergillus niger Wat Res* **39**:533–540

Parsons JG Gardea-Torresdey JL Tiemann KJ and Gamez G (2003) Investigation of trace level binding of $PtCl6^{2-}$ and $PtCl4^{2-}$ to alfalfa biomass (*Medicago sativa*) using Zeeman graphite furnace atomic absorption spectrometry. *Anal. Chim. Acta.* **478**: 139–145.

Pavasant P Apiratikul R Sungkhum V Suthiparinyanont P Wattanachira S and Marhaba TF (2006) Biosorption of Cu^{2+}, Cd^{2+}, Pb^{2+}, and Zn^{2+} using dried marine green macroalga *Caulerpa lentillifera Bioresour Technol* **97**: 2321–2329

Pérez Silva RM Rodriguez AA G_omez Montes De Oca JM Cantero Moreno D (2009) Biosorption of chromium, copper, manganese and zinc by Pseudomonas aeruginosa AT18 isolated from a site contaminated with petroleum. *Bioresour Technol* **100:** 1533-1538

Pethkar AV and Panikar KM (1998) Recovery of gold from solutions using *Cladosporium cladosporioides* biomass beads *J Biotechnol* **63**:121–136

Pillichshammer M Pumple T Eller K Klima J and Schinner F (1995) Biosorption of chromium to fungi *Biometals* **8**: 117-121

Plette A Benedetti M and Van Rimsdijk (1996) Comparative binding of protons, calcium, cadmium and zinc to isolated cell walls of a Gram-positive soil bacteria. *Environ Sci Technol* **30**: 1902-1910

Prasad M NV Sajwan K S and Naidu R (2006) Trace elements in the environment CRC Press Boca Racton

Preetha B and Viruthagiri T (2007) Batch and continuous biosorption of chromium(VI) by *Rhizopus arrhizus Sep Purif Technol* **57:** 126–133

Preetha B and Viruthagiri T (2005) Biosorption of zinc(II) by *Rhizopus arrhizus*: equilibrium and kinetic modeling, *Afr J Biotechnol* **4**: 506–508

Puranik PR and Paknikar KM (1997) Biosorption of lead and zinc from solutions using *Streptoverticillium cinnamoneum* waste biomass, *J Biotechnol* **55**: 113–124

Puranik PR Chabukswar NS and Paknikar KM (1995) Cadmium biosorption by *Streptomyces pimprina* waste biomass *Appl Microbiol Biotechnol* **43:** 1118–1121

Puranik PR and Paknikar KM (1999) Biosorption of lead, cadmium and zinc by *Citrobacter* strain MCM B-181: characterization studies. *Biotechnol Progress* **15**: 228-237

Qin JJ Wai MN Oo MH Lee H (2004) A pilot study for reclamation of a combined rinse from a nickel plating operation using a dual-membrane UF/RO process, ***Desalination*** **161:** 155–167

Raraz AG (1995) Biological and biotechnological waste management in materials processing. *JOM* 56–62

Robles LC Feo JC and Aller AJ (2000) Selective preconcentration of phenyl-mercury by living *Escherichia coli* and its determination by cold vapour atomic absorption spectrometry. *Anal Chim Acta* **423:** 255–263

Ruiz Manriquez A Magana PI López V Guzmán R (1997) Biosorption of Cu by *Thiobacillus ferrooxidans Bioprocess Eng* **18**:113–118

Sag Y and Kutsal T (1995) Biosorption of heavy metals by *Zoogloea ramigera*: use of adsorption isotherms and comparison of biosorption characteristics. *Chem Eng J* **60**:181–188

Sar P Kazy SK Asthana RK and Singh SP (1999) Metal adsorption and desorption by lyophilized *Pseudomonas aeruginosa Int Biodeterior Biodegradation* **44**: 101–110

Sar P and D'Souza S F (2001) Biosorptive uranium uptake by a *Pseudomonas* strain: Characterization and equilibrium studies. *J Chem Technol Biotechnol* **76**: 1286-1294

Savvaidis I Hughes M N Poole R K (*2003*) Copper *biosorption* by *Pseudomonas cepacia* and other strains. *World J Microbiol Biot* **19**: 117-121

Savvaidis I Hughes MN and Poole RK (1992) Differential pulse polarography: a method of directly measuring uptake of metal ions by live bacteria without separation of biomass and medium. *FEMS Microbiol Lett* **92**: 181-186

Seki H Suzuki A and Maruyama H (2005) Biosorption of Cr (VI) and As (V) onto methylated yeast biomass. *J Colloid Interf Sci* **281**: 261–266

Seki H Suzuki A and Mitsueda SI (1998) Biosorption of Heavy Metal Ions on *Rhodobacter sphaeroides* and *Alcaligenes eutrophus H16 J Colloid Interf Sci*, **197**: 185-190

Selatnia A Boukazoula A Kechid N Bakhti MZ Chergui A Kerchich Y (2004) Biosorption of lead (II) from aqueous solution by a bacterial dead *Streptomyces rimosus* biomass. *Biochem Eng J* **19**: 127–135

Senthilkumar R Vijayaraghavan K Thilakavathi M Iyer PVR and Velan M (2007) Application of seaweeds for the removal of lead from aqueous solution. *Biochem Eng J* **33**: 211–216

Sheng PX Ting YP Chen JP and Hong L (2004) Sorption of lead, copper, cadmium, zinc, and nickel by marine algal biomass: characterization of biosorptive capacity and investigation of mechanisms. *J Colloid Interf Sci* **275**: 131–140

Shumate H Strandberg G and Parrott J (1978) Biological removal of metal ions from aqueous process streams. *Biotechnol Bioeng Symp* **8**: 12–20

Shuttleworth K L and Unz R F (1993) Sorption of heavy metals to *Thiothrix* strain A1 *Appl Environ Microbiol* **59**:1274–1282

Sips R (1948) On the structure of a catalyst surface. *J Chem Phys* **16**: 490-495

Srinath T Verma T Ramteke PW and Garg SK (2002) Chromium (VI) biosorption and bioaccumulation by chromate resistant bacteria. *Chemosphere* **48**: 427–435

Tan T Cheng P (2003) Biosorption of metal ions with Penicillium chrysogenum *Appl Biochem Biotechnol* **104**:119-128

Tewari N Vasudevan P Guha BK (2005) Study on biosorption of Cr(VI) by Mucor hiemalis. *Biochem Eng J* **23**:185–192

Tong Mi (2007) Polymer modified biomass of baker's yeast for treating simulated wastewater containing nickel and lead. *Polymers for Advanced Technologies* **18**: 829 – 834

Tsezos M (2001) Biosorption of metals. The experience accumulated and the outlook for technology development. *Hydrometallurgy* **59**:241–243

Tsezos M Remoudaki E and Angelatou V (1995) Systematic study on equilibrium and kinetics of biosorptive accumulation. The case of Ag and Ni. Int. *Biodet Biodeg* **35**: 331-332

Tunali S and Akar T (2006) Zn (II) biosorption properties of *Botrytis cinerea* biomass. *J Hazard Mater* **131**: 137-45

Tunali S Akar T Safi Ozcan A Kiran I and Ozcan A (2006) Equilibrium and kinetics of biosorption of lead (II) from aqueous solutions by *Cephalosporium aphidicola*. *Sep Purif Technol* **47**:105–112

Tuzen M Saygi KO Usta C Soylak M (2008) *Pseudomonas aeruginosa* immobilized multiwalled carbon nanotubes as biosorbent for heavy metal ions. *Bioresour Technol* **99**:1563-1570

Uslu G and Tanyol M (2006) Equilibrium and thermodynamic parameters of single and binary mixture biosorption of lead (II) and copper (II) ions onto *Pseudomonas putida*: Effect of temperature *J Hazard Mater* **135**: 3187-3193

Veglio F and Beolchini F (1997) Removal of metals by biosorption: a review. *Hydrometallurgy* **44**: 301–316.

Vegliò F Beolchini F and Gasbarro A (1997) Biosorption of toxic metals: an equilibrium study using free cells of *Arthrobacter* sp. *Process Biochem* **2**: 99-105

Vijayaraghavan K and Yun Y-S (2008) Bacterial biosorbents and biosorption. *Biotechnol Adv* **26**:266-291

Vijayaraghavan K Padmesh TVN Palanivelu K and Velan M (2006) Biosorption of nickel(II) ions onto *Sargassum wightii*: application of two-parameter and three-parameter isotherm models. *J Hazard Mater* **133**: 304–308

Vilar VJP Botelho CMB and Boaventura RAR (2006) Equilibrium and kinetic modelling of Cd(II) biosorption by algae *Gelidium* and agar extraction algal waste. *Wat Res* **40**: 291–302

Vilar VJP Botelho CMS and Boaventura RAR (2007) Modeling equilibrium and kinetics of metal uptake by algal biomass in continuous stirred and packed bed adsorbers. *Adsorption* **13**: 587–601

Volesky B Naja G (2005) Biosorption: application strategies. 16th Internat. Biotechnol. Symp.Compress Co., Cape Town, South Africa.

Volesky B (1990) Biosorption by fungal biomass. In: Volesky B, Biosorption of heavy metals. Florida: CRC press, (140–171).

Volesky B (1994) Advances in biosorption of metals: selection of biomass types. *FEMS Microbiol Rev* **14**: 291-302

Volesky B (2007) Biosorption and me. *Water Res* **41**:4017–29

Volesky B and Holan ZR (1995) Biosorption of heavy metals. *Biotechnol Prog* **11**: 235-250

Volesky B (2003) Sorption and biosorption BV Sorbex, Inc. Montreal-St. Lambert, Quebec, Canda

Wang J and Chen C (2006) Biosorption of heavy metals by *Saccharomyces cerevisiae*: a review, *Biotechnol Adv* **24**: 427–451

Wang CL Michels S PC Sawson C Kitisakkul S Baross J A Keasling J D and Clark DS (1997) Cadmium removal by a new strain of *Pseudomonas aeruginosa* in aerobic culture. *Appl Environ Microbiol* **63**: 4075-407

Whistler R and Daniel TR (1985) Carbohydrates. In Owen R Fennema (Ed.) Food Chemistry. 96-105

Wood JM and Wang H-K (1985) Microbial resistance to heavy metals. In: Irgolic KJ, editor: Environmental inorganic chemistry (proceedings). Deerfield Beach, Florida7 VCH Publishers; (pp. 487–512).

Yan G and Viraraghavan T (2000) Effect of pretreatment on the bioadsorption of heavy metals on *Mucor rouxii, Water SA* **26**: 119–124

Yan G and Viraraghavan T (2006) Effect of pretreatment on the bioadsorption of heavy metals on *Mucor rouxii Water SA* **26**: 119–124

Yilmaz EI Ensari NY (2005) Cadmium biosorption by Bacillus circulans strain EB1. *World J Microbiol Biotechnol* **21**:777–779

Zhou M Liu Y Zeng G Li X Xu W and Fan T (2007) Kinetic and equilibrium studies of Cr(VI) biosorption by dead *Bacillus licheniformis* biomass. *World J MicrobiolBiotechnol* **23**: 43–48

Ziagova M Dimitriadis G Aslanidou D Papaioannou X Tzannetaki EL and Liakopoulou-Kyriakides M (2007) Comparative study of Cd(II) and Cr(VI) biosorption on *Staphylococcus xylosus* and *Pseudomonas* sp. in single and binary mixtures. *Bioresour Technol* **98**: 2859–2865

5

INFLUENCE OF HERBICIDES ON BIOLOGICAL PROCESSES IN GRAIN LEGUMES

VIRENDER SARDANA AND PUSHP SHARMA

INTRODUCTION

Nitrogen is the most important nutrient required for plant growth. Its manufacturing requires non-renewable fossil fuel and thus its application in the form of chemical fertilizers is expensive, has low use efficiency and its use causes environmental pollution (Keeney 1982, Newbould 1989). Legumes are the integral part of cropping systems. Legumes globally play an important role in nitrogen cycling. Nitrogen fixation is carried out by legume-*rhizobia* symbiosis. Only *prokaryotes* have the ability to fix nitrogen biologically. Nitrogen fixed by various legumes in association with different species of *rhizobia* through the process of biological nitrogen fixation (BNF) is better alternative as it results in increased soil fertility, meets most of the N requirements of the crop and leaves substantial amount of N_2 in soil for utilization of subsequent crops in rotation (Peoples and Herridge 1990). These contribute about 25 per cent of the total nitrogen in grain legumes/pulses, 32 per cent in permanent grasslands and 29 per cent in forest ecosystems (Suhas *et al.* 2007). It is estimated that legumes contribute about 35×10^6 tonnes of biologically fixed nitrogen per annum (Burns and Hardy 1975, Paul and Clark 1996). Legumes vary in their ability to fix N_2 depending upon several factors (Giller and Wilson 1991, Peoples and Crasswell 1992, Wani *et al.* 1995).

Grain legumes or pulse crops are grown both in *kharif* and *rabi* seasons. The important *kharif* season pulses are pigeonpea (*Cajanus cajan* L. Millsp.), blackgram/ urdbean (*Phaseolus mungo* L. Hepper), greengram/moong (*Phaseolus radiatus* L. Wilczek), horsegram (*Macrotyloma uniflorum* L. Verdc), mothbean (*Phaseolus aconitifolia* J. Marechal) and cowpea (*Vigna unguiculata* L. Walp). The *rabi* pulses include chickpea (*Cicer arietinum* L.) lentil (*Lens culinaris* Medic), pea (*Pisum sativum* L), lathyrus (*Lathyrus sativus*

L.) and frenchbean (*Phaseolus vulgaris* L.). Greengram and blackgram are also grown in *spring/summer* season in northern India after harvest of sugarcane, wheat, potato, mustard etc. and during *rabi* season in the rice fallows in south-eastern and southern region.

Weeds constitute an important factor in crop production and their presence causes significant reductions in yields besides impairing quality of produce due to competition and allelopathic effects. At the global level, the losses caused by weeds exceed the losses from any other category of pests. Of the total losses to agricultural produce, weeds account for about 45 per cent in comparison to 30 per cent by insects, 20 per cent by diseases and 5 per cent by other organisms (Rao 1987). Infestation of weeds in pulses is more in the *kharif / rainy* season than during *rabi*. Important weeds associated with pulse crops in different seasons are listed in Table 1.

TABLE 1 : Major weeds of pulse crops

Classification of weeds	*Predominant weeds*
Kharif	
Annual grasses and sedges	*Echinochloa colonum, Echinochloa crus-galli, Eleusine aegiptiacum, Eleusine indica, Eragrostis sp., Dactyloctenium aegiptium, Digitaria sanguinalis, Panicum maximum, Setaria glauca,*
Broadleaf weeds	*Amaranthus viridis/spinosus, Celosia argentia, Cleome viscosa, Commelina benghalensis, Digera arvensis, Phyllanthus niruri, Physallis minima, Trianthema monogyna/portulacastrum, Tribulus terrestris,*
Perennials	*Cynodon dactylon, Cyperus rotundus, Saccharum spontaneum, Sorghum halepanse,*
Rabi	
Annual grasses and sedges	*Avena fatua/ludoviciana, Phalaris minor, Poa annua*
Broadleaf weeds	*Anagallis arvensis, Asphodelus tenuifolius, Chenopodium album, Cuscuta spp., Fumaria parviflora, Lathyrus aphaca, Medicago denticulata, Melilotus spp., Rumex dentatus/spinosus, Vicia spp.,*
Perennials	*Circium arvenses, Cynodon dactylon, Cyperus rotundus*

While the losses caused by insects-pests and diseases to standing crop and stored grains are visible, weeds invisibly rob the crop of essential and limited nutrients and compete for space, light, moisture etc. Use of herbicides for weed control (Tables 2 and 3) is an essential input in crop management and their role is gaining importance due to rising costs and non-availability of labour to carry out weeding. Presently, herbicides form the largest class of agro-chemicals.

Application of any agro-chemical including pesticides on crops may alter the equilibrium by direct toxic action on micro-organisms inhabiting the soil or may cause selective toxicity to a group of micro-organisms. Herbicides used to control weeds may adversely affect crop plants through their effect on growth, nodulation, nitrogen fixation, other physiological processes or economic yield by virtue of their toxic nature. Herbicide use may alter the biological equilibrium in soil (Grossbard 1976). As with other biocides, biological effect of herbicides often extends to non-target organisms in the environment. These unintended effects may inhibit or stimulate the activities of harmful or beneficial non-target organisms. These depressive or stimulating effects depend on the kind of chemicals, their concentration and environmental conditions.

TABLE 2 : Commonly used herbicides in pulse crops in India

Common name	*Chemical name*	*Trade name*	*Legumes in which mainly used*
Alachlor	2-chloro-N-(2,6-diethylphenyl) N-(methoxymethyl)acetamide	Lasso	Pigeonpea, green gram, black gram, cowpea, peas
Dinitramine	N^3-N^3-diethyl-2,4-dinitro-6-(trifluoromethyl)-1-3-benzenediamine	Cobex	-
Fluchloralin	*N*-(2-chloroethyl)-2,6-dinitro-*N*-propyl-4-(trifluoromethyl)benzenamine	Basalin	Pigeonpea, green gram, black gram, peas, lentil chickpea
Fluorodifen	2-nitro-1-(4-nitrophenoxy)-4-trifluoromethylbenzene	Soyex, Preforan	-
Isoproturon	*N,N*-dimethyl-*N*-(4-{1-methylethyl}phenyl)urea	Isopro-turon	-
Linuron	*N*-(3,4-dichlorophenyl)-*N*-methoxy-*N*-methylurea	Afalon, Lorex	Chickpea
Nitrofen	2,4-dichloro-1-(nitrophenoxy)benzene	TOK	-
Metolachlor	2-chloro-N-(2-ethyl-6-methylphenyl)-N-(2-methoxy-1-methylethyl) acetamide	Dual	Soybean
Oxadiazon	3-[2,4-dichloro-5-(1-methylethoxy) phenyl]-5-(1,1-dimethylethyl)-1,3,4-oxadiazole-2-(3 *H*)-one	TOK E 25	Pigeonpea, soybean, peas, chickpea, lentil
Oxyfluorfen	2-chloro-1-(3-ethoxy-4 nitrophenoxy)-4-(trifluoromethyl) benzene	Goal	Chickpea, lentil
Sethoxydim	2-[1-(ethoxyimino)butyl]-5-[2-(ethylthio)propyl]-3-hydroxy-2-cyclohexen-1-one	-	Green gram, chickpea

Common name	*Chemical name*	*Trade name*	*Legumes in which mainly used*
Pendimethalin	(N-(1-ethylpropyl)-3,4-dimethyl-2,6-dinitrobenzamine)	Stomp, Prowl	Soybean, green gram, black gram, chickpea, peas
Prometryn	*N,N*-bis(1-methyl)-6(methylthio)-1,3,5-triazine-2,4-diamine	Caparol, Gessagard	Peas
Diallate	S-(2,3-dichloroallyl-diisopropylthiocarbamate	Avadex	
Triallate	S-(2,3,3-trichloro-2-propenyl) bis(1-methylethyl)carbamothioate	Avadex, Far Go	
EPTC	S-ethyl dipropylthiocarbamate	Eptam	Pigeonpea, peas, chickpea
Terbutryn	*N*-(1,1-dimethylethyl)-N-ethyl-6 (methylthio)-1,3,5-triazine-2,4-diamine	Igran, Prebane	Pigeonpea, peas, chickpea
Imazaquin	2-[4,5-dihydro-4-methyl-4-(1-methylethyl)-5-oxo-1 *H*-imidazol-2-yl]-3-quinolinecarboxylic acid	-	Soybean
Imazamox	2-[4,5-dihydro-4-methyl-4-(1-methylethyl)-5-oxo-1 *H*-imidazol-2-yl]-5-methyl-3-pyridinecarboxylic acid	-	Peas
Imazethpyr	2-[4,5-dihydro-4-methyl-4-(1-methylethyl)-5-oxo-1 *H*-imidazol-2-yl]-5-ethyl-3-pyridinecarboxylic acid	-	Soybean, black gram, peas, lentil, faba bean
Bentazone	3-(1-methylethyl)-(1*H*)-2,1,3-benzo thiadiazin-4(3*H*)-one 2,2-dioxide	Basagran	Peas
Carbetamide	*N*-ethyl-2-[1{phenylamino}carbonyl) oxy]propanamide(R)-isomer	Legurame	
MCPA	(4-chloro-2-methylphenoxy) acetic acid	Chiotox, Rhomene, DOW MCP	
MCPB	4-(4-chloro-2-methylphenoxy) butanoic acid	Can-Trol, Thistrol	
Metribuzin	4-amino-6-(1,1-dimethylethyl)-3(methylthio)-1,2,4-triazin-5(4H)-one	Sencor	
Dalapon	2,2-dichloropropionic acid	Dowpon	
Dinoseb	2-(1-methylpropyl)-4,6-dinitrophenol	DNBP, Premerge, Basanite	

Common name	*Chemical name*	*Trade name*	*Legumes in which mainly used*
Methabenzthia zuron	*N*-(2-benzothiazolyl)-*N,N*-dimethylurea	Dosanex	
Chloroxuron	*N*-[4-(4-chlorophenoxy) phenyl]-*N,N*-dimethylurea	Tenoran, Norex	Soybean, peas
Fluazifop butyl	(R)-2-(4-[{5-(trifluoromethyl)-2-pyridinyl}oxy)phenoxy]propionic acid	-	Soybean, greengram, lentil
Fomesafen	5-[2-chloro-4(trifluoromethyl) phenoxy]-N-(methylsulfonyl)-2-nitrobenzamide	-	Green gram
Trifluralin	2,6-dintro-*N,N*-dipropyl-4-(trifluromethyl)benzenamine	Treflan	Pigeonpea, green gram, black gram, peas, chickpea

TABLE 3 : Commonly used herbicides and their time of application in pulse crops

Crop	*Herbicide (Time of application)*
Soybean	Pendimethalin (pre-em), Imazaquin (pre-plant), Imazethpyr (pre-plant, pre-em, post–em)), Metolachlor (pre-em), Trifluralin (pre-plant, pre-em), Oxadiazon (pre-em), Sethoxydim (pre-em), Chloroxuron (post-em), Fluazifop butyl (post-em)
Pigeonpea	EPTC, Trifluralin (pre-plant), fluchloralin (pre-em), Oxadiazon (pre-em), Terbutryn (pre-em), Bentazon (post-em)
Green gram	Fluchloralin (pre-plant, pre-em), Trifluralin (pre-plant, pre-em), Pendimethalin (pre-em), Fluazifop butyl (post-em), Fomesafen (post-em), Sethoxydim (post-em)
Black gram	Fluchloralin (pre-plant, pre-em), Trifluralin (pre-plant, pre-em), Pendimethalin (pre-em), Imazethpyr (post-em)
Peas	Diallate (pre-plant), Triallate (pre-plant, pre-em), EPTC (pre-plant) Fluchloralin (pre-plant, pre-em), Trifluralin (pre-plant, pre-em), Oxadiazon (pre-em), Pendimethalin (pre-em), Prometryn (pre-em), Terbutryn (pre-em), Bentazon (post-em), Carbetamide (post-em), Chloroxuron (post-em), Cyanazine (post-em), Dalapon (post-em), Dinoseb (post-em), Imazamox (post em), Imazethpyr (post-em), MCPA (post-em), MCPB (post-em), Metribuzin (post-em)
Chickpea	EPTC (pre-plant), Fluchloralin (pre-plant, pre-em), Trifluralin (pre-plant, pre-em), Oxadiazon (pre-em), Pendimethalin (pre-em), Terbutryn (pre-em), Oxyfluorfen (pre-em), Sethoxydim (post-em)
Lentil	Fluchloralin (pre-plant, pre-em), Oxadiazon (pre-em), Oxyfluorfen (pre-em), Fluazifop butyl (post-em), Imazethpyr (post-em)
Fababean	Imazethpyr (post-em)

EFFECT ON LEGUME CROPS

Germination and emergence

Pre-plant application of fluchloralin and pre-emergence application of pendimethalin or metolachlor @ 1.0, 1.5 and 2.0 kg/ha did not influence germination of chickpea, though higher dose reduced it marginally (Yadav *et al.* 2006). Pre–emergence application of oxyflourfen @ 1.68–6.72 kg/ha caused significant reduction (20%) in germination of pigeonpea (Semidey and Alimodovar 1987). Pahwa *et al* (1988) recorded inhibition in germination of pigeonpea with application of fluchloralin (0.75, 1.0 and 1.25 kg/ha), pendimethalin (1.0, 1.5 and 2.0 kg/ha) and fluazifop-p-butyl (0.5, 1.0 and 1.5 kg/ha) compared to control. Pendimethalin and trifluralin (1.1, 1.7 and 2.2 kg/ha) delayed emergence and caused severe injury to soybean seedlings at concentration of 1.7 and 2.2 kg/ha (Bollich *et al.* 1988).

Growth

Swelling of root tip is universally recognized morphological effect of dinitroaniline herbicides. The toxic effect occurs between times of radical and shoot emergence of seedlings from soil (Kolhe *et al.* 1984).

Root growth of soybean was retarded more than the top growth with application of trifluralin and threshold concentration for root growth was recorded to be 0.75 ppm (Ronald 1965). Bertholet and Clark (1985) also reported reduction in root length of soybean with trifluralin. Primary root length of pigeonpea was inhibited by fluchloralin, pendimethalin and fluazifop-p-butyl up to 45 DAS which recovered at later stages (Pahwa *et al.* 1988). Fluazifop-p-butyl caused maximum reduction in dry weight of roots per plant in pigeonpea at all concentrations in comparison to untreated control at 60 DAS (Pahwa *et al.* 1988).

Salvo *et al* (1991) observed that root and shoot growth of chickpea seedlings remained either unaffected or little affected by application of fluchloralin (0.5, 1.0 and 1.5 kg/ha), metribuzin (0.25, 0.5 and 0.75 kg/ha) and pendimethalin (0.75, 1.0 and 1.25 kg/ha). Pahwa *et al.* (1988) observed increase in fresh and dry weight of roots of pigeonpea at 45 DAS at lower concentration of fluchloralin and pendimethalin which decreased at higher concentration of herbicides.

Pahwa and Prakash (1992) reported that higher concentration of pendimethalin reduced root and shoot development of chickpea. Stem length and number of leaves of pigeonpea were reduced with application of herbicide, particularly fluazifop-p-butyl (Pahwa *et al.* 1988). Fresh and dry weights of shoots were significantly reduced with all the herbicides at initial stges, which regained at later stages particularly where fluchloralin was applied (Pahwa *et al.* 1988).

Pahwa (1972) reported reduction in length of roots and chlorophyll content of shoots of peas with pre-emergence application of dinitroaniline herbicides. Pre-emergence application of methabenzthiazuron and pendimethalin @ 1.2 kg/ha, diclofopmethyl @ 0.6-1.2 kg/ha, fluchloralin @ 0.5-1.0 kg/ha and oxyfluorfen @ 0.1-0.2 kg/ha decreased root length but not shoot length in peas (Prakash and Pahwa 1984). Reduction in fresh and dry weight of shoot of cowpea at initial stages and recovery at later stages on with fluchloralin application was also noticed by Prakash and Pahwa 1985.

Herbicides have been found to affect growth of crops like mungbean and peas (Prakash and Pahwa 1982, Pahwa and Prakash 1988). Nadasy *et al.* (2000) reported reduction in leaf area of soybean with application of metolachlor. Chaurasia (2001) and Murti *et al.* (2004) reported reduction in number of branches per plant and dry matter production in urdbean with application of pendimethalin and fluchloralin. Fluchloralin, pendimethalin and metolachlor @ 2.0 kg/ha caused significant reduction in number of branches per plant of chickpea than their lower dose (1.0 kg/ha) but was at par with untreated control (Yadav *et al.* 2006). All these herbicides except fluchloralin @ 1.0 kg/ha reduced leaf area of plants at 30 DAS compared to untreated plants (Yadav *et al.* 2006).

Lower doses of fluchloralin and metribuzin increased number of leaves per plant in blackgram (Pahwa and Prakash 1988). Majority of the herbicides have been observed to cause no inhibitory effect on bacterial population and activity at normal dosages of their application (Mohammad 1984, Tanski *et al.* 1984). However, dalapon (Kruglov *et al.* 1973) and diclobenil (Grossbard 1976) inhibited bacterial population at field dosages.

BIOLOGICAL NITROGEN FIXATION (BNF)

Herbicides applied for weed control may influence the process of BNF by affecting plant growth or by influencing N_2 fixing bacteria. These may also influence physiological processes of nodulation of *rhizobia* (Eberbach and Douglas 1989, Martensson 1992, Madhavi *et al.* 1994). Different strains of *rhizobia* can show differential response to herbicides. (Mallik and Tesfai 1983, Olson *et al.* 1984, Van Rensburg and Strijdom 1984, Kishinevsky *et al.* 1988, Mahmood and Mohammed 1991, Martensson 1992) due to their differential ability to detoxify the herbicide. The effect of pesticides including herbicides on nitrogen fixation in legumes has earlier been reviewed by Gaur (1980; Lal 1988 and Roberts 1992). Effects of herbicides on biological N_2 fixation have been reported in soybean (Rennie and Dubetz 1984, Mallik and Tesfai 1985, Bollich *et al.* 1985, 1988, Ozair and Moshier *et al.* 1988, Ozair *et al.* 1990, Yueh and Hensley 1993), fababean (Bertholet and Clark 1985, Haider *et al.*1991, Vidal *et al.* 1992, Martinez *et al.* 1996), kidneybean (Bethlenfalvay *et al.* 1979, Schenelle and Hensley 1989 1990), lentil (Sandhu *et al.* 1991, Sprout *et al.* 1992), chickpea (Kumar *et al.* 1981, Khokhar and Malik 1988, Pahwa and Praksah1992), lupins (De Felipe *et al.* 1987), groundnut (Kishinevsky *et al.* 1988), clovers (Brock 1972, Lindstrom *et al.* 1985, Martensson and Nilsson 1989, Eberbach and Douglas 1989, 1991, Clark and Mohanti 1991), peas (Singh *et al.* 1994, Gonzalez *et al.* 1996), *Trifolium repens sylvestre* (Fletcher *et al.* 1956, 1957).

MICROBIAL ACTIVITY AND NODULATION

Rhizobium

No effect: Fletcher (1960) observed no adverse effect of 2,4-D, MCPA, 2,4,5-DB or MCPB on growth of *Rhizobium trifolii* up to 25 ppm. Rhizobium population was not affected by trifluralin applied @1.0 kg/ha to soil (Olson *et al.* 1984). Methabenzthiazuron (1.0 and 1.5 kg/ha), pendimethalin (1.0 and 1.5 kg/ha) and terbutryn (0.6 and 1.2 kg/ha) caused no adverse effect on *rhizobial* growth in chickpea (Kumar and Kolar 1988) and lentil (Kumar and Kolar 1989). However, higher concentration was deleterious. Sandhu (1984) found no adverse effect of methabenzthiazuron and terbutryn on nodulation and rhizobial activity in lentil. Lewis and Hamner (1946) observed that 2, 4-D upto 1000 ppm failed to influence the activity of *Rhizobium leguminosarum*. N-serve application of 1-10 ppm caused little effect on *rhizobium* survival and nodulation in lucerne but higher rates caused morphological changes (McKell and Whalley 1964).

Dunigan *et al* (1972) observed that pre-emergence application of alachlor, chloramben, DCPA or post-emergence application of chloroxuron, dinoseb, linuron and prometryne at their recommended rates did not affect nodulation in soybean. Bentazone application had little or no effect on number of nodules, weight of nodules, root and shoot and also nitrogenase activity in soybean (Ozair and Moshier 1988, Ozair *et al* 1990). Terbutryn (0.6 and 1.2 kg/ha) caused no adverse effect on number of nodules in chickpea (Kumar and Kolar 1988), lentil (Kumar and Kolar 1989) and peas (Van Rensburg and Strijdom 1984). Renne and Dubetz (1984) found no adverse effect of recommended rates of chloramben, linuron, metribuzin and trifluralin on N_2 fixation in soybean. Similarly recommended doses of sethoxydim, alachlor, fluazifop butyl and metolachlor caused no adverse effect on N_2 fixation in soybean (Kucey *et al.* 1988). No marked differences in number of nodules per plant with application of pendimethalin and metolachlor @ 1.0 kg/ha were observed in peas (Singh and Wright 1999) and chickpea (Yadav *et al.* 2006). Ammonification and sulphur oxidation were slightly reduced by diquat dibromide, paraquat, dichloride and paraquat dimethyl sulphate at 0.25-0.50 ppm concentration in silty clay loam soil (Tu 1966).

Adverse effect: The antagonistic effects of herbicides reported by several workers in differerent legumes, *rhizobial* growth and nodule activities have been vividly summarised in Tables 4 and 5. Alachlor and metribuzin (Mallik and Tesfai 1985), paraquat (Clark and Mahanty 1991), trifluralin (Mallik and Tesfai 1985, Bollich *et al.* 1988) decreased nodulation when applied at their recommended rates. Even at recommended rate of application, some herbicides decreased the process of BNF (Bollich *et al.* 1988, Martensson and Nilsson 1989, Haider *et al.* 1991, Sprout *et al.* 1992) or N_2 fixation (Bethlenfalvay *et al* 1979, Cardina *et al.* 1986, Madhavi *et al.* 1993, 1994) possibly by influencing internal processes of plant even without any visual symptoms of phytotoxicity.

In symbiosis with legumes, as the *rhizobia* obtain their energy from the host plant, any adverse effect of a herbicide on plant may decrease the supply of carbohydrates. Herbicide may directly inhibit growth of *rhizobia* (Van Rensburg and Strijdom 1984, Martensson 1992) or may reduce their nitrogen fixation capacity.

Herbicides generally affect nodulation and N_2 fixation by influencing plant growth rather than directly affecting rhizobia (Brock 1972, Bethlenfalvay *et al.* 1979, Lindstrom *et al.* 1985, Martensson and Nilsson 1989, Sprout *et al.* 1992, Gonzalez *et al.* 1996). Paraquat significantly reduced the amount of N_2 fixed in soybean (Kucey *et al.* 1988).

Trifluralin and pendimethalin up to 2.2 kg/ha were reported to inhibit N_2 fixation by affecting acetylene reduction (Bollich *et al.* 1988). In another study, inhibitory effect of higher doses of herbicides on nodulation has been reported by Prakash and Pahwa 1982). Pahwa *et al.* (1988) observed increase in fresh and dry weight of roots, number, fresh and dry weights of nodules and leghaemoglobin content of nodules of pigeonpea at 45 DAS at lower concentration of fluchloralin and pendimethalin which decreased at higher concentration of herbicides.

Herbicides have been found to affect nodulation in crops like mungbean and peas (Prakash and Pahwa 1982, Pahwa and Prakash 1988, Singh and Wright 2000a). Earlier, Szabo (1964) and Avrov (1966) observed reduction in number and size of nodules of peas due to prometryne. Fluazifop-p-butyl caused maximum reduction in number of nodules per plant at all concentrations in comparison to untreated control at 60 DAS (Pahwa *et al.* 1988). Fluchloralin, pendimethalin and metolachlor @ 2.0 kg/ha caused significant reduction in number and dry weight of nodules per plant of chickpea than their lower dose (1.0 kg/ha) but was at par with untreated control (Yadav *et al.* 2006).

Trifluralin applied at normal rates reduced nodulation in soybean, gram and othe legumes (Dunigan *et al.* 1970, Brock 1972, Misra and Gaur 1974). Nodulation decreased in soybean with application of trifluralin (Yueh and Hensley 1993). Application of simazine (1.6-3.2 kg/ha) to chickpea (Kumar *et al.* 1981), simazine (2.0 kg/ha) to fababean (Haider *et al.* 1995) and prometryn (0.75 kg/ha) to kidneybean (Fischer and Tasistro 1981) decreased dry weight of roots and shoots as well as number and weight of nodules. Prometryn applied to peas @ 6g/kg soil decreased nodule weight and nitrogenase activity (Paromenskaya and Lebskii 1986, Singh and Wright 2000b). Misra and Gaur (1975) reported severe reduction in root nodulation, growth, yield and nitrogen uptake by groundnut due to simazine application. Terbutryn also decreased number and weight of nodules in groundnut (Van Rensburg and Strijdom 1984, Kishinevsky *et al.* 1988) and lentil (Sandhu *et al.* 1991).

Adverse effect of trifluralin (Udoh and Nelson 1986) and alachlor (Kao and Wang 1981) on nodulation in soybean has also been well documented. Luthra *et al.* (1983) reported reduction in N_2 fixation in pigeonpea with application of herbicides. Bollich *et al.* (1988) found inhibitory effect of triflurlin and pendimethalin on nodulation and N_2 fixation, whereas Akhtar *et al.* (1990) reported reduction in nodulation in soybean with increasing concentration of pendimethalin and fluazifop butyl. Similarly, the number and dry weight of nodules were reduced with herbicide application in soybean (Jain *et al.* 1990). Though lower concentration of herbicides did not cause marked reduction in nodulation, pre-plant application of pendimethalin (1.25 kg/ha) and trifluralin (1.08 kg/ha) and pre-emergence application of alachlor (2.88 kg/ha) and oxadiazon (0.875 kg/ha) reduced the period of nodulation and N_2 fixation in chickpea (Salvo *et al.* 1991). In drybeans, there was differential response of herbicides on nodulation (Schenella and Hensley 1990, Ahmad and Rashad 1996).

Specific nitrogenase activity (SNA) in drybeans was depressed with post-em application of bentazone @ 2.2 kg/ha without any adverse effect on number and weight of nodules even at three times higher concentration (Schnelle and Hensley 1990). Bentazone affected infection, nodule development and plant dry matter production but not bacteria population (Martensson 1992). Number and weight of nodules in birdsfoot tremoil (*Lotus corniculatus*) remained unaffected by dalapon but were reduced with application of 2, 4-DB (Garcia and Jordan 1967). Drastic reduction in plant dry weight as well as nitrogenase activity in lucerne (*Medicago sativa*), white clover (*Trifolium repens*) and peas (*Pisum sativum*) were also observed with increasing rate of bentazone from 0 to 10 ppm (Torstensson 1975). Simazine applied @ 0.1 ppm markedly reduced nodulation of legumes such as clover, lucerne and peas (Hauke Pacewiczowa 1969, 1970). Nodule formation in peas (*Pisum sativum*), beans (*Phaseolus lunatus*), red clover (*Trifolium pratense*), alfalfa (*Medicago sativa*) and kidney bean (*Phaseolus vulgaris*) was prevented with application of 2,4-D particularly sodium and amine salts (Carlyle and Thorpe 1947, Payne and Fults 1947). Reduced nodulation in white clovers (*Trifolium repens*) following application of MCPA was observed by Payne and Fults (1947) and Anderson and Baker (1950). Choudhri and Rajgopal (1958) observed reduction in nodule formation even at 10 ppm of 2,4,5-T, IPC and MH and the adverse effect increased with increasing concentration of herbicides. 2,4-DB alone or in combination with dalapon reduced N_2 fixation, number and dry weight of nodules of bird foot trefoil (Gracia and Jordan 1967). Dalapon alone reduced dry weight rather than number of nodules whereas 2,4-DB affected host plant rather than nodules.

The aniline, trifluralin and nitralin appear to have largely adverse effect on *rhizobia* and nodulation. Most of the triazines appear inhibitory even at the recommended rates of application. Hamdi and Tewfik (1969) observed inhibition in growth and nodulation of cowpea with recommended rates of trifluralin applied on the day of planting but stimulated when herbicide was applied 4 weeks before planting.

TABLE 4 : Effect of herbicides on *Rhizobial* growth

Crop	*Herbicide*		*Effect*	*Reported by*
	Name	*Dose*		
Peas	Linuron	0.75 kg/ha	Reduced population of soil bacteria	Singh *et al* (1994)
Peas	Methabenz thiazuron	1.31 kg/ha	Reduced population of soil bacteria	Singh *et al* (1994)
Peas	Pendimethalin	0.75 l/ha	Reduced soil bacterial population	Singh *et al* (1994)
Peas	Prometryn	1-2 kg/ha	Toxicity to rhizoshere Rhizobia	Bhalla and Chourasia (1985)
Peas	Simazine	0.9-1.5 kg/ha	Toxicity to rhizoshere Rhizobia	Bhalla and Chourasia (1985)

Crop	*Herbicide*		*Effect*	*Reported by*
	Name	*Dose*		
Peas	Methabenz thiazuron	1.05 kg/ha	Reduced rhizobial population	Bhalla and Chourasia (1985)
Pigeonpea	Simazine		Reduced rhizobium population of *Rhizobium* spp.	Madhavi *et al* (1994)
Pigeonpea	Oxyfluorfen		Reduced population of *Rhizobium* spp.	Madhavi *et al* (1994)
Lentil	Oxyfluorfen		Reduced population of *Rhizobium leguminosarum*	Madhavi *et al* (1994)
Lentil	Simazine		Reduced population of *Rhizobium leguminosarum*	Madhavi *et al* (1994)
Colver, lucerne, pea	Simazine	0.1 ppm	Reduced nodulation and yield. Higher rates decreased amino acid in protein and modified levels of free amino acid	Hauke-Pacewiczowa (1969, 1970)
Alfalfa	Simazine		Reduced population of *Rhizobium meliloti*	Madhavi *et al* (1994)
Alfalfa	Oxyfluorfen		Reduced population of *Rhizobium meliloti*	Madhavi *et al* (1994)
Groundnut	Simazine	2 ppm	Reduced nodulation, growth and yield	Misra and Gaur (1975)
Not specified	Simazine	0.1-1.0 ppm	Morphological changes in nodules. Decreased nodulation	Thomas and Hammond (1971) Dunigan *et al.* (1970)
Not specified	Prometryne	2.5-5.0 ppm	Initially decrease and subsequently increase in nodulation in light chestnut soils	Shkola (1970)
	2,-D	50 ppm	Suppression *R. leguminosarum*	Gaur and Misra (1972)
	2, 4-D	250 ppm	Reduction in population of *R. meliloti* and *R japonicum*	Gaur and Misra (1972)
	2,-D	1000 ppm	No effect on *R. leguminosarum*	Lewis and Hamre (1946)
	2,-D	400 ppm	Complete inhibition of *R. leguminosarum*	Carlyla and Thorpe (1947)

TABLE 5 : Effect of herbicides on nodulation and nodule activity in different crops

Crop	*Herbicide*		*Effect*	*Reported by*
	Name	*Rate*		
Soybean (*Glycine max*)	2,4-DB	0.23 kg/ha	Reduced number and weight of nodules and SNA	Mallik and Tesfai (1985)
	Alachlor	1.7 l/ha	Significantly reduced number and weight of nodules and SNA	Mallik and Tesfai (1985)
	Bentazone	1.1 kg/ha	Marginal reduction in number and weight of nodules and ARA and SNA	Ozair and Moshier (1988), Ozair *et al* (1990)
	Bentazone	2.8 ml/l	No effect on nodule weight but number and SNA decreased	Yueh and Hensley (1993)
	Glyphosate	1.12 l/ha	Reduced number and weight of nodules and SNA	Mallik and Tesfai (1985)
	Metolachlor	1.92 l/ha	No effect on number and weight of nodules	Greenfield (1991)
	Metribuzin	0.34 - 1.7 kg/ha	Significant reduction in number and weight of nodules and SNA	Mallik and Tesfai (1985)
	Prometryn	1.6-3.2 kg/ha	Reduced number and weight of nodules; increased SNA	Kumar *et al* (1981)
	Sethoxydim	29.4 ml/l	No effect on number and weight of nodules and SNA	Yueh and Hensley (1993)
	Trifluralin	0.96 l/ha	No effect on number and weight of nodules	Greenfield (1991)
	Trifluralin	0.56-2.8 l/ha	Reduced number and weight of nodules and SNA	Mallik and Tesfai (1985)
	Trifluralin	0.92-2.8 l/ha	Drastically reduced nodulation but increased SNA	Yueh and Hensley (1993)
Kidney bean (*Phaseolus vulgaris*)	Acifluorfen	0.44 kg/ha	Reduced nodule weight	Fischer and Tasistro (1981)
	Alachlor	13.3 l/ha	No effect on number and weight of nodules or SNA	Schenelle and Hensley (1990)

Crop	*Herbicide*		*Effect*	*Reported by*
	Name	*Rate*		
	Alachlor	1.92 l/ha	Reduced weight of nodules	Fischer and Tasistro (1981)
	Bentazone	2.2 kg/ha	No effect on number and weight of nodules but SNA declined	Schenelle and Hensley (1990)
	Bentazone	0.96 kg/ha	Nodule weight reduced by 86% over control	Fischer and Tasistro (1981)
	Bentazone	1.8 kg/ha	SNA declined	Bethlenfalvay *et al* (1979)
	EPTC	25.5 kg/ha	No effect on number and weight of nodules; slightly increased SNA	Schenelle and Hensley (1990)
	Linuron	0.75 l/ha	Reduced nodule weight by two third	Fischer and Tasistro (1981)
	Prometryn	0.75 l/ha	Drastically reduced weight of nodules	Fischer and Tasistro (1981)
	Trifluralin	6.7 l/ha	No effect on number and weight of nodules and SNA	Schenelle and Hensley (1990)
Lupins (*Lupines albus/ angushfolius*)	Cyanazine	1.75 kg/ha	Increase in SNA	Pozuelo *et al* (1989)
	Cyanazine	1.38 kg/ha	Drastically reduced SNA	De Felipe *et al* (1987)
	Linuron	1.38 l/ha	Drastically reduced SNA	De Felipe *et al* (1987)
	Metribuzin	0.5 kg/ha	Increase in SNA	Pozuelo *et al* (1989)
	Simazine	0.75 kg/ha	Enhanced SNA	Pozuelo *et al* (1989)
	Simazine	1.38 kg/ha	Drastically reduced SNA	De Felipe *et al* (1987)
	Trifluralin	0.85 l/ha	Increased SNA	Pozuelo *et al* (1989)
Red clover (*Trifolium pretense*)	Bentazone	25-200 mg/l	Reduced number of nodules	Martensson (1992)
White clover (*Trifolium repens*)	Bentazone	1.21 kg/ha	Reduced number of nodules	Clark and Mahanty (1991)

Crop	Herbicide		Effect	Reported by
	Name	Rate		
Alfalfa/ Lucerne (*Medicago sativa*)	Chlorsulfu ron	2 g or more	Significant reduction in nodulation and SNA	Martensson and Nilsson (1989)
Lentil (*Lens culinaris*)	Fluchloralin	1.5 l/ha	Marginal to no effect on number and dry weight of nodules or SNA	Yadav *et al* (1990)
	Linuron	0.5-0.75 kg/ha	Reduced number and weight of nodules and SNA	Sandhu *et al* (1991)
	Methabenz thiazuron	1.05 kg/ha	No effect on number and weight of nodules; Slight reduction in SNA	Sandhu *et al* (1991)
	Methabenz thiazuron	1.0-1.5 kg/ha	No effect on number of nodules	Kumar and Kolar (1989)
	Metribuzin	0.5-0.75 kg/ha	Reduced number and weight of nodules and SNA	Sandhu *et al.* (1991)
	Metribuzin	2.2-4.4 ppm	Reduced number of nodules and SNA	Sprout *et al* (1992)
	Oxadiazon	0.5 l/ha	Marked reduction in number and weight of nodules and SNA	Sandhu *et al* (1991)
	Oxyfluorfen	0.15-0.20 l/ha	Reduced number and weight of nodules and SNA	Sandhu *et al* (1991)
	Pendimetha lin	1.0 l/ha	No effect on number and dry weight of nodules and SNA	Yadav *et al* (1990)
	Pendimetha lin	1.0-1.5 l/ha	No effect on number of nodules	Kumar and Kolar (1989)
	Terbutryn	0.6 kg/ha	No effect on number of nodules but weight of nodules decreased	Sandhu *et al* (1991)
	Terbutryn	0.6-1.2 kg/ha	No effect on nodules	Kumar and Kolar (1989)
Chickpea (*Cicer arietinum*)	Fluchloralin	0.5-1.5 l/ha	Reduction in number and weight of nodules and SNA at higher rates	Pahwa and Prakash (1992)
	Methabenz thiazuron	1.0-1.5 kg/ha	No effect on number of nodules	Kumar and Kolar (1988)

Crop	Herbicide		Effect	Reported by
	Name	Rate		
	Methabenz thiazuron	2.35-3.5 kg/ha	No effect on number and weight of nodules; reduction in SNA	Khokhar and Malik (1988)
	Metribuzin	0.25-0.75 ppm	Reduced number and weight of nodules and SNA at higher rates	Pahwa and Prakash (1992)
	Pendimethalin	1.0-1.5 l/ha	No effect on number of nodules	Kumar and Kolar (1988)
	Pendimethalin	0.75-1.25 l/ha	Reduced number and dry weight of nodules at higher rates; SNA increased	Pahwa and Prakash (1992)
	Simazine	1.6-3.2 kg/ha	Drastically reduced number and weight of nodules and nitrogenase activity	Kumar *et al* (1981)
	Terbutryn	0.6-1.2 kg/ha	No effect on nodule population	Kumar and Kolar (1988)
	Terbutryn	2-4 kg/ha	No effect on number of nodules but reduced weight of nodules and SNA	Khokhar and Malik (1988)
Peas (*Pisum sativum*)	Linuron	0.75 kg/ha	Reduced number and weight of nodules and SNA	Singh *et al* (1994)
	Imazethapyr	0.35-6.91 uM	Reduced number and weight of nodules	Gonzalez *et al* (1996)
	Methabenz-thiazuron	1.31 kg/ha	Reduced number and weight of nodules and SNA	Singh *et al* (1994)
	Prometryn	6 g/kg soil	Reduced nodule weight and nitrogenase activity significantly	Paromenskaya and Lebskii (1986)
	Pendimethalin	0.75 l/ha	Reduced number and dry weight of nodules and SNA	Singh *et al* (1994)
Faba bean (*Vicia faba*)	Linuron	1-4 kg/ha	Reduced weight of nodules and ARA	Haider *et al* (1991)
	Methabenz-thiazuron	2.5-4.0 kg/ha	Reduced nodule weight and ARA	Martinez *et al* (1996)
	Simazine	0.5-2.0 kg/ha	Reduced nodule weight and ARA at higher rates	Haider *et al* (1995)

ARA = Acetylene reduction activity, SNA = Specific nitrogenase activity

Favourable effect

Herbicide application may, sometimes, stimulate the metabolic and physiological processes (Berger and Heitefuss 1991). Nutman (1948) found enhanced rate of nodulation in dalapon inoculated red clover. At normal/recommended rates of application, prometryne increased number of nodules of *Phaseolus* roots (Rankov *et al.* 1966). Lower dose of simazine (0.75 kg/ha) increased specific nitrogenase activity in lupins (Pozuelo *et al.* 1989). Nitrogenase activity of soybean plants was 1.8 times higher with trifluralin application than unsprayed plants (Yueh and Hensley 1993).

Praksah and Pahwa (1984) reported increase in number of nodules per plant of pea with methabenzthiazuron (1.0 and 1.5 kg/ha), diclofopmethyl (0.6 kg/ha), fluchloralin (1.0 kg/ha), pendimethalin (1.0 and 1.5 kg/ha) and oxyfluorfen (0.1 kg/ha) in comparison to control. Prakash and Pahwa (1985) reported increase in N_2 fixation efficiency of nodules in cowpea with lower concentration of herbicides.

Application of fluchloralin (0.75, 1.0 and 1.25 kg/ha), pendimethalin (1.0, 1.5 and 2.0 kg/ha) and fluazifop butyl (0.5, 1.0 and 2.0 kg/ha) increased number, fresh and dry weight, leghaemoglobin content and N_2 fixation efficiency of nodules in pigeonpea at lower doses but decreased at higher doses (Pahwa *et al.* 1988). Pre-emergence application of pendimethalin @ 0.25 kg/ha to groundnut increased number of nodules per plant at 45 DAS but caused reduction at 1.0 and 1.25 kg/ha (Pahwa and Praksah 1990)

Azolla

Inoculation time of azolla in rice coincides with pre-emergence application of herbicides. It is apprehended that herbicides used for weed control at this stage may prove to be deleterious to Azolla under transplanted conditions (Moody and Janiya 1992; Gopalaswamy *et al.* 1994). Monuron applied as low as 1 ppm proved toxic to soil algae, *Stichococcus bacillaris* (Raud *et al.* 1959). Butyric acid form of MCPA was highly toxic to green algae whereas 2,4-D, 2,4,5-T and acetic acid form of MCPA up to 500 ppm were harmless (Shennan and Fletcher 1965). Janiya and Moody (1981, 1986) reported deleterious effect of herbicides on biomass production of azolla. Thiobencarb, propanil, butachlor, 2,4-D and piperopos caused significant reduction in fresh weight of azolla at 30 DAT; 2,4-D caused least suppression, propanil was more deleterious while thiobencarb was the most deleterious. Mixture of molinate-simetryn-MCPB was also highly damaging to azolla (Jainiy and Moody 1981). In an earlier study, reduction in growth and N_2 fixation in Azolla due to herbicide application was reported by Moody and Janiya 1972). Azolla is susceptible to butachlor, oxadiazon and thiobencarb but not to 2, 4-D (Srinivasan *et al.* 1993). Butachlor was highly toxic causing 65.0 per cent reduction while EPTC S-ethyl dipropyl thiocarbamate applied alone or in combination with 2,4-D ester caused reduction of 21.0-23.6 per cent. Hoist *et al.* (1982) reported reduction in growth and N_2 fixation of azolla with application of triazine, dinitroanaline and substituted urea herbicides at concentration of 0.1–10 ppm. Relative growth rate of azolla was significantly reduced by anilophos concentration above 1.0 ppm, whereas *in vitro* biomass production of azolla hybrids and wild cultures decreased with increasing concentration of anilophos with 1.5 and 2.5 ppm adversely affecting growth and biomass production while tolerance limit for butachlor and thiobencarb was found to be only 0.5 ppm (Gopalasamy and Kannaiyan 1997). High sensitivity to azolla at low concentration of herbicides under *in vitro* conditions might be due to its direct contact with herbicides.

Anderson and Domsch (1980) and Kumar *et al.* (1987) reported initial set back to microbial population and restitution after certain period of time. Magnitude of herbicide toxicity is governed by soil organic carbon status and amino acid contents in plants (Andreae and Good 1957). Such restitution can be ascribed to addition of organic matter to soil through decaying azolla fronds in advanced stages of growth resulting in activation of soil microflora. However, N content of azolla was not influenced by any of the herbicides viz; butachlor, EPTC, molinate, thiobencarb (Srinivasan *et al.* 1993).

EPTC either alone or in combination with 2,4-D ester resulted in significantly higher biomass and relative growth rate compared to control and thus can be safely applied in azolla dual cultured rice fields provided azolla inoculation is made 3 days after herbicide application (Srinivasan *et al.* 1993). Inoculation of rice crop with azolla at 12 days after herbicide (butachlor or anilophos) application resulted in higher biomass production than that inoculated 7 days after application of herbicide (Gopalasamy *et al.* 1994) ***Azotobacter*** : Sharma and Saxena (1974) reported that application of 2,4-D sodium salt (4-5 ppm) increased *Azotobacter* population in soil whereas higher concentration (.5 ppm) reduced it. Nair *et al.* (1957) concluded that growth of *Azotobacter* can be stimulated at lower concentration of 2,4-D. Piter (1981) observed stimulation in the growth of Azotobacter in the presence of pyramin (20 and 100 ppm). Diuron and 2, 4-D sodium salt stimulated *Azotobacter* population and 2, 4-D amine, atrazine, Destum and Sinbar showed good growth but asulox 40, Gramoxone, TotaCol, Paracot and Sirmate retarded growth of *Azotobacter* (Jadhav *et al.* 1988). MCPA application @ 2 kg/ha to lime rich soil caused no adverse effect on *Azotobacter* population while soils poor in lime, population of *Azotobacter* following MCPA application declined. Lenhard (1959) postulated that though higher concentration of herbicides than that normally used for weed control inhibits growth and respiratory metabolism in *Azotobacter*, the actual nitrification process may be affected at much lower concentration. Higher concentration of phenoxy herbicides can markedly inhibit growth and respiration of *Azotobacter* species (Colmer 1953). Herbicides such as monuron, neburon, diuron, dinoben, chlorazine, alipur, pebulate, 2, 4-DES, chloramben, pentanochlor and diphenamide caused greater inhibition on growth of *Azotobacter chroococcum* than *A. galophilium*.

Soil microbes

Soil fertility depends on the delicate balance of various types of micro-organisms. Mineral nutrient status of soil depends on activities of microbes in soil whose role in nutrient transformation is well established. The use of herbicides for weed in crops is well documented but the effect of these herbicides on soil fauna and beneficial microorganisms is less understood (Cullimore *et al.* 1977). Herbicide may disturb and alter the biological equilibrium in soil (Grossbard 1976) by influencing biological activities in soil (Gopalasamy *et al.* 1994, Singh and Dudeja 2007). Inhibitory or stimulatory effects depend on kind of herbicide, its concentration and period of incubation (Amakiri 1980). Some herbicides may show differential effect in different types of soils (Bollich *et al.* 1985). Dinitro aniline herbicides showed highest persistence in silty clay soil under normal tillage conditions and least persistence in sandy loam soils (Flom and Miller 1978).

McCall *et al.* (1962) observed reduction in O_2 uptake by microbes at higher rates of 2,4-D and PBA. EPTC temporarily depressed soil respiration (Chandra *et al.* 1968) while chlorpropham increased it (Stotsky *et al.* 1956, Teater *et al.* 1958). Soils low in organic matter is known to absorb less metribuzin resulting in greater availability for uptake by plants (Savage 1976, Sharom and Stephenson 1976, Lafleur 1979). Soybean injury to metribuzin increased with increase in moisture content (Coble and Schrader 1973). Bertholet and Clark (1985) also reported that metribuzin caused greater injury to faba bean in soils low in organic matter and in high rainfall areas.

Skrdleta *et al* (1964) reported inhibition in population of *Rhizobium* sp. with 2, 4-D and MCPA. Similarly Siduron suppressed growth of soil bacteria and *Azotobacter* but did not affect actinomycetes, filamentous fungi, soil ciliates and algae (Fields and Hemphill 1968). Avrov (1966) observed stimulatory effect of simazine on lupine nodule bacteria. Microbial population decreased soon after application of pendimethalin, increased after some time and then again decreased (Ponnuswamy *et al.* 1997). *Trichoderma harzianum* is used as a biocontrol agent. Since this fungus is inoculated with soil, it is likely to interact with herbicides which are applied to soil. Application of pendimethalin and alachlor significantly reduced the growth, survival and competitive ability of *Trichoderma harzianum* (Jayaraj and Radhakrishnan 2000). Effect of pendimethalin was more pronounced than alachlor. The fungus tends to regain its multiplication in later stages. Many workers have reported this kind of initial reduction in microbial population and its restitution after a period of time Anderson and Domsch 1980, Kumar *et al.* 1987, Gopalasamy *et al.* 1994). This phenomenon of restitution of growth might be attributed to gradual decomposition of the toxicant residues by microbial action and decomposition through physico-chemical reactions that take place in soil. Vintikova *et al* (1965) concluded that bacterial sensitivity was a property of microbial strains and was independent of plant species.

Sidhu *et al* (1985) reported suppressing effect of fluchloralin on soil microflora up to 7 days after application. Similarly, Patel and Patel (1998) observed significant reduction in bacterial as well as fungal population up to 10 days after application of fluchloralin @ 1.5 and 3.0 kg/ha. Pre-emergence application of pendimethalin @ 1.0 kg/ha significantly reduced population of bacteria, fungi and actinomycetes one day after treatment, whereas, glyphosate applied @ 1.0 kg/ha as post–emergence caused reduction of only fungal colony at one day after treatement (Nalayini *et al.* 2000). However, 30 days after treatment, there was no difference in the population of these microbes with pendimethalin and glyphosate. Kumar *et al* (1987), Nalayini and Sankaran (1992) and Selvamani and Sankaran (1993) reported similar facts. Since pendimethalin was applied directly to soil resulting in more amount of it reaching the soil than glyphosate which was applied on foliage of weeds; hence the inhibitory effects vary.

Actinomycetes were stimulated in the presence of herbicides 2, 4-D, Simazine, Bladex and NaTA (Desmukh and Shrikhande 1974). Application of 2, 4-D, Bladex and NaTA depessed bacterial and fungal populations in red clay loam soil whereas similar reduction bacterial population in black clay soil was observed with 2, 4-D only whereas fungal population was reduced with application of 2, 4-D, NaTA and Bladex (Desmukh and Shrikhande 1974). Dalapon is reported to have slight stimulatory effect on soil population (Hale *et al.* 1957, Worsham and Giddens 1957). Cole (1976) did not observe any adverse effect of long term application of atrazine on soil microbes. Giardina *et al.* (1979) observed no change in microbial population with application of 1% ammonium sulphate or 30 ppm atrazine. More harmful effect of herbicides on these organisms in red soils compared to black soil could be ascribed to the adsorption of herbicides on mineral colloids in the black soil. Earlier Magee and Colmer (1960) and Steinbrenner *et al* (1969) observed stimulatory effect of herbicides on population of actinomycetes. Fungal population was affected more than the bacterial population (Patel and Patel 1998). Nalayini *et al.* (2000) also observed that bacteria were influenced to lesser extent than actinomycetes.

BIOCHEMICAL PROCESSES

Application of trifluralin significantly decreased the amount of reserve protein *arginine* in the hypocotyls of mungbean (Mercado and Baltazar 1971). Dukes and Biswas (1967) recorded increase in carbohydrate content of peanut plants treated with trifluralin @ 5 and 10 ppm followed by gradual decrease with its increasing concentration. Total soluble carbohydrates and reducing sugars in cotyledons and embryo axis of treated seeds of different legume crops increased with application of pendimethalin @ 5, 10 and 25 ppm (Dhall 1992). Trifluralin did not cause marked reduction in protein synthesis (Mann *et al.* 1965). Higher rates of simazine decreased amino acid in the protein and modified the levels of free amino acids (Hauke Pacewiczowa 1969, 1970). Moreland *et al* (1969) reported that diuron inhibited RNA and protein synthesis in excised tissue of soybean.

There was significant reduction in activity of RNase and DNase enzymes during germination of pigeonpea with increase in dose (5, 10, 15 and 20 ppm) of trifluralin and pendimethalin (Pahwa and Bajaj 2000). Pendimethalin was more toxic than trifluralin.

DISEASES MANAGEMENT

Several pathogens survive on annual and perennial weeds. The capacity of a potential pathogen to induce disease may be reduced if the herbicide applied to combat weeds in crop fields proves toxic to that pathogen or put it at competitive disadvantage (Altman and Campbell 1977). Application of herbicides significantly increased germination of *Rhizobium bataticola* inoculated seeds of linseed (Ahmad and Vyas 1988) probably due to fungitoxicity of these herbicides (Ahmad 1982). Fields and Hemphill (1967) recorded depression in growth of certain bacteria with siduron. Fungi are more resistant to phenoxy herbicides as compared to bacteria. Paraquat was more fungitoxic to a range of organisms than other herbicides (Wilkinson and Lucas 1969). While

Houseworth and Tweedy (1972) reported fungitoxic potential of alachlor against *Fusarium oxysporum* and *Diplodia maydis*, Altman and Campbell (1977) observed disease reduction with 2,4-D. Gopalaswamy *et al.* (1994) observed reduction in population of bacteria, fungi and actinomycetes up to 20 days after application of herbicide. Paraquat, butachlor, alachlor and anilophos caused significantly more inhibition (18.2, 22.7, 32.9 and 36.2%) of fungal growth as compared to pendimethalin (49.4%) and 2,4-D (63.3%) when applied @ 25 ppm and control (85.0%) and the fungal growth decreased with increasing concentration of herbicides (Rai *et al.* 2000).

Stimulating effect of phenoxy herbicides on fungi has been reported (Anderson and Baker 1950, Roberts and Bollen 1955, Newman and Downing 1958). Thiobencarb (1.0 kg/ha) slightly stimulated bacterial population whereas population of fungi was markedly reduced with all the herbicides up to 20 days after applications followed by restitution after 30th day. Anilophos and butachlor though reduced bacterial population up to 20 days after application, the restitution occurred on 30th day (Gopalaswamy *et al.*1994).

Aspergillus and *Penicillium* were little affected by simazine and atrazine up to 1000 mg/kg except for slight delay in spore formation. Proliferation of *Aspergillus niger* was reduced by 2,4-D at 10-50 ppm whereas, picloram up to 50 ppm did not reduce fungal growth. Damping off of pea seedlings was less in autoclaved soil subsequently inoculated with *Pythium debaryanum* and to which DMPA @ 100 ppm was added than in inoculated soil without DMPA (Fields and Hemphill 1967). Addition of FYM @ 5t/ha significantly increased fungal population up to 30 days after application of fluchloralin as compared to control (Patel 1990, Patel and Patel 1998).

REFERENCES

Ahmad M (1982) *Studies on contol of plant diseases by herbicdes*. MSc (Agri.) thesis. Jawaharlal Nehru Vishwavidyalaya, Jabalpur (MP).

Ahmad M and Vyas SC (1988) Effect of herbicides on seed borne nature of fungi. *Pestology* **12(6)** : 8-9.

Ahmed AHH and Rashad MH (1996) Comparative studies on the effect of fusillade in corn and soybean plants. Bulletin of the Faculty of Agriculture, Univeristy of Cairo. pp. 570-610.

Akhtar M, Afghan S, Mahmood T, Abbas G (1990) Effect of pre and post emergence weedicides application on nodulation, growth and yield of soybean at different fertility levels. *Journal of Agricultural Research*, Lahore **28** : 29-37.

Altman J and Campbell CL (1977) Effect of herbicides on plant diseases. *Annual Review of Phytopathology* **15** : 361-385.

Amakiri MA (1980) Effect of herbicides on microbial population and activities in soil under teak (*Tectona grandis* LF) plantation. *African Agricultural and Forestry Journal* **42** : 420-426.

Anderson GR and Baker GO (1950) Some effects of 2,4-D in representative Idaho soils. *Agronomy Journal* **42** : 456-458.

Anderson JPE and Domsch KH (1980) Influence of selected pesticides on the microbial degradation of ^{14}C-Triallate and ^{14}C-Diallate in soil. *Archives of Environment Contamination Toxicology* **9** : 115-123.

Andreae WA and Good ME (1957) Studies on 3-indole acetic acid metabolism. IV. Conjugation with aspartic acid ammonia as processes in the metabolism of carbonic acids. *Plant Physiology* **32** : 566-572.

Andus IJ. (1976) Herbicide: Physiology, Biochemistry and Ecology. Vol. I. Academic Press, New York.

Avrov OE (1966) Effect of herbicides on nodule bacteria and nodule formation in legumes. Dokl Vses Akad Sel Khuz nauk No. **3** : 16-19 (*cf Soils and Fertilizers 29 : 454*).

Bakalivanov D and Bajchev A (1979) Effect of the herbicide Dalapon on soil micro organisms and its decomposition in a smolnitza. *Soils and Fertilizer* **44(1)** : 1073.

Berger A. and Heitefuss R (1991).Use of isoproturon alone and in combination with other compounds post-emergence in winter and spring on wheat and barley. II. Side effects on dehydrogenase activity, nitrogen formation and straw decomposition. *Weed Research* 31 : 1-18.

Bertholet J and Clark KW (1985) Effect of trifluralin and metribuzin on faba bean growth, development and symbiotic nitrogen fixation. *Canadian Journal of Plant Science* **65** : 9-21.

Bethlenfalvay GJ, Norris RF and Phillips DA (1979) Effect of bentazon, a Hill reaction inhibitor, on symbiotic nitrogen-fixing capability and apparent photosynthesis. *Plant Physiology* **63** : 213-215.

Bhalla PL and Chourasia KK (1985) Effect of herbicidal treatments on the growth and seed yield of garden pea (*Pisum sativum subsp. hortense* L). *Seeds & Farms (September)* : 32-42.

Bockman OC (1997) Fertilizers and biological nitrogen fixation as sources of plant nutrients: Perspectives for future agriculture. *Plant and Soil* **194** : 11-14.

Bollich PK, Dunigan EP and Jadi AWM (1985) Effects of seven herbicides on $N_2(C_2H_2)$ fixation by soybeans. *Weed Science* **33** : 427-430.

Bollich PK, Dunigan EP, Kitchen LM and Taylor V (1988) The influence of trifluralin and pendimethalin on nodulation, N_2 (C_2H_2) fixation and seed yield of field grown soybean. *Weed Science* **36** : 15-19.

Brock JL (1972) Effects of the herbicides trifluralin and carbetamide on nodulation and growth of legume seedlings. *Weed Research* 12 : 150-154.

Burns RC and Hardy RWF (1975) *Nitrogen Fixation in Bacteria and Higher Plants*. Springer-Verlag, Berlin.

Cardina J, Hartwig NL and Lukezic FL (1986) Herbicidal effects on crownvetch rhizobia and nodule activity. *Weed Science* **34** : 338-343.

Carlyle RE and Thorpe JD (1947) *Journal of American Society of Agronomy* **39** : 929 -936.

Chandra P, Furtick WR and Bollen WB (1968) The effects of four herbicides on micro organisms in nine Oregon soils. *Weeds* **8** : 589-598.

Chaurasia RM (2001) Effect of pendimethalin and fluchloralin on growth, nodulation and yield of urdbean. MSc (Ag) thesis. Narender Dev University of Agriculture and Technology, Kumarganj, Faizabad.

Choudhri RS and Rajgopal R (1958) Modern Development in Plant Physiology, Delhi University, India.

Clark SA and Mahanti HK (1991) Influence of herbicides on growth and nodulation of white clover, *Trifolium repens*. *Soil Biology and Biochemistry* **23** : 725-730.

Coble HD and Schrader JW (1973) Soybean tolerance to metribuzin. *Weed Science* **21** : 308-309.

Cole MA (1976). Effect of long term atrazine application on soil microbial activity. *Weed Science* **24** : 473-476.

Colmer AR (1953) Applied Microbiology **1** : 184-187.

De Felipe MR, Fernandez-Pascual M and Pozuelo JM (1987) Effects of the herbicdes lindex and simazine on chloroplast and nodule development, nodule activity, and grain yield in *Lupinus albus* L. *Plant and Soil* **101** : 99-105.

Deshmukh VA and Shrikhande JG (1974). Effect of some herbicides on soil microflora. *Journal of Indian Society of Soil Science* 22 : 300-303.

Dhall V (1992) Effect of pendimethalin on germination, early seedling growth and metabolism of pigeonpea, lentil, *Trianthema portulacastrum* and *Melilotus indica*. M Sc. Thesis, CCS Haryana Agricultural University, Hisar.

Domsch KH, Jagnow G and Anderson TH (1983) An ecological concept for the assessment of side effects of agrochemicals on soil microorganisms. *Residue Review* **86** : 66-103.

Dukes IE and Biswas PK (1967) The effect of trifluralin on the nucleic acid and carbohydrate content of sweat potato and peanuts. Abstract Weed Science Society of America. p. 59.

Dunigan EP, Allen LD and Frey JP (1970) *Losangles Agric* 13 : 6-7.

Dunigan EP, Frey JP, Allen Jr LD and McMahon A (1972) Herbicidal effects on the nodulation of glycine max (l) Merrill. *Agronomy Journal* 64 : 806-808..

Eberbach PL and Douglas LA (1989) Herbicide effects on the growth and nodulation potential of *Rhizobium trifolii* with *Trifolium subterraneum* L. *Plant and Soil* **119** : 15-23.

Eberbach PL and Douglas LA (1991) Effect of herbicide residues in a sandy loam on the growth, nodulation and nitrogenase activity (C_2H_2/C_2H_4) of *Trifolium subterraneum* L. *Plant and Soil* **131** : 67-76.

Fields ML and Hemphill DD (1967) The influence of DMPA on damping off of peas by *Pythium debaryanum* ATCC 9998. *Weed Science* **15** : 281-282.

Fields MI and Hemphill DD (1968) Effect of siduron and its degradation products on soil microflora. *Weed Science* **16** : 417-420.

Fischer A and Tasistro A (1981) Effect of certain herbicides upon the *Rhizobium phaseoli-Phaseolus vulgaris* symbiosis. In: *Theory and Practice of the Use of Soil Applied Herbicides*, Proceedings of the EWRS Symposium, Versailles, pp. 120-128.

Fletcher WW (1960) The effect of herbicides on soil microorganisms. In: *Herbicides and the Soil* (EK Woodford and GR Sagar, Eds), pp. 20-62. Blackwell Scientific Publications, Oxford.

Fletcher WW, Dickenson PB and Raymond JC (1956) The effect of certain hormone herbicides on the growth and nodulation of *Trifolium repens sylvestre* in aseptic culture. *Phyton* **7** : 121-130.

Fletcher WW, Dickenson PB, Forrest JD and Raymond JC (1957) The effect of soil application of certain substituted phenoxyacetic and phenoxybutyric acids on the growth and nodulation of *Trifolium repens sylvestre. Phyton* **9** : 41-47.

Flom DG, Miller SD (1978) Persistence of dinitroanaline herbicides. Proceedings of North Central Weed Control Conference **33** : 35-36.

Garcia M and Jordan DC (1967) Effect of herbicides on nodulation and N-fixation in birdfoor trefoil. *Bacteria Proc.* p. 8.

Gaur AC (1980) Effect of pesticides on symbiotic nitrogen fixation by legumes. *Indian Journal of Microbiology.* **20** : 362-370.

Gaur AC and Misra KC (1972) *Indian Journal of Microbiology* **12** : 45-46.

Geller IA and Khariton EG (1961) *Microbiologiya* **30** : 494-499.

Giardina MC, Giardi MT and Buffone R (1979). Soil enrichment studies with atrazine long term effects on degradation and microbiological composition. *Soils and Fertilizers* **44(3)** : 275.

Giller KE and Wilson KT (1991) *Nitrogen Fixation in Tropical Cropping Systems*. CAB International Wallingford.

Gonzalez A, Gonzalez-Murua C and Royuela M (1996) Influence of imazethapyr on *Rhizobium* growth and its symbiosis with pea (*Pisum sativum*). *Weed Science* **44** : 31-37.

Gopalaswamy G, Anthoni R and Abulkareem A (1994) Interaction of herbicides with *azolla* and soil microbes. *Indian Journal of Weed Science* **26(2)** : 28-34.

Gopalaswamy G and Kannaiyan S (1997) Influence of herbicides on the growth and biomass production of *Azolla* hybrids. *Indian Journal of Weed Science* **29(1&2)** : 53-57.

Gracia M and Jordan DC (1967) *Soils and Fertilizer* **30** : 476.

Grau CR (1977) Effect of dinitramine and trifluralin on growth, reproduction and activity of *Aphanomyces entelches*, *Phytopathology* **67** : 551-556.

Greenfield PL (1991) The influence of method of inoculation and certain herbicides on nodulation and seed yield of soybean. *South African Journal of Plant and Soil* **8** : 119-122.

Grossbard E. (1976) Effect of herbicides on the soil microflora. In: Herbicides – Physiology, Biochemistry, Ecology. Andus JJ (ed.). Academic Press, London. pp. 99.

Haider J, Fendrik I and Niemann EG (1991) Effect of linuron and dinosebacetate on *Vicia faba-Rhizobium leguminosarum* symbiosis. *Soil Science and Plant Nutrition* **37** : 601-609.

Haider J, Fendrik I and Niemann EG (1995) Effect of linuron and simazine on growth and N_2 fixation by *Vicia faba* L in a sandy soil. *Annals of Bangladesh Agriculture* **5** : 1-8.

Hale MG, Hulcher FH and Chappel WE (1957) The effects of several herbicides on nitrification in a field soil under laboratory conditions. *Weeds* **5** : 331-341.

Hamdi YA and Tewfik MS (1969) *Acta Microbiology* Pol 18 : 53-67.

Hauke-Pacewiczowa T (1969) Pam Pulawaki 37 : 241-259. (*cf Soils and Fertilizers 33(6) : 1970*).

Hauke-Pacewiczowa T (1969) Meded Rijksfae Landb Wet Gent 35 : 497-503.Pam Pulawaki 37 : 241-259. (*cf Soils and Fertilizers 34(5) : 1971*).

Hoist RW, Yopp JH and Kapusta G (1982) Effect of several pesticides on the growth and nitrogen assimilation of the *Azolla-Anabaena* symbiosis. *Weed Science* **30** : 54- 58.

Houseworth LD and Tweedy BG (1972) Effect of herbicides on soil borne pathogens. *Plant Disease Reporter* **40** : 52-56.

Jadhav JJ, Jadhav SK and Hapase DG (1988). Compatability of weedicides with *Azotobacter chroococcum*. *Pestology* **12(6)** : 24-26.

Jain HC, Tiwari JP and Jain NK (1990). Influence of herbicides on nodulation under different row spacings and seeding rates in soybean. Proceedings of Biennial Conference. Indian Society of Weed Science. p. 198.

Janiya JD and Moody K (1981) Effect of herbicides on azolla. *International Rice Research Newletter* **6(5)** : 23.

Janiya JD and Moody K (1986) Herbicides reduce azolla growth. *International Rice Research Newletter* **11(1)** : 28.

Jayaraj J. and Radhakrishnan NV (2000) Survival of the biocontrol agent *Trichoderma harzianum* in soil as influenced by certain herbicides. *Indian Journal of Weed Science* **32(1&2)** : 116-117.

Jordan TN, Baber RS and Barrentina WL (1978) Comparative toxicity of several dinitroaniline herbicides. *Weed Science* 26 : 72-75.

Kao TC and Wang CC (1981) Study on effect of herbicides on growth of *rhizobia* and development of root nodules. Journal of Agricultural Association, China, *Microbial Abstract* **17(2)** : 12-15.

Keeney DR (1982) Nitrogen management for maximum efficiency and minimum pollution. In: *Nitrogen in Agricultural Soils* (FJ Stevenson, Ed), pp. 605-649. American Society of Agronomy, Crop Science Society of America, and Soil Science Society of America, Madison, Wisconsin.

Khokhar SN and Malik BA (1988) Effect of herbicides on nodulation and nitrogenase activity of chickpea. *Pakistan Journal of Agricultural Research* **9** : 493-497.

Kishinevsky B, Lobel R, Lifshitz N and Gurfel D (1988) Effects of some commercial herbicides on rhizobia and their symbiosis with peanuts. *Weed Research* **28** : 291-296.

Kolhe SS, Mitra BN, Meur SK and Bhadauria SS (1984) Influence of pre-emergence herbicide on germination and growth parameters of paddy. *Indian Journal of Weed Science* **16** : 207-209.

Kruglov YuV, Gersh NB and Bui Benko NV (1973) The effects of metribuzin (a methyl urea) on soil biology. *Khim Sel'sk khoz* **11** : 294-299.

Kucey RMN, Chaiwanakupt P, Arayangkool T, Snitwongse P, Siripaibool C, Wadisirisul P and Boonkerd N (1988) Nitrogen fixation (^{15}N dilution) with soybean under Thai field conditions II. Effect of herbicides and water application schedule. *Plant and Soil* **108** : 87-92.

Kumar K and Kolar JS (1988) Relative performance of terbutryn, methabenzthiazuron and pendimethalin for weed control in chickpea (*Cicer arietinum* L.). *Journal of Research, Punjab Agricultural University* **25** : 534-538.

Kumar K and Kolar JS (1989) Effect of chemical weed control and Rhizobium inoculation on the yield of lentil. *Journal of Research, Punjab Agricultural University* **36** : 19-24.

Kumar K, Kolar JS and Bhandari SC (1987) Effect of terbutryn, methabenzthiazuron and pendimethalin on soil microflora in Bengal gram (*Cicer arietinum* L.) and lentil (*Lens culinaris*). *Indian Journal of Ecology* **14** : 52-58.

Kumar S, Pahwa SK, Promila and Sharma HR (1981) Effect of simazine and prometryn on the growth and nodulation of chickpea (*Cicer arietinum* L.). *Journal of Agriculttural Science* (Cambridge) **97** : 663-668.

Lafleur KS (1979) Sorption of metribuzin by model soils and agronomic soils: rates and equilibrium. *Soil Science* **127** : 51-55.

Lal S (1988) Effects of pesticides on Rhizobium-legume symbiosis. In: *Pesticides and Nitrogen Cycle,* Vol III (R Lal and S Lal, Eds), pp. 47-128. CRC Press, Inc., Boca Raton, Florida.

Lenhard GS (1959) African *Journal of Agricultural Science* **2** : 487-497.

Lewis RW and Hamner Cl (1946) Bulletin Michigan Experimental Station **29** : 112.

Lindstrom K, Polkunen J and Kansanen P (1985) Effect of dinoseb on nitrogen fixation of red clover (*Trifolium pretense*). *Soil Biology & Biochemistry* **17** : 865-869.

Lund ZF, Pearson RW and Buchanan GA (1970) An implanted soil mass technique to study the herbicide effects on root growth. *Weed Science* **18** : 279-281.

Luthra YP, Sheoran IS and Singh R (1983) Ontogenetic interactions between photosynthesis and symbiotic nitrogen fixation in pigeonpea. *Annals of Applied Biology* **103** : 549-556.

Madhavi B, Anand CS, Bharathi A and Polasa H (1993) Effects of pesticides on growth of *rhizobia* and their host plants during symbiosis. *Biomedical and Environmental Sciences* **6** : 89-94.

Madhavi B, Anand CS, Bharathi A and Polasa H (1994) Biotoxic effects of pesticides on symbiotic properties of rhizobial species. *Bulletin of Environmental* Contamination and Toxicology **52** : 87-94.

Magee LA (1958) Dissertation Abstract **19** : 413-414.

Magee LA and Colmer AR (1960) *Sugar Bulletin* 38 : 242.

Mahmood SM and Mohammed SH (1991) Effect of certain herbicides and insecticides on the growth of some *Rhizobium* species. *Assiut Journal of Agricultural Sciences* **22** : 243-259.

Mallik MAB and Tesfai K (1983) Compatibility of *Rhizobium japonicum* with commercial pesticides in vitro. *Bulletin of Environmental Contamination and Toxicology* **31** : 432-437.

Mallik MAB and Tesfai K (1985) Pesticidal effect on soybean-rhizobia symbiosis. *Plant and Soil* **85**: 33-41.

Mann JD, Jordan LS and Day BE (1965) A survey of herbicides for their effect upon protein synthesis. *Plant Physiology* **40** : 840-843.

Martensson AM (1992) Effects of agrochemicals and heavy metals on fast growing rhizobia and their symbiosis with small seeded legumes. *Soil Biology & Biochemistry* **24** : 435-445.

Martensson AM and Nilsson AK (1989) Effects of chlorsulfuron on *Rhizobium* grown in pure culture and in symbiosis with alfalfa (*Medicago sativa*) and red clover (*Trifolium pretense*). *Weed Science* **37** : 445-450.

Martinez J, Vidal D and Simon E (1996) Nitrogenase activity, nodulation and seed production in *Vicia faba* as affected by methabenzthiazuron. *Journal of Agricultural Science* (Cambridge) **127** : 319-324.

McCall TM, Arm TJ and Wiess AF (1962) Comparison of the effects of chemical and sweep tillage methods of summer fallow on some properties of Pullman silty clay loam. *Agronomy Journal* **54** : 404-407.

McKell CM and Whalley RDB (1964) Compatibilityof 2-Chloro-6(trichloromethyl) pyridine with *Medicago sativa* L inoculated with *Rhizobium meliloti*. *Agronomy Journal* **56** : 26-28.

Mercado BL and Baltazar AM (1971) Morphological and physiological responses of crops and weeds to trofluralin. VI Amino acids-trifluralin interaction in mungbean (Phaseolus aureus). *Phillipine Agriculture* **54** : 384-389.

Misra KC and Gaur AC (1974) Influence of simazine, lindane and ceresin on different parameters of nitrogen fixation by groundnut. *Indian Journal of Agricultural Sciences* **44** : 837-840.

Misra KC and Gaur AC (1974) Zbl Bakt Abt II 130 : 598-602.

Mohammed B (1984) Influence of alachlor and fluchloralin on the rhizosphere fungi of soybeans. *Journal of Maharastra Agricultural Universities* **9** : 303.

Moreland DE, Malhotra SS, Gruenhagen RD and Shokran EH (1969) Effects of herbicides on RNA and protein synthesis. *Weed Science* **17** : 556-563.

Moody K and Janiya JD (1972) The role of Azolla in weed control in rice. *Phillipines Journal of Weed Science* **19** : 79-102.

Murti R, Khan AK, Vaishya RD and Yadav PK (2004) Effect of some dinitroaniline herbicides on growth, nodulation, chlorophyll content and nitrate reductase activity of urdbean (*Vigna mungo* (L) Hepper). *Indian Journal of Weed Science* **36(1&2)** :124-126.

Nadasy E, Lehoczky E, Lucas P and Adam P (2000) Influence of different pre-emergence herbicides on the growth of soybean varieties. *Zeitschrift-furpflanzekrankheiten-Und-Pflanzenschutz, Sanderh* (Germany) **17** : 635-639.

Nair KS, Menon KMK and Mariakulandai A (1957) *Madras Agriculural Journal* **44** : 667-668.

Nalayini P and Sankaran S (1992) Effect of pre-emergence herbicides on soil microorganisms. *Indian Journal of Agronomy* **37** : 625-626.

Nalayini P, Kandasamy OS and Balasubramaniyan A (2000) Effect of pre-emergence and post emergence herbicides on soil microgflora. *Pestology* **14(10)** : 22-23.

Nelson DR, Belliville RJ and Porter CR (1984) Role of nitrogen assimilation in seed development of soybean. Plant Physiology **74** : 128-133.

Newbould P (1989) The use of nitrogen fertilizer in agriculture. Where do we go practically and ecologically? *Plant and Soil* **115** : 297-311.

Newman AS and Downing CE (1958) Journal of Agriculture and Food Chemistry **6** : 352-363.

Nutman PS (1948) *Annals of Botany* **12** : 81.

Olson BM, McKercher RB and Germida JJ (1984) Microbial populations in trifluralin-treated soil. *Plant and Soil* **76** : 379-387.

Ozair CA and Moshier IJ (1988) Effect of post emergence herbicides on nodulation and nitrogen fixation in soybean (*Glycine max*). *Applied Agricultural Research* **3** : 214-219.

Ozair CA, Moshier IJ and Chaudhry AH (1990) Green house evaluation of selected herbicides for their effect on soybean (*Glycine max*) growth, nodulation and nitrogenase activity. *Pakistan Journal of Botany* **22** : 197-203.

Pahwa SK (1972) Effect of some pre-emergence herbicides on weed control and physiology of pea (*Pisum sativum* L.). MSc thesis, Haryana Agricultural University, Hisar.

Pahwa SK and Bajaj K (2000) Effect of pre-emergence herbicides on the activity of ribonuclease and deoxyribonuclease enzyme during germination in pigeonpea and carpetweed. *Indian Journal of Weed Science* **32 (1&2)** : 21-24.

Pahwa SK and Prakash J (1988) Effect of some herbicides on growth, nodulation, leghaemoglobin and nitrogen content of soybean. Abstract in the Biennial Conference of Indian Society of Weed Science. pp. 45.

Pahwa SK and Prakash J (1990) Effect of pre-emergence herbicides on the growth and nodulation in groundnut (*Arachis hypogaea* L.). Proceedings of the Biennial Conference. Indian Society of Weed Science. p. 1.

Pahwa SK and Prakash J (1992) Effect of some herbicides on the growth, nodulation and nitrogen fixation in chickpea (*Cicer arietinum* L.). *Indian Journal of Plant Physiology* **35** : 207-212.

Pahwa SK, Prakash J and Sharma HR (1988) Effect of herbicides on the growth, nodulation and symbiotic nitrogen fixation in pigeonpea (*Cajanus cajan* (L.) Millsp.) *Crop Research* **1** : 131-140.

Paromenskaya LN and Lebskii VK (1986) Influence of the herbicide prometryn on symbiotic nitrogen fixation and morphology of pea nodules. *Microbiology* **54** : 655-658.

Patel PR (1990) Studies on the effect of herbicides on plant pathogenic and beneficial microorganisms in groundnut. MSc Thesis, Gujarat Agricultural University, Sardar Krushinagar.

Patel RB and Patel BK (1998) Effect of fluchloralin and farm yard manure on soil microflora. *Journal of Research, Gujarat Agricultural University* **24(1)** : 15-18.

Paul EA and Clark FE (1996) Soil Microbiology and Biochemistry, 2nd Edition. New York: Academic Press.

Payne MG and Fults JL (1947) *Agronomy Journal* **39** : 52-55.

Peoples MB and Crasswell ET (1992) Biological nitrogen fixation: Investments, expectations and actual contributions to agriculture. *Plant and Soil* **141** : 13-39.

Peoples MB and Herridge DF (1990) Nitrogen fixation by legumes in tropical and sub-tropical agriculture. *Advances in Agronomy* **44** : 155-223.

Piter SJ (1981) Herbicidal effect on some metabolic processes of *Azotobacter crococcum. Soils and Fertilizer* **44(11)** : 1055.

Pochon J, Tardieux P and Charpentier M (1960) *Compt Rend* **250** : 115-156.

Ponnuswamy K, Santhi P, Gopalasamy G and Mohammed Ali A (1997) Interaction of herbicides with soil microbes in cassava (*Manihot esculenta* Crantz). *Indian Journal of Weed Science* **29(3&4)** : 147-150.

Pozuelo JM, Fernandez-Pascual M, Lucas MM and De Felipe MR (1989) Effect of eight herbicides from five different chemical groups on nitrogen fixation and grain yield in *Lupinus albus* L. grown in semi-arid zones. *Weed Research* **29** : 419-425.

Prakash J and Pahwa SK (1982). Effect of some new herbicides on growth and chlorophyll content of mungbean. Abstract in the Biennial Conference of Indian Society of Weed Science. pp. 42-43.

Prakash J and Pahwa SK (1984) Effect of some new herbicides on growth of pea plants. *Pesticides* **18** : 56-58.

Prakash J and Pahwa SK (1985). Effect of some herbicides on growth, nodulation and chlorophyll content of cowpea (*Vigna unguiculata* (L) Walp) plants. *Haryana Journal of Agronomy* **1** : 39-41.

Raghuwanshi MS, Jain HC and Tiwari JP (1998) Influence of herbicides on nodulation under different seeding rates and row spacings in soybean. *Pestology* **22(11)** : 35-37.

Rai JP, Dubey KS and Sinha A (2000) Effect of some herbicides on the growth of *Rhizoctonia solani* Kuhn *in vitro*. *Indian Journal of Weed Science* **32(1&2)** : 112-113.

Rankov V, Elenkov E and Surelkov P (1966) Effect of some herbicides on development of nitrogen fixing bacteria. *Agrokhumisja* **4** : 115-120 (*cf Soils and Fertilizers* **29** *: 452*)

Rao VS (1987) *Principles of Weed Science*. Oxford and IBH Oxford Publishing Company Pvt Ltd, New Delhi, India.

Raud G, Tyssett C and Vacher b (1959) *Annals of Luise Pasteur* **96** : 242-244.

Rennie RJ and Dubetz S (1984) Effect of fungicides and herbicides on nodulation and N_2 fixation in soybean fields lacking indigenous *Rhizobium japonicum*. *Agronomy Journal* **76** : 451-454.

Rethinam p, Kesavan R and Purushothaman (1974) *Indian Journal of Microbiology* 14 : 149-151.

Roberts JE and Bollen WB (1955) *Annals of Applied Microbiology* **3** : 190-194.

Roberts JMF (1992) Effects of pest management systems on biological nitrogen fixation. In: *Biological Nitrogen Fixation and Sustainability of Tropical Agriculture* (K Mulongoy, M Gueye and DSC Spencer, Eds), pp. 97-108. Johan Wiley & Sons, Chichester.

Ronald ET (1965) Effect of trifluralin on soybean root development. Proceedings South Asia Weed Conference **18** : 652.

Salvo-Soares N, Souza-Paes M-Do-C-del, Cruz L, Martina ALM, Nagar V, Oliverira-Lombardi MLC, Lopes ES, Ambrosio LA, Novo M-do-C-de, Guz LSP, Lombardi MLC de O (1991). Effect of different herbicides on nodulation and nitrogenase activity in groundnuts. *Bragantia* **50** : 139-151.

Sandhu PS (1984) *Effects of sowing dates, phosphorus levels and herbicides on response of rhizobium in lentil*. Ph D Thesis, Punjab Agricultural University, Ludhiana.

Sandhu PS, Dhingra KK, Bhandari SC and Gupta RP (1991) Effects of hand hoeing and application of herbicides on nodulation, nodule activity and grain yield of *Lens culinaris* Med. *Plant and Soil* **135** : 293-296.

Savage KE (1976) Adsorption and mobility of metribuzin in soil. *Weed Science* **24** : 525-528.

Schenelle MA and Hensley LE (1986) Effects of pesticides upon nitrogen fixation and nodulation by dry beans. *Pesticide Science* **28** : 83-88.

Schenelle MA and Hensley DL (1989) Mode of action of bentazon in depressing nitrogen fixation by dry bean (Phaseous vulgaris). *Transactions of the Kansas Academy of Science* **92** : 149-154.

Schenelle MA and Hensley DL (1990) Effects of pesticides upon nitrogen fixation and nodulation by dry bean. *Pesticide Science* **28** : 83-88.

Selvamani S and Sankaran S (1993) Soil microbial population as affected by herbicides. *Madras Agricultural Journal* **80** : 397-399.

Semidey N and Alimodovar L (1987) Oxyflourfen : Canadian herbicides for weed control in pigeonpea. *Journal of Agricultural University, Puerto Rico* **71** : 277-285.

Shaheid SI (1971) Cytological and biochemical effects of the herbicide trifluralin on plant roots. Dissertation Abstarct B **31** : 4457.

Sharma LN and Saxena SN (1974). Influence of 2,4-D on soil micro organisms with special reference to *Azotobacter. Journal of Indian Society of Soil Science* **22** : 168-174.

Sharom MS and Stephenson GR (1976) Behaviour and fate of metribuzin in eight Ontario soils. *Weed Science* **24** : 153-160.

Shennan JL and Fletcher WW (1965) *Weed Research* **5** : 256-274.

Shkola AI (1970) Vest Sel Kho Nauk Am=lma-Ata 13 : 34-37.

Sidhu PS, Bhandari SC, Mehla SP and Gill HS (1985) Effect of fluchloralin, oxadiazon and pendimethalin on soil microflora and nodulation in groundnut. *Journal of Research, Punjab Agricultural University, Ludhiana* **22(1)** : 33-38.

Singh M and Dudeja SS (2007) Soil Plant Microbe interaction for ecologicl sustainability. *In: National Symposium on Legumes for Ecological Sustainability: Emerging Challenges and Opportunities*, held at Indian Institute of Pulses Research, Kanpur from November 35-5. p. 34.

Singh G and Wright D (1999) Effect of herbicides on nodulation, symbiotic nitrogen fixation, growth and yield of pea (*Pisum sativum* L). *Journal of Agricultural Science* **133** : 21-30.

Singh G and Wright D (2000a) Effects of herbicides on nodulation and growth of two varieties of peas (*Pisum sativum*). *Acta Agronomica Hungarica* **50**: 337-348.

Singh G and Wright D (2000b) Effects of prometryn and trietazine/simazine on weeds, nodulation, nitrogenase activity and yield of pea (*Pisum sativum*). *Annals of Biology* **18** : 1-18.

Singh G, Kundra HC, Brar LS and Gupta RP (1994) Effect of herbicides on soil microorganism dynamics, *Rhizobium*-legume symbiosis and grain yield of pea (*Pisum sativum* L.). *Annals of Agricultural Research* **15** : 22-26.

Skrdleta V, Vintikova H and Srogl M (1964) Susceptibility of rhizobia to some seed killers. *Rostl Vyroba* **37** : 827-834 (*cf Soils and Fertilizers* **28** *: 50*).

Sprout SL, Nelson LM and Germida JJ (1992) Influence of metribuzin on the *Rhizobium leguminosarum*-lentil (*Lens culinaris*) symbiosis. *Canadian Journall of Microbiology* **38** : 343-349.

Srinivasan G, Pothiraj P and Choudhry GK (1993) Response of azolla (*Azolla pinnata*) to rice (*Oryza sativa*) herbcides. *Indian Journal of Agronomy* **38** : 156-158.

Steinbrenner K, Naglitsch AF and Schlight I (1960) Albrecht-Thaer-Arch 8 : 611.

Stotsky G, Marlin WP and Mortensen JL (1956) *Proceedings Soil Science Society of America* **20** : 392-396.

Suhas PW, Singh P, Sreedevi TK, Thang NV and Bangong T (2007) Increased resource conservation and improved livelihood through legume based cropping systems. *In: National Symposium on Legumes for Ecological Sustainability: Emerging Challenges and Opportunities,* held at Indian Institute of Pulses Research, Kanpur from November 35-5. p. 29.

Szabo I (1964) The effect of two herbicides on the soil microflora and root nodule formation in pea. Mosomagy, Agrartud Foisk Kosl **7** : Kem Talajt 26 : 33 (*cf Soils and Fertilizers* **28** *: 250*)

Tanski HH, Montonece L, Ervio LR, Junnila S and Ressala B (1984) The effects of chlorsulfuron, glyphosate, metribuzin and TCA on soil nitrification capacity and dehydrogenase activity in compartment. INRA publication, Paris No. 31.

Teater RW, Mortensen JL and Pratt PF (1958) Effect of certain herbicides on the rate of nitrification and carbon dioxide evolution in soil. *Journal of Agriculture and Food Chemistry* **6** : 214-216.

Thomas MA and Hammond HD (1971) *Ohio Journal of Science* 71 : 21-29.

Tortensson L (1975) Effects of bentazon and dinoseb on soil microorganisms and on the *Rhizobium-Leguminosae* symbiosis. *Swedish Journal of Agricultural Research* **5** : 177-183.

Tripathi B, Kumar CPC, Rangarajan T, Chandurkar PS and Vijayalakshmi (1993) Pendimethalin, metolachlor, residues in soil and their effect on soil microflora, weed spectrum and green peas *(Pisum* sativum L.). In: *Integrated Weed Management for Sustainable Agriculture*, Vol II, pp. 98-100. Proceedings of the International Symposium. Indian Society of Weed Science, Hisar, India.

Tu CM (1966) PhD thesis. Oregon State University. pp. 125.

Udoh DJ and Nelson LE (1986) Trifluralin induces chlorosis in soybean grown on clayey high ph soil. *Plant and Soil* **96** : 175-184.

Van Rensburg HJ and Strijdom BW (1984) Effect of herbicides on survival of rhizobia and nodulation of peas, groundnuts and Lucerne. *South African Journal of Plant and Soil* **91** : 135-138.

Vidal D, Martinez J, Bergareche C, Miranda AM and Simon E (1992) Effect of methabenzthiazuron on growth and nitrogenase activity in *Vicia faba*. *Plant and Soil* 144 : 235-245.

Vintikova H, Skrdleta V and Srogl M (1965) The sensitivity of nodule bacteria to herbicides. p. 264-268 (*In: Plant microbe relationships. Procedings Symposium Prague 1963 (cf Soils and Fertilizers* **28** *: 570).*

Wani SP, Rupela O Lee KK (1995) Sustainable agriculture in the semi-arid tropics through biological nitrogen fixation in grain legumes. *Plant and Soil* **174** : 29-49.

Wilcke E Sonderh Z (1968) *Pfla path Pfefraukh Pflschuts* **4** : 163-167.

Wilkinson V and Lucas RL (1969) Effect of herbicides on the growth of soil fungi. *New Phytology* **68** : 709-719.

Worsham AD and Giddens J (1957) Some effects of 2,2-dichloro propionic acid on soil microorganisms. *Weeds* **5** : 316-320.

Yadav K, Prasad V, Rai R and Ahmad N (1990) Effect of some herbicides on the nodulation and grain yield of lentil. *Journal of the Indian Society of Soil Science* **38** : 749-752.

Yadav PK, Khan AH, Murti R and Upadhyay RK (2006) Effect of herbicides on germination, growth and nodulation in chickpea (*Cicer arietinum*). *Indian Journal of Agricultural Sciences* **76** : 682-684.

Yueh LY and Hensley DL (1993) Pesticide effet on acetylene reduction and nodulation by soybean and lima bean. *Journal of the American Society for Horticultural Science* **118** : 73-76.

6

SCIENCE OF OMICS

RITU MAHAJAN

Cells are the fundamental working units of every living system. All the instructions needed to direct their activities are contained within the chemical called DNA (deoxyribonucleic acid). DNA from all organisms is made up of the same chemical and physical components. The DNA sequence is the particular side-by-side arrangement of bases along the DNA strand (e.g., ATTCCGGA). This order spells out the exact instructions required to create a particular organism with its own unique traits. The genome is an organism's complete set of DNA. Genomes vary widely in size: The smallest known genome for a free-living organism (a bacterium) contains about 600,000 DNA base pairs, while human genome has about 3.2 billion base pairs.

The sequencing of the first genome of a higher plant *Arabidopsis,* was available to public domain in 2001 and with new generations using more efficient sequencing machines (Marshall and Pennisi, 1998), the sequence of several other plant genomes may become available in the coming decade. In the wake of these sequencing efforts, plant research enters an exciting period in which genome-wide approaches are becoming an integral part of plant biology (Somerville and Somerville, 1999). This is reflected by the development of tools to exploit and further expand this wealth of information.

The term functional genomics can be referred to as the "development and application of global (genome-wide or system-wide) experimental approaches to assess gene function by making use of the information and reagents provided by structural genomics" (Hieter and Boguski, 1997). It involves the use of high-throughput methods for the study of large numbers of genes (ideally the entire set) in parallel. Gene "function" can be considered from several points of view: it can be mean biochemical function (e.g. protein kinase), cellular function (e.g. a role in a signal transduction pathway), developmental function (e.g. a role in pattern formation), or adaptive function (the contribution of the gene product to the fitness of the organism). Having identified a new sequence, the comparison with sequence databases is the simplest way to obtain (essentially biochemical) functional information.

Currently, about 50% of newly identified genes show sequence similarity to previously described genes. However, computerized analyses are generally not sufficient to define gene function with a high level of confidence, and experimental confirmation is needed in most cases. Indirect information on cellular or developmental function can be obtained from spatial and temporal expression patterns; for example, the presence of mRNA and/or protein in different cell types, during development, during pathogen infection, or in different environments. The sub cellular localization and post-translational modifications of proteins can be informative as well. Knocking out or over expressing the gene permits the gene sequence to be linked to a phenotype from which a cellular role or a role in development may be deduced. Finally, the fitness of plants carrying mutations or natural variants for the gene can be compared in different environments with wild-type plants to study the adaptive function.

SINGLE NUCLEOTIDE POLYMORPHISM (SNPs)

SNPs along with EST markers are the type of third generation markers, which are extensively used now days in various genomic studies. ESTs basically are used in functional genomics while SNPs are used in individual genotyping.

Single nucleotide polymorphisms or SNPs (pronounced "snips") are DNA sequence variations that occur when a single nucleotide (A, T, C, or G) in the genome sequence is altered (Gupta *et al*, 1999). For example a SNP might change the DNA sequence AAGGCTAA to ATGGCTAA.

So SNPs are the sequence based molecular markers.

This variation is due to **point mutation** at a given location. The mutation can be of two types:

1. **Transition:** In this case there is subtitution of a purine with other purine and a pyrimidine with other pyrimidine. This type of transition occurs basically at **CpG** islands. The methylated Cytosine gets deaminated to form Thymine base pair. This leads to the occurrence of SNP at a given location.
2. **Transversion:** Here in this case purine base gets substituted with a pyrimidine base and leads to occurrence of SNP.

Some important features of SNPs

- Two of every three SNPs involve the replacement of cytosine (C) with thymine (T).
- SNPs can occur in both coding (gene) and noncoding regions of the genome. Many SNPs have no effect on cell function and they do not form a mutant phenotype.
- When they are present in coding regions they may present at the third position of the codon and there is hence no effect on the phenotype. But when a nucleotide is

present at first and second position it they may lead to change in sequence of amino acid and hence leads to the mutant phenotype.

➢ When the SNP is present in coding region they may show 100% transmission with the trait. But when they are away from the gene or are present in the non-coding region they do not show 100% transmission with the trait.

The SNPs show intraspecific diversity all individuals of a same species have same DNA sequence but the differences occur at certain loci due to SNPs. In case of human beings 99.9% DNA is same but the difference is due to 0.1% SNPs. These SNPs make us individuals rather than identical clones.

Methods used for discovery and identification of new SNPs

Some of the methods available for the discovery of these SNPs are described in this section, although rarely one or more methods described later for SNP genotyping can also be used for SNP discovery.

These methods involve:

- **Locus specific PCR amplification**
- **Alignment among available genomic sequences**
- **Whole genome shotgun sequences**

Locus specific-PCR amplification

While using this method, locus specific PCR primers are synthesized from the genomic sequences available in the data banks and PCR amplification is undertaken using DNA samples from several individuals for which we want to find SNPs. PCR amplified products are sequenced and sequence differences are used for discovery of new SNPs. This method is not commonly used as it requires prior sequence information so this method can be used only for genomic regions with known sequences.

Alignment among available genomic sequences

Alignment of genomic and EST sequences is the most convenient method for discovery of SNPs, if genomic/EST sequences from a heterozygote or from more than one individuals of the same species are available in the databases. There are various data bases having full genomic as well as EST sequences. The alignment of sequences is automated through computer software, and will allow detection of SNPs. This alignment can be pairwise if two sequences are to be compared or multiple alignment if more sequences are there. However, a comparison of genomic sequences will detect SNPs, both in coding and non-coding regions, while that of ESTs will detect SNPs only in the coding region.

Whole genome shotgun sequences

In this approach, random clones from the genomic library prepared from a mixture of DNA from several individuals are sequenced. SNPs can be detected by alignment of sequences belonging to the same locus.

Methods used for genotyping individuals at SNP Loci

These are basically two terms. SNP discovery is the which involves the determination of new SNP at a given location while the term SNP genotyping involves the genotyping of the individuals for the SNPs which have already been discovered.

A number of methods for SNP genotyping are now available, although all of them may not be equally useful.

There are basically two classes:

- **Gel based methods:** These methods using agarose gel electrophores is done for genotyping.
- **Non gel based methods:** These methods do not involve the use of agarose gel electrophoresis rather fluorescent dyes are used here.

GEL based methods for SNP detection:

1. **RFLP and AFLP-based assay***:* The presence of SNP Scan be detected by RFLP or AFLP conducted on PCR products, whenever such SNP generates or destroys a specific restriction site for an enzyme. The PCR product in this method is subjected to restriction digestion with individual enzymes. The detection is done by RFLP or AFLP to find the differences in patterns, which will be due to changes in the restriction sites.
2. **Single-strand conformation polymorphism:** Single strand conformation polymorphism (SSCP) technology detects allows detection of polymorphism which is to differences of one or more base pairs in the PCR products and is therefore suitable for SNP detection. The technique relies on the secondary structure being different for single strands derived from PCR products that differ by one or more nucleotides at an internal site The technique has been utilized for detection of SNPs in the genus *Picea* (spruce).
3. **Allele-specific amplification for SNP genotyping**: SNP Scan also be detected by designing allele-specific primers for individual SNP sites. Different fluorochromes are attached at the 5′ ends and the polymorphic nucleotides are attached at the 3′ ends of the two primers. These allele-specific primers, when used for PCR with pre amplified DNA as the template; the amplified product will be allele-specific and could be identified either on polyacryamide gel or on an automatic sequencer. This approach has been utilized for SNP genotyping in barley.

Non-gel based methods

1. **TaqMan assay***:* In TaqMan assay an oligonucleotide TaqMan probe is labelled with a fluorescent reporter molecule (e.g. FAM or TET) at the 5′end and a quencher (e.g. TAMRA) at the 3′ end. The probe on hybridization to the template DNA is degraded at its 5′ end due to exonuclease activity of *Taq* polymerase (TaqMan), so that the reporter is released leading to a rise in fluorescence signal.

However, when due to the presence of a SNP, the probe mismatches with the template leading to failure in duplex formation. So no such degradation at the 5′ end of the probe is possible and there is no rise in fluorescence signal.

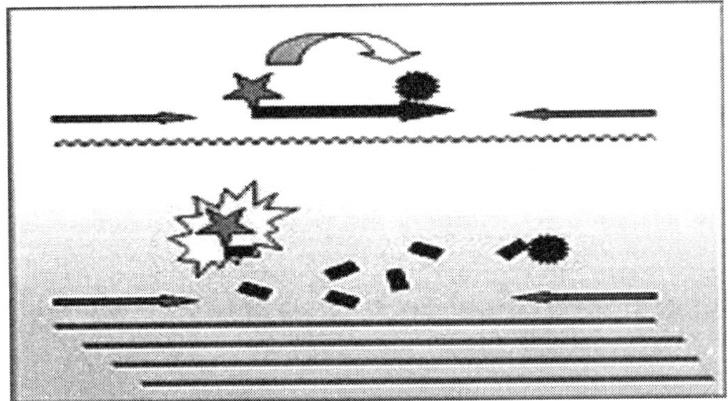

Fig. 1. TaqMan Assay

2. **Molecular beacons**: These are the hairpin shaped, oligonucleotide, hybridization probes which have quencher dye at 3′ end and reporter dye at 5′ end. When the beacon is hair pin shaped it will not give the signal but when the beacon does not find the SNP it will hybridize and give the signals.

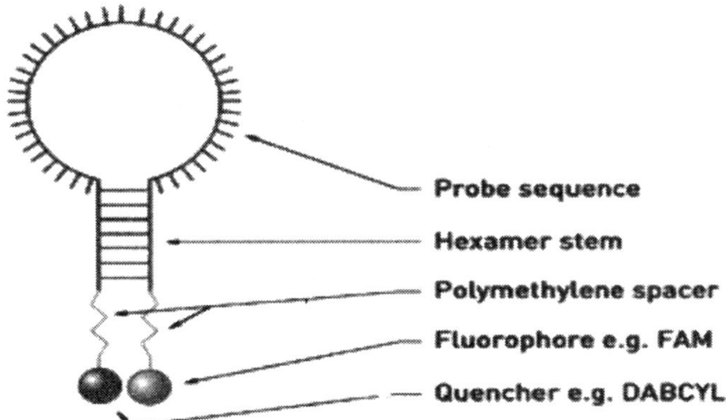

Fig. 2. Molecular Beacons

3. **Oligonucleotide ligation assay**: In this approach, two independent probes (one is 5′ biotinylated and the other is 3′ fluorescent-labelled) are used for hybridization with PCR product, so that when the probe matches the product then these two probes anneal with the PCR product and undergo ligation resulting in an oligonucleotide which is biotinylated at the 5′ end and fluorescent-labelled at the 3′ end. The ligation product, which is fluorescent labelled at the 3′ end, is captured on a solid streptavidin-coated matrix due to biotinylation at its 5′ end, and the signal is detected by autoradiography.

The SNPs are called **mother of all markers** as they have certain uses over others. These are:

- These are the most abundant type of molecular markers present and are spread on whole of the genome. and can be used to create high density maps and also can be used to tag the gene of interest
- These are codominant markers and can be used to distinguish heterozygotes from the homozygotes.

Limitations

- These are biallelic in nature, detects the population where only two alleles are present
- These are population specific and can be used only on one population.

Applications

- **Cultivar identification and evaluation of diversity**: SNP are used to identify level of intervarietal variation for each genome. In wheat, one SNP per 20 bp and in the maize genome, one SNP per 70 bp have been recorded in certain regions of these genomes.
- **Functional proteomics**: To identify the SNPs that modify proteins and protein active sites.
- **Gene therapy**: It involves the insertion of normal genes into germ cells or fertilized cells so as to create a beneficial genetic change.

DNA MICROARRAYS

DNA microarray technology has given rise to the study of functional genomics (Brown and Botstein, 1999; Lockhart and Winzeler, 2000). The entire set of genes of an organism can be microarrayed on an area as small as a fingernail and the expression levels of thousands of genes are simultaneously studied in a single experiment. DNA microarray technology allows comparisons of gene expression levels on a genomic scale in all kinds of combinations of samples derived from normal and diseased tissues, treated and nontreated time courses, and different stages of differentiation or development. Further computational analysis of microarray data allows the classification of known or unknown genes by their mRNA expression patterns. Global gene expression profiles in cells or tissues will provide us with a better understanding of the molecular basis of phenotype, pathology, or treatment.

DNA arrays or gene chips

- DNA arrays consists of large number of DNA molecules spotted in systemic order on a solid surface.
- Depending on the size of each spot on the array, DNA array can be categorized as microarrays (dia < 250μm and macroarrays (dia > 300 μm).
- When the solid substrate used is small in size, arrays are also referred as DNA chips or gene chips.

- ➢ DNA chips can be used as Variant Detector Array (VDAs) to look for DNA sequences that differ by single nucleotide polymorphism ("SNPs")

Based on type of probes

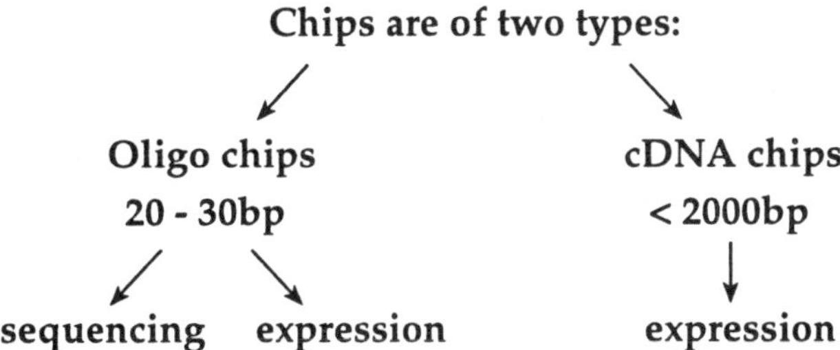

On a glass surface, complementary DNAs (cDNAs) or gene-specific oligonucleotides can be spotted. Oligonucleotides can also be synthesized *in situ*.

***in situ* synthesis of oligonucleotide arrays**

Photolithography

This method was developed by Fodor *et al* (1991) and it is now referred to as the Affymetrix GeneChip system. In this method oligonucleotide arrays, DNA oligonucleotides are synthesized in situ onto the DNA chip using photolabile protecting groups and photolithographic masks are used to add the selective sequences of nucleotides (Lipshutz *et al.* 1999). This technique use phosphoramides, surface linker chemistry and combinatorial synthesis strategies.

Synthesis support: The synthesis support consists of a glass substrate derivatised with polylysine or aminosilane.

Masking: Photolithographic masks having apertures are used to cover the regions, where a particular nucleotide is not desired to be added.

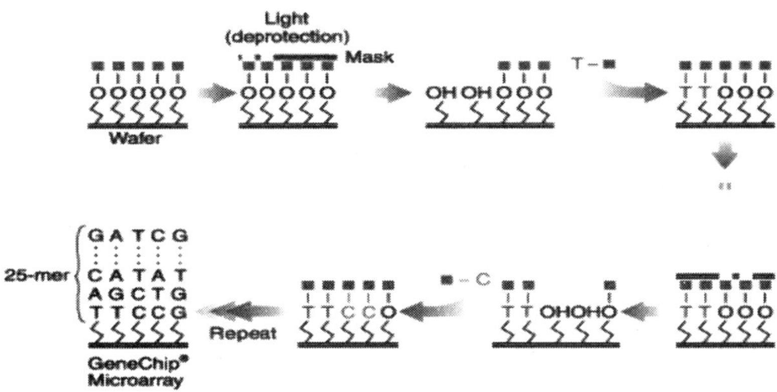

Fig. 3. Photolithography (Fodor *et al* 1991)

Illumination: The surface of solid support is illuminated, by illumination (wavelength more than 280nm) through micro apertures of the photolithographic masks, so that specific regions of the support are activated for synthesis.

Coupling: A 3′-O-phosphoramidite activated deoxynucleoside is protected with a group present to the surface that was illuminated and coupling occurs in the sites that were exposed to the light.

The selective photodeprotection and coupling cycles are repeated until the desired sets of products are available.

Ink jet

Ink jet is another technology can be used for *in situ* production of oligonucleotide microarrays.

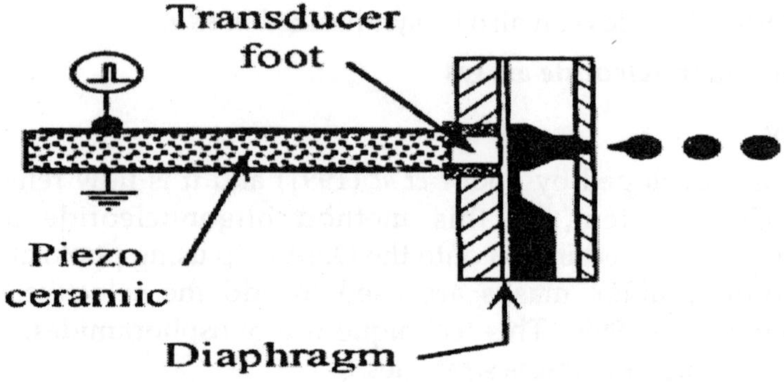

Fig. 4. Piezoelectric delivery method for microspotting

It involves direct delivery of tiny droplets of phosphoramidite solutions to precise locations on a surface rather than using lithographic masks. Piezoelectric delivery method is recently used for the manufacture of high-density oligonucleotide microarrays.

cDNA deposited arrays

In deposited arrays, which were developed by Pat Brown lab in Stanford University (Schena *et al* 1995) a high-speed spotting robot is used to affix PCR-amplified cDNA or synthesized oligonucleotides onto a chemically modified glass slide. Polylysine or amino silane is commonly used as a substrate for coating the surface of these glass slides. Mechanical microspotting allows transfer of premade substances onto a solid surface. This is done by capillaries, needles and tweezers. The inkjet printing method produces more uniform spots, which had been difficult to achieve with available pin spotting techniques.

cDNA vs. oligonucleotide as a target DNA

For reference, the term "probe" refers to the fluorescently labeled DNAs or RNAs and the term "target" refers to the DNAs on the slide. The definitions are reversed in some articles, especially when the Affymetrix GeneChip system is mentioned.

The oligonucleotide targets require only the sequence information of genes, and thereby can maximally exploit the genome sequences of the organism. The precise fold that changes from cDNA or oligonucleotide microarrays are not identical, but the general trends of changes and the fold differences are similar to each other when the same genes are compared in both array experiments. In oligonucleotide arrays, the design of oligonucleotides is very critical to the results and they should be designed very carefully. In cDNA arrays, there may be potentially confounding effects of cross-hybridization due to sequence homologies among members of a gene family. When oligonucleotides are used, they can be specifically designed to differentiate between highly homologous members of a gene family as well as alternatively spliced forms of the same gene (exon-specific). Oligonucleotide arrays can also be designed to allow detection of mutations and SNPs.

Hybridization

In two-color hybridization, fluorescently labeled cDNA probes, directly reverse-transcribed from two separate mRNA populations, are hybridized onto the target DNA fixed on the slide. One cDNA sample is labeled with Cy3 fluorescent dye, and the other sample with Cy5. Cy3 has advantage over Cy5 as Cy5 has strong fluorescent background and is sensitive to photobleaching. After hybridization and washing on a single array, the relative level in mRNA transcript for each gene is measured by the Cy5/Cy3 color ratio using a confocal laser scanner.

Applications

- **Detection of SNPs**: Use of SNPs offer great promise for rapid and highly automated gene sequencing and are extensively used as markers of choice.
- **Functional genomics**: Microarrays for gene expression analysis provide an integrated platform for functional genomics. Samples of mRNA from a variety of cells and tissues are used for microarray analysis that would yield information about specific changes in gene expression patterns. Novel expression profiles were identified for many sequences. DNA chips can also be used to distinguish the transcripts from individual members of multigene families.
- **Diagnostics and gene mapping**: Chips have been prepared to detect mutant alleles in CFTR (cystic fibrosis), BRCA 1 (cancer susceptible genes) and beta globulin genes.
- **DNA chips and agricultural biotechnology**: DNA chips with ESTs can also be used to collect data on expression in an agricultural crop under different conditions. This information can be of practical utility in agricultural biotechnology. Transgenic plants can be rapidly analyzed using microarrays on DNA chips, and expression pattern under different environmental conditions can be predicted at the gene level.
- **Proteomics**: Like genomics, proteomics is the study of protein-protein interactions. Protein linkage maps can be generated using genomic sequence

information. Protein-protein interactions can be studied using the yeast-two hybrid systems.

Limitations

- Post-transcriptional regulatory events cannot be detected.
- The expression patterns of individual members of closely related gene families may not be reliably detected.
- Sensitivity of detection of transcripts expressed at low levels.
- Most arrays are based on known genes and ESTs, which are currently biased towards most abundant genes.

CARBOHYDRATE MICROARRAYS

Carbohydrates, as components of glycoprotein, glycolipids and proteoglycans, play an important biological role as recognition markers through carbohydrate-protein interactions. For the most part, biophysical and biochemical methods have been used to analyze these bimolecular interactions. In contrast, less attention has been given to the development of high-throughput procedures to elucidate carbohydrate protein recognition events. Recently, carbohydrate arrays were developed and employed as a novel highthroughput analytic tool for monitoring carbohydrate-protein interactions. This technique has been used to profile protein binding and enzymatic activity. Carbohydrate microarray technologies are new developments at the frontiers of glycomics (Ratner, 2005). While conventional DNA and protein microarray technologies have been a boon to industrial and academic research, investigators interested in studying the biological role-played by glycans.

Carbohydrates and glycoconjugates are known to be involved in inflammation, cell-cell interactions, pathogen-host adhesion and signal transduction. With half or more of all proteins containing some carbohydrate modification, the challenges associated with their study have been a considerable hindrance to the field of glycobiology and to understanding of biology as a whole.

Carbohydrate recognition often consists of low affinity interactions between glycans and their associated binding partners. Therefore a successful microarray platform must be able to probe protein-carbohydrate, carbohydrate-carbohydrate and RNA-carbohydrate interaction with sufficient specificity to detect weak binding. To, achieve this glycan moiety should be displayed in fashion that is readily accessed by the analyte. Carbohydrate-modified surfaces are ideal for interrogating these weak interactions, as monolayers of sugars on the surface of chip to the multivalent display of glycan frequently found on the cell surface.

Factors for consideration

- Several factors must be taken into consideration while designing a microarray.
- A suitable surface must be selected for displaying the glycan library. This can include microtitre plates, glass slides, nitrocellulose and gold surfaces.

- A linking chemistry whether covalent or non-covalent binding must be adopted. For e.g. lipid modified glycans are required for non-covalent binding and amino and maleimido derivatised sugars for covalent attachment.
- A protocol must be developed for blocking potential sites for non-specific binding of analyte to surface.
- A method must be developed to detect weak binding events on the surface of the array.
- Probe is fluorescently labeled prior to hybridization and then the signal is detected.

Early microarray platforms

- **Wang and Feizi** (2003) utilized nitrocellulose surfaces for immobilization of carbohydrates and neoglycolipids to study microbial polysaccharides. The surface was blocked by BSA (Bovine Serum Albumin) to reduce non-specific binding.
- **Mrkish and his colleagues** (2003) covalently attached synthetically defined sugars to an array. They used linker-modified saccharides bound to derivatised self-assembled monolayers of alkanethiolate on gold surface. The arrays were used for on-chip enzymatic glycosylations and direct analysis of protein binding partners by MALDI-TOF MS.

The Consortium for Functional Glycomics currently possesses a library of more than 1,000 synthetic chips for microarray fabrication.

Biological applications

- Arrays of carbohydrates have been used to identify the glycan specificity of proteins, nucleic acids and intact cells.
- Glycochip has been employed to examine human serum for carbohydrate reactivity and to profile glycan-dependent enzymes and enzyme-inhibitors.
- The arrays have application in the study of HIV glycobiology as the virus surface has protein glycoprotein gp120, which plays a vital role in entry of host surface into the cell.
- Amino glycoside chips have been prepared to study the antibiotic resistance, antibiotic toxicity and to detect RNA binding.
- Efforts are there to develop glycochips against malaria, HIV and cancer.

Carbohydrate-binding proteins of the immune system-antibodies, selecting, a cytokine and a chemokine - and several plant lectins indicate that microarrays of carbohydrates (glycoconjugates, oligosaccharides and monosaccharides) will greatly facilitate not only surveys of proteins for carbohydrate-binding activities but also elucidation of their ligands. It is predicted that both naturally occurring and synthetic carbohydrates will be required for the fabrication of microarrays that are sufficiently comprehensive and representative of entire glycomes (Shin *et al*, 2004).

SERIAL ANALYSIS OF GENE EXPRESSION (SAGE)

With the advent of the human genome project, a vast amount of information about genes and gene structure is suddenly at our fingertips. But this information is limited. Every cell within an organism has the same genetic composition (with the exception of its gametes), and yet, obviously skin tissue is very different from nervous tissue. The DNA sequence cannot provide information about these differences, which represent the next level of complexity and organization within an organism: DNA expression. A method to efficiently count large numbers of mRNA transcripts by sequencing short tags, usually 10 bp in length. It uses a new analytical method of reliably matching SAGE tags to known genes. SAGE is a technique that allows rapid, detailed analysis of thousands of transcripts in a cell.

Principles

The basic concept of SAGE rests on two principles:

- A short nucleotide sequence tag (9-10bp) containing sufficient information to uniquely identify a transcript provided it is isolated from a defined position within the transcript.
- Concatenation of short sequence tags allows the efficient analysis of transcript in a serial manner by sequencing of multiple tags within a single clone. The number of times that a sequence of a given gene is found within the isolated clones represents the frequency of the transcript .

Serial Analysis of Gene Expression (SAGE), originally described by Velculescu *et al* (1995) counts polyadenylated transcripts by sequencing a short 14 bp tag at the genes 3'end, adjacent to the last restriction site (Polyak *et al*, 2001).

Basic steps in the SAGE protocol:

- Select cells for profiling.
- Isolate mRNA using magnetic beads and synthesize cDNA.
- Cut transcript with anchoring enzyme (NlaIII).
- Release 14 bp tags from the 3' end, adjacent to the last restriction site with tagging enzyme BsmF1.
- Pair and ligate tags to form ditags, and concatenate ditags into plasmids for automtic sequencing.
- Extract and count tags to calculate expression level of each transcript.

Once the cDNA has been created, it is then cleaved using an anchoring enzyme. The anchoring enzyme is a restriction endonuclease that recognizes and cuts specific 4 base pair (bp) DNA sequences. Since this enzyme requires only 4 specific nucleotides, it cleaves DNA molecules often, resulting in every cDNA that has been generated being cut at least once. The cut cDNA is then bound to streptavidin beads by virtue of its multiple thymidine ('T's) at its 3' end, thereby immobilizing it. At this point, we have multiple cDNAs bound to beads that are all of varying lengths, each with the same cut site and 5' overhang, because the anchoring enzyme recognizes the same 4 bp in all transcripts, but these 4 bp occur at different points within each transcript.

The sample of bound cDNAs is then divided in half and ligated to either linker A or B. These linkers are designed to contain a type IIS restriction site. Type IIS restriction endonucleases cut at a defined distance up to 20 base pairs away from their recognition sites. The type IIS restrictive endonuclease, also called the tagging enzyme, cleaves the cDNA to release it from its bound bead. Blunt ends are then created, so that neither the 3' nor 5' end has overhanging single-stranded sequences. Once this is achieved, the cDNA tags bound to linker A and B are ligated to each other to create ditags. These ditags have linker A on one end, linker B on the other, and both transcript tags are adjacent to one another in the middle. These ditags are then amplified by PCR, using primers that are complimentary to sequence in either linker.

Once the ditags have been amplified, they are then cleaved using the anchoring enzyme again. This has two effects: first, it releases the linkers from either end of the ditag, leaving only sequence from the two tags. Secondly, it creates sticky ends, or 3' and 5' ends that have overhanging, single-stranded DNA that can complimentary base pair with single-stranded DNA of another ditag. In this way, all of the ditags generated are linked, or concatenated to produce one long string of tags. This collection of tags is then introduced into a vector to be cloned and sequenced.

Problems with Gene Tags

There are several problems:

- The tags generated during SAGE are extremely short (13 - 14 bp). If the tag derived is from an unknown gene, it is difficult to investigate its potential function using such a short sequence. In cancer research, for example, the most attractive feature of SAGE is its ability to evaluate the expression pattern of thousands of genes in a quantitative manner without prior sequence information. This has lead to analysis of differential gene expression of cancer cells and their normal counterparts, and identification of several novel genes that could potentially play a role in tumourigenesis.
- Another problem with tags is tag specificity. There are instances in which multiple genes share the same tag, as there is an overlap in sequence between the two genes; using longer tag sequences can eliminate this problem. Increasing tag length to 18 bp in an attempt to increase tag specificity has been reported to yield a better representation of DNA expression.
- Another downfall of the SAGE technique is that typeIIS restriction enzymes (typically *Bsm*FI) do not always yield the same length of fragments. *Bsm*FI should yield exact 14 bp tags, but, depending on the temperature, the length of fragments produced varies. Since two tags are ligated tail to tail, it is hard to ensure each tag is 14 bp long in a ditag of 28 bp - the ditag could be consist of a 12 bp and 16 bp tag, a 13 bp and 15 bp tag, or any variation thereof. Maintaining the temperature at 65° C can minimize this problem.

Applications

- SAGE technique allows the rapid and detailed analysis including the relative abundance of all the transcripts in a given sample (cell or tissue).
- SAGE has been widely used in biological, medical, and pharmaceutical areas of research (Wakatsuki *et al*, 2001).
- SAGE has been used successfully to compare expression profiles between normal and diseased cells (Tuteja and Tuteja, 2004) or any two differentially treated samples.

EXPRESSED SEQUENCE TAGS (ESTs)

As one can imagine, finding a gene that codes for a protein, or proteins, is not easy. Traditionally, scientists would start their search by defining a biological problem and developing a strategy for researching the problem. Oftentimes, a search of the scientific literature provided various clues about how to proceed (Newman *et al*, 1994; Cooke *et al*, 1996; Yamamoto and Sasaki, 1997). Researchers would then work to isolate that protein, determine its function, and locate the gene that coded for the protein. Alternatively, scientists could conduct what is referred to as linkage studies to determine the chromosomal location of a particular gene. Once the chromosomal location was determined, they would use biochemical methods to isolate the gene and its corresponding protein.

An Expressed Sequence Tag is a tiny portion of an entire gene that can be used to help identify unknown genes and to map their positions within a genome.

ESTs are small pieces of DNA sequence (usually 200 to 500 nucleotides long) that are generated by sequencing either one or both ends of an expressed gene. The idea is to sequence bits of DNA that represent genes expressed in certain cells, tissues, or organs from different organisms and use these "tags" to fish a gene out of a portion of chromosomal DNA by matching base pairs. The challenge associated with identifying genes from genomic sequences varies among organisms and is dependent upon genome size as well as the presence or absence of introns, the intervening DNA sequences interrupting the protein coding sequence of a gene.

Gene identification is very difficult because most of our genome is composed of introns interspersed with a relative few DNA coding sequences, or genes. These genes are expressed as proteins, a complex process composed of two main two steps. Each gene (DNA) must be converted, or transcribed, into messenger RNA (mRNA), RNA that serves as a template for protein synthesis. The resulting mRNA then guides the synthesis of a protein through a process called translation. Interestingly, mRNAs in a cell neither contain sequences from the regions between genes, nor from the non-coding introns that are present within many genes. Therefore, isolating mRNA is key to finding expressed genes in the vast expanse of the human genome.

The problem, however, is that mRNA is very unstable outside of a cell; therefore, scientists use special enzymes to convert it to complementary DNA (cDNA). cDNA is a much more stable compound and, importantly, because it was generated from a mRNA in which the introns have been removed and cDNA represents only expressed DNA sequence. cDNA is a form of DNA prepared in the laboratory using an enzyme called reverse transcriptase. cDNA production is the reverse of the usual process of transcription in cells because the procedure uses mRNA as a template rather than DNA. Unlike genomic DNA, cDNA contains only expressed DNA sequences, or exons.

From cDNAs to ESTs

A "gene family" is a group of closely related genes that produces similar protein products. Once cDNA representing an expressed gene has been isolated, scientists can then sequence a few hundred nucleotides from either end of the molecule to create two different kinds of ESTs. Sequencing only the beginning portion of the cDNA produces what is called a 5' EST. A 5' EST is obtained from the portion of a transcript that usually codes for a protein. These regions tend to be conserved across species and do not change much within a gene family. Sequencing the ending portion of the cDNA molecule produces a 3' EST. Because these ESTs are generated from the 3' end of a transcript, they are likely to fall within non-coding or untranslated regions (UTRs), and therefore tend to exhibit less cross-species conservation than do coding sequences.

A UTR is that part of a gene that is not translated into protein.

ESTs are generated by sequencing cDNA, which itself is synthesized from the mRNA molecules in a cell. The mRNAs in a cell are copies of the genes that are being expressed. mRNA does not contain sequences from the regions between genes, nor from the non-coding introns that are present within many interesting parts of the genome.

ESTs as Genome Landmarks

Just as a person driving a car may need a map to find a destination, scientists searching for genes also need genome maps to help them to navigate through the billions of nucleotides that make up the human genome. For a map to make navigational sense, it must include reliable landmarks or "markers". Currently, the most powerful mapping technique, and one that has been used to generate many genome maps, relies on Sequence Tagged Site (STS) mapping. An STS is a short DNA sequence that is easily recognizable and occurs only once in a genome (or chromosome). The 3' ESTs serve as a common source of STSs because of their likelihood of being unique to a particular species and provide the additional feature of pointing directly to an expressed gene.

Limitations

- Low copy number genes
- Rare tissues
- Enrichment of 3′ ends
- Incomplete coverage of genes

PROTEOMICS

The term proteomics was coined in 1995 by an Australian postdoctoral fellow Marc Wilkins as a complement of genomics, the study of genes (Wilkins *et al* 1996). While the genome is a rather constant entity, the proteome differs from cell to cell and is constantly changing through its biochemical interactions with the genome and the environment In simplest term, it is the study of proteome, that refers to the collection of total set of proteins expressed in a cell at a given time, in an organism or Proteomics is the study of protein expression, regulation, modification, and function in living systems for understanding how living systems use proteins. Using a variety of techniques, proteomics can be used to study how proteins interact within a system or how protein expression changes in different parts of body, in different stages of its life cycle and in different environmental conditions as every individual has one genome and many proteomes.

The needs for proteomics research is necessary as in a cell there are certain genes that encode proteins with specific functions. The large increase in protein diversity is thought to be due to alternative splicing and about 200 different post-translational modifications of the proteins. This discrepancy implies that protein diversity cannot be fully characterized by gene expression analysis alone, making proteomics a useful tool for characterizing cells and tissues of interest. The proteome in any individual has a collection of 30-50% different gene products which are expressed at low levels, and its only proteomics which can convert proteins to peptides, followed by their conversion to amino acids which can be further determined and identified. Moreover the genome is static and the proteome reflects the cellular processes.

The study of proteomics is to obtain a global integrated view of:

- measurement of protein composition of organelles (spliceosome, phagosome, speckles etc.)
- normal and abnormal cellular processes
- protein-protein interactions
- protein-drug interactions
- determination of protein structure (NMR, X-ray, in silico etc.)
- identification of how and where proteins are modified (Post translational modifications)

Methods of studying proteins

1. Two-dimensional gels and mass spectrometry: Developed in late 1970s. Proteins from a sample are separated on the basis of charge and mass, and then analyzed by spectrometry (Henzel *et al* 1993). This method of detection is mainly used for cellular Proteomics. This is one of the established technology (can resolve 10000 spots per gel) used for studying proteins expressed at different levels. MS instruments consist of three modules: an ion source, which splits the sample molecules into ions; a mass analyzer, which sorts the ions by their masses by applying electromagnetic fields; and a detector, which measures the value of an indicator quantity and thus provides data for calculating the abundances of each ion present. The technique has both qualitative and quantitative uses. MS can determine the isotopic composition of elements in a molecule, the structure of a compound by observing its fragmentation, can quantify the amount of a compound in a sample or study the fundamentals of gas phase ion chemistry. It is biased at detecting very small, very large and nuclear bound proteins and at the same time it is time consuming.

Similarly, MALDI, or Matrix-assisted laser desorption/ionization, is a soft ionization technique used in mass spectrometry. This technique allows the analysis of biomolecules (proteins, peptides, and sugars) and larger organic molecules (polymers and dendrimers). MALDI is used to ionize these molecules because they are very fragile and will usually fragment when ionized by other, more conventional ionization methods. A laser beam is used for ionization, and the biomolecule is protected by a matrix to prevent it from being destroyed, and to allow for optimal vaporization and ionization.

2. **Protein microarrays**: Protein microarrays are tools that can be used in many different areas of research, including basic and translational research. Proteomic arrays are typically high-density arrays (> 1000 elements/array) that are used to identify novel proteins or protein-protein interactions, to identify the substrates of protein kinases, or to identify the targets of biologically active small molecules (MacBeath and Schreiber, 2000). Silicon chips are coated with specific surface chemistries or known protein ligands, and proteins from a sample are fractionated on the chip, and then analyzed by mass spectrometry or chips are prepared with different surface chemistries, much like traditional protein chromatography columns, but in a flat array. To detect proteins that are bound to the array, the samples must be labeled directly with a fluorophore or a hapten. Alternatively, in some applications antibodies can be used to detect binding events, where antibodies are spotted onto the protein chip and are used as capture molecules to detect proteins from cell lysate solutions. For immunoproteomics approaches, antigen microarrays are used to detect disease associated changes in the autoantibody profiles.

The ultimate goal of proteomics is to the study biochemical activities of every protein encoded by an organism or proteome. Zhu *et al* (2001) made a landmark where they prepared the first proteome chip by cloning ~94% (>5800 of 6200) of the yeast open reading frames in a yeast expression vector which expressed the proteins as N-terminal GST-His x6 double tagged fusions. A high-throughput yeast protein purification method was developed to individually purify proteins. 80% of yeast proteins were full length and of sufficient quantity to be detectable by most assay types. The proteins were then purified using the GST tags and were then attached to Ni-NTA-coated glass slides using the HisX6 tags. In addition to identifying known interactions, 33 novel binding proteins were detected and 150 novel lipid-binding proteins were also identified. This study demonstrated that an entire proteome can be immobilized on a glass surface to directly screen for interactions with proteins and small molecules.

3. Automated yeast two-hybrid screen: Pioneered by Fields and Song in 1989, the technique was originally designed to detect protein-protein interactions using the GAL4 transcriptional activator of the yeast Saccharomyces cerevisiae. The GAL4 protein activated transcription of a protein involved in galactose utilization which formed the basis of selection (Hurt *et al* 2003). Since then, the same principle has been adapted to describe many alternative methods including some that detect protein-DNA interactions, DNA-DNA interactions. Joung *et al* (2000) use E. coli instead of yeast which facilitates the rapid analysis of larger libraries (due to the higher transformation efficiency and faster growth rate). Such methods also might be faster than phage display, which is an enrichment technique requiring multiple rounds of affinity purification and amplification and also allows studies of sequences (of much larger proteins) that are not readily displayed on a phage surface.

In general, in any two-hybrid experiment a protein of interest is fused to a DNA-binding domain and transfected in a yeast host cell bearing a reporter gene controlling this DNA-binding domain. When this fusion protein cannot activate transcription on its own, it can be used as "bait" or as a "target" to screen a library of cDNA clones that are fused to an activation domain. The cDNA clones within the library that encode proteins capable of forming protein-protein interactions with the bait are identified by virtue of their ability to cause activation of the reporter gene. So the yeast two-hybrid system is devised to identify genes encoding proteins that are physically associated with a given protein in vivo. Since the emergence of the two-hybrid approach in 1989, a number of improvements have been incorporated that have increased its applicability. A laboratory robot performs bioassay to find whether pairs or groups of interact (Albers *et al* 2004). A successful technique for rapid results in global proteomics but provides information only on binding partners and not on biochemical function. It can yield false-positive and false- negative results with no structural information.

4. High-throughput crystallography: The high-throughput protein crystallography beamline is a dedicated facility for determining the structure of protein crystals and undertaking initial assessment of more complex crystals. Single crystals are analysed using multiple wavelength anomalous dispersion (MAD). Hypothetical proteins from a sequenced genome are expressed, purified, and crystallized to obtain three-dimensional structure, then tested for their biochemical functions (Kim *et al* 2003). This technique is important to determine the three-dimensional structures of nonmembrane proteins for which sequence-based methods have been unable to predict a function. Membrane proteins were excluded because it is technologically challenging to produce them on a high throughput scale, so a subset of protein structures is designed that may be useful in elucidating protein function and that will contribute to our understanding of the sequence/structure relationship. This technique can be useful to uncover certain protein functions and also provide their structural information.

5. **Shotgun proteomics**: Shotgun proteomics using liquid chromatography-tandem mass spectrometry (LC-MS/MS) provides the most powerful analytical platform for global inventory of complex proteomes. It is a method of identifying proteins in complex mixtures using a combination of high performance liquid chromatography combined with mass spectrometry. In this method, the proteins in the mixture are digested and the resulting peptides are separated by liquid chromatography. Tandem mass spectrometry is then used to identify the peptides.

6. **Bottom up proteomics**: Bottom-up proteomics is the most mature and most widely used approach for protein identification and characterization. A method to identify proteins, characterize their amino acid sequences and post-translational modifications by proteolytic digestion of proteins prior to analysis by mass spectrometry. The proteins may first be purified by gel electrophoresis. Then the crude protein extract is digested directly, followed by one or more dimensions of separation of the peptides by liquid chromatography coupled to mass spectrometry. By comparing the masses of the proteolytic peptides or their tandem mass spectra with those predicted from a sequence database, peptides can be identified and multiple peptide identifications assembled into protein identification. Reversed-phase HPLC provides high-resolution separations of peptide digests with solvents that are compatible with ESI.

7. **Top-down proteomics**: In top-down proteomics, intact protein molecular ions generated by ESI are introduced into the mass analyzer and are subjected to gas-phase fragmentation. It is a method of protein identification that uses an ion trapping mass spectrometer so as to store an isolated protein ion for mass measurement and tandem mass spectrometry analysis. An obstacle to this approach is the determination of product ion masses from multiply charged product ions

Applications

Protein as a diagnostic tool: One of the most promising developments is the interaction between human genes and proteins resulting in identification of new drugs for the treatment of diseases. If a certain protein is implicated in a disease, its 3D structure provides the information to design drugs which then interfere with the action of the protein. A molecule that fits the active site of an enzyme, but cannot be released by the enzyme, will inactivate the enzyme. The identification of new drugs to target and inactivate the enzyme HIV-1 protease that cleaves a very large HIV protein into smaller, functional proteins. The virus cannot survive without this enzyme; therefore, it is one of the most effective protein targets for killing HIV (Sullivan *et al* 1998). In Alzheimer's disease, targeting the enzyme beta secretase (which causes plaque to build up in the patient's brain), slows the progression of the disease (Cole and Vassar 2008). Specific proteins can be used as biomarkers to diagnose disease. A number of techniques like western blot, immunohistochemical staining, enzyme linked immunosorbent assay (ELISA) or mass spectrometry allow us to test for proteins produced during a particular disease, which helps to diagnose the disease quickly.

Protein chips: Chip-based separation of proteins (David *et al* 2007) provide methods that are faster, automated and more convenient than conventional gel electrophoresis. A chip that held thousands of immobilized proteins could be used to detect protein-protein interactions quickly and easily (Kersten and Feilner, 2007). Zyomyx developed a high-density protein chips made of silicon, which is microfabricated to posts to create a 3-D substrate. This technology offers profiling biochipTM for high throughput, low sample volume protein studies. A human cytokine chip was the first product to be commercialized, while various other products and technologies are under development. The chips are expensive.

Determining the existence of proteins in complex mixtures: Earlier antibodies to particular proteins or to their modified forms have been used in biochemistry and cell biology studies or ELISA was used for quantative determination of protein amounts, but now more recent techniques such as Matrix-assisted laser desorption/ionization have been employed for rapid determination of proteins in particular mixtures.

Differential expression of proteins: Proteins expressed under normal conditions are compared to those under stressed conditions. The approach is quite similar to microarray analysis where the expression levels of proteins are significantly changed, indicating the potential role of the protein to the causation of the stress.

Proteins as messengers: The binding of one protein to other protein receptors so as to send a signal to the cell results in forming structural connections between cells and thus leading to study of interaction between the proteins.

In vivo isotopic labeling: This is a common method used in comparative proteomics (Schwender and Ohlrogge, 2002). Here one set of samples is grown on a natural nitrogen source while the comparative sample is grown in the presence of the heavy isotope. The isotopic label can be introduced either as an amino acid (termed stable isotopic labeling in cell culture, SILAC) or by ^{15}N as the sole nitrogen source, typically in the form of $K^{15}NO_3$. Due to difference in masses of peptides from the two or three populations a direct comparison of MS peak intensities can be made for the two samples.

Proteomics is an extremely powerful technology but few limitations are:

- 2-D gels only work for hydrophilic (soluble) proteins. So membrane proteins simply don't run into the gel.
- 2-D gels are resolution sensitive. It's possible to resolve only hundreds to a few thousand spots on each gel whereas low abundance proteins may be masked by nearby abundant proteins on the gel.
- No two gels are exactly the same. So computers stretch the gel images (rubber-sheeting) to try to make them look as similar as possible before comparing them but now in DiGE (difference gel electrophoresis) (Unlu *et al* 1997; Tonge *et al* 2001) two protein samples are labelled with different fluorescent dyes, then mixed and run on the same gel. The gel is then viewed by fluorescence at two wavelengths. Though this technique is simple but very expensive.

Protein databases

A variety of protein sequence databases exist which play an important role as central comprehensive resources of protein information (Pruess *et al* 2006)

- UniProt (The UniProt Consortium comprises the European Bioinformatics
- Institute (EBI), the Swiss Institute of Bioinformatics (SIB), and the Protein Information Resource (PIR) and is the world's most comprehensive resource on protein information..
- PIR (Protein Information Resource to support genomic and proteomic research).
- Swiss-Prot (manually curated biological database of protein sequences) (Junker *et al* 2000)
- TrEMBL- computer-annotated supplement of Swiss-Prot that contains all the translations of EMBL nucleotide sequence entries not yet integrated in Swiss-Prot.
- PDB (Protein Data Bank is a repository for 3-D structural data of proteins and nucleic acids)
- NCBI (National Center for Biotechnology Information houses genome sequencing data in GenBank and an index of biomedical research articles in PubMed Central and PubMed, and information relevant to biotechnology)
- Human Protein Reference Database (manually curated scientific information pertaining to the biology of most human proteins).
- Proteopedia The 3D encyclopedia of proteins and other molecules.

METABOLOMICS

Metabolomics is the study of small metabolites present in biological system. It is species independent and can be applied on diverse array of plants. However, despite the immense technical challenges associated with analyzing a wide range of metabolites of varying concentrations, different techniques have been developed for their detection. These techniques are high throughput and are cheap on a per sample basis, making them ideal as rapid screening tools. The combination of high throughput and breath will help metabolomics in becoming a key tool in variety of fields in the years to come.

The history of metabolomics traces back to 2000-1500 B.C. when Chinese doctors use body fluids to determine the biological condition of patients in order to predict disease. In 1614 Santorio Sanctorius whose work was first to obtain physical data and provide quantitative basis of pathology, is considered to be the founding father of metabolic studies. Thomas Willis performed the first analysis of urine in 1674 and he found that people with diabetes mellitus and diabetes insipidus could be distinguished based solely upon the sweetness of their urine which was confirmed by Matthew Dobson in 1776.

The term metabolomics was coined by Oliver *et al* in 1998 and in 2006 a new approach called "Sniper" metabolomics was introduced which uses both chromatography and mass spectrometry in order to analyze metabolites. Metabolism and metabolomics share as their root the ancient Greek word, metabol, which means change. Metabolomics is an emerging field in analytical biochemistry and can be regarded as the end point of the "omics" cascade. Thus metabolomics is the quantitative measurement of the metabolic profiles of model organisms or metabolites to characterize their phenotype or phenotypic response to genetic or nutritional perturbations whereas metabonomics is the quantitative measurement of "total" metabolic response of vertebrates to pathophysiological situations (nutritional, xenobiotic, surgical or toxic stimuli).

Where genomics deals with the analysis of the complete genome in order to understand the function of single genes, the majority of functional genomics studies are currently based on the analysis of gene expression (transcriptomics) and comprehensive protein analysis (proteomics). However, changes in the metabolome are the ultimate answer of an organism to genetic alterations, disease, or environmental influences. Even the genome of a large number of plants has been sequenced but the function of majority of their genes is not cleared, because these genes are believed to involve in plant metabolism. By studying their sequence homology one can assign putative functions but cannot describe their biochemical functions. It is likely that at least some of metabolites fulfill important biological functions such as helping plants to survive in specialized ecological niches. To study function of such metabolites metabolite profiling is required. Metabolite profiling is an analytical method for relative quantization of a number of metabolites from biological samples (a specific tissue, a part of a tissue of interest or a mixture of different organs). The requirements and current strategies for the comprehensive and quantitative analysis of the molecular phenotype have been recently reviewed for all three levels of gene products (mRNA, proteins and metabolites), including thoughts on database requirements and informatics tools (Fiehn 2001).

Metabolomics techniques

The main aim of metabolomics is the quantification and identification of all metabolites of a given biological system under defined conditions. There are variety of methods that facilitates the study the metabolic alterations on comprehensive scale. No single technique or protocol can be used for metabolic analysis due to difference in metabolite concentrations. Thus the two dominant detection systems are nuclear magnetic resonance (NMR) and mass spectrometry (MS). An advantage of NMR is the linearity of quantitative responses on increasing metabolite concentrations, almost irrespective of the chemical compound class. Large signals in NMR can directly be interpreted as high level concentrations, whereas in MS quantitative responses strongly rely on the ionization potential of each metabolite.

Therefore, quantization in MS is limited to relative abundances of a given metabolite between samples, or requires calibration curves if absolute comparisons of different metabolites are needed. Thus MS is generally more favored than NMR due to four reasons: (i) MS is advantageous with respect to the capability to resolve complex mixtures of compounds, (ii) for most compounds, MS is far more sensitive than NMR,

(iii) due to the need for high end magnets, NMR instruments are usually far more expensive than most MS instruments and (iv) coupling of NMR to separation techniques such as liquid chromatography is far from straightforward. This leads to the main use of NMR as a tool for 'metabolite fingerprinting' in which we classify samples according to the origin or their biological relevance in order to maintain a high throughput (Fiehn 2002).

The most frequently used tool for profiling primary metabolites is Gas Chromatography coupled to mass spectrometry (GC-MS). Compounds screened by GC-MS profiling cover large parts of primary metabolism that are conserved across species, facilitating comparative studies between genetically and environmentally perturbed potato tuber systems (Roessner *et al* 2001). The most versatile method of separating compounds with wide range of polarity is LC. Semi-polar compounds, alkaloids, flavonoids can be separated by using reverse phase columns whereas hydrophilic columns are used for polar compounds like sugars, amino acids, vitamins, nucleotides etc. The mass detection of LC-MS is conditioned by the capacity of an analyte to ionize while being a part of complex mixture as only ions can be measured by MS.

Other elegant method of obtaining metabolite information is the use of 13C material (Mashego *et al* 2004). Nowadays, in stabilized (LC or GC)-MS setups, retention-time variation can be relatively low, allowing direct comparisons of chromatograms and the construction of metabolite databases (Moco *et al* 2006). The resolving power of much used Quadrupole (Q) MS instruments is 4 times less then Time-of- Flight (TOF)-MS and that of Fourier transform (FS)-ion cyclotron (ICR)-MS is 400 times greater than Q-MS (Moco *et al* 2007). Seifar *et al* (2008) developed a rapid, sensitive and selective ion-pair reversed-phase liquid chromatography–electrospray ionization isotope dilution tandem mass spectrometry (IP-LC–ESI-ID-MS/MS) for quantitative analysis of free intracellular metabolites in cell cultures. No recovery check of metabolites in the cell extraction step was needed because 13C-labeled extract was added to the cells before the extraction of intracellular metabolites. This effectively minimizes the extraction differences between the labeled and unlabeled analytes.

Advantages of Metabolomics:

- Identification of target organ, severity, onset, duration and reversal of the effects (time-course)
- Classify sample as "normal" vs "abnormal"
- Determine mechanisms of action within the organ
- Potential for identifying novel biomarkers of toxic effect
- Non-invasive
- No a priori decisions about samples need be made
- No sample processing necessary other than cold collection
- Complete time course data can readily be obtained
- Minimization of compound requirements

- Relatively fast analysis (200-300 samples/day)
- Useful tool for modeling physiological variation and exposure conditions

Applications

As a diagnostic aid: To determine the mode of action of various herbicides using GC-MS. The analysis of metabolic profiles from seedlings treated with median lethal (LC50) doses of acetyl CoA carboxylase (ACC) and acetolactate synthase (ALS) inhibitors, alongside untreated control seedlings, resulted in the visual discrimination of metabolite profiles of the different plants (Sauter *et al* 1991). Similarly used in tomato to catalogue the metabolite composition of developing tomato fruit where the influence of Hexose Phosphorylation diminishes during fruit development (Roessner *et al* 2003). Ma and Gang, 2006 studied difference in chemical curcuminoids between micropropagated and green house grown turmeric plants using GC-MS method.

Metabolic response to stress: An interesting area in plant system is the metabolic response of plants to various biotic and abiotic stresses. Thimm *et al* (2004) analyzed the effect of carbon starvation by exposing Arabidopsis plants to an extended night by metabolic profiling. Similarly Cook *et al* (2004) compared Ws-2 and Cvi-1 ecotypes to Arabidopsis lines over expressing CBF transcription factor genes, and this CBF over expression configured the metabolome of Arabidopsis in a way that resembles cold acclimation treatments.

Gene function: The technique has been used to explore the function of genes in plants, yeast and experimental animals. Metabolomic strategies have also been used to define gene function in the yeast. Allen *et al* (2003) studied extracelluler metabolites in *S. cerevisiae* to distinguish between the wild-type and mutant strains of yeast during different growth phases. Similarly, Nakagawa *et al* (2004) studied that in transgenic petunia plants over expression of a cDNA for a zinc-finger protein (LIF), alters cytokinin metabolism and plant forms.

Food Science: Plants are a fabulously rich source of diverse functional biochemicals and metabolomics is also already proving valuable in an applied context. By characterizing their biochemical composition, metabolomics will greatly assist in identifying the unexploited biodiversity available today (Kussmann *et al* 2007). Food quality is easily and rapidly deteriorated by a number of different pests, leading to major losses. Vikram *et al* (2004) studied the response of four different types of fungi applied to apples to study the effect of different metabolic responses during food storage.

Databases

There are different tools for visualization or databases that can be used to display the coupling of different "omics" data (e.g. KEGG (www.genome.jp/kegg), MAPMAN (gabi.rzpd.de/projects/ MapMan), KappaView (kpv.kazusa.or.jp/kappaview) and DOME (a database of OMEs). AraC is a tool to visualize biochemical pathways of Arabidopsis (Mueller *et al* 2003). MetaCyc (http://metacyc.org/) provides metabolic pathways and citations to 150 different organisms. MetNetDB is a metabolic and regulatory networkmap that contains a growing map of Arabidopsis entities (genes, RNAs, polypeptides, protein complexes and metabolites) and the catalytic and regulatory interactions between them. For identifying GS-MS signal an important databases is AMDIS (Automated mass spectral deconvolution and identification system) (www.amdis.net) and for MS and NMR Advanced Chemistry Development Labs (ACDLabs) is mainly used.

Limitations of Metabolomics:

- Requirement of specialized equipments
- Extensive expertise is required
- Information is limited to time- and dose-points taken
- High risk of false positive data
- Difficulty in separation of physiological (adaptive) and toxicological (adverse) effects
- Sensitivity of the assay
- Advisability of sub-fractionating metabolites (highly charged, neutral)
- Supplemental MS or other tests for final identifications (e.g., fructose and glucose)

CONCLUSION

Given the speed with which functional genomics, proteomics and metabolomics has advanced in the last decade, there are potential developments around the corner that will make much of the current technology redundant. There are some more advanced approaches currently being developed.

Together with the completion of the sequence of different plant genomes, a magnificent tool kit is becoming available that will dramatically improve the quality of research in all disciplines of plant biology (Bouchez and Hofte, 1998). This tool kit comprises large collections of knockout mutants in model organisms for systematic gene function searches and methods for the study of the regulation of entire sets of genes at the mRNA (transcriptomics) or protein level. A challenge in the next decade will be to build integrated databases combining information on such things as sequence, map position, mRNA and protein expression, mutant phenotypes, metabolism, and allelic variation. Also, intelligent software tools will be required for the efficient mining of this wealth of information and to construct intelligible models describing the complex molecular interactions that constitute the regulatory networks. EST sequencing and mapping or genomic sequencing of a few strategic crop species may suffice to facilitate the transfer of information from model species to the majority of crop plants, to be used for crop improvement in the future (Leung and An, 2004).

The science of omics has fundamentally altered the way in which we view plant biology, by providing the possibility of taking a global view of cellular processes. Sequencing programs are rapidly documenting the expressed genes of many species. Several functional genomic, proteomics approaches discussed in this chapter, address plant gene function on a large scale. Plants are combinatorial chemists par excellence, and understanding the principles that relate enzyme structure to function will open up unlimited possibilities for the rational design of new enzymes to generate novel biologically active natural products. Knowledge of the molecular genetics of plant natural product pathways will also facilitate the engineering of these pathways for plant improvement and human benefit. It is highly probable that the field of omics will continue to be driven by technological developments for the foreseeable future.

REFERENCES

Albers M, Kranz H, Kober I, Kaiser C, Klink M, Suckow J, Kern R and Koegl M 2004 Automated yeast two-hybrid screening for nuclear receptor-interacting proteins. *Mol Cell Proteomics* **4**: 205-213.

Allen J, Davey H M, Broadhurst D, Heald J K, Rowland J J, Oliver S G and Kell D B 2003 High-throughput classification of yeast mutants for functional genomics using metabolic footprinting. *Nat Biotechnol* **21**: 692–696.

Bouchez D and Hpfte H 1998 Functional genomics in plants. *Plant Physiol* **118**: 725-732.

Brown PO and Botstein D 1999 Exploring the new world of the genome with DNA microarrays. *Nature Genet* **21**: 33-37.

Cagney G, Amiri S, Premawaradena T, Lindo M and Emili A 2003 In silico proteome analysis to facilitate proteomics experiments using mass spectrometry. *Proteome Science* **1**: 5-7.

Cole L S and Vassar R 2007 The Alzheimer's disease _-secretase enzyme, BACE1. *Mol Neurodegeneration* **2**: 22-26.

Cook D, Fowler S, Fiehn O and Tomashow M F 2004 A prominent role for the CBF cold response pathway in configuring the low-temperature metabolome of Arabidopsis. *PNAS* **101**: 15243–15248.

Cooke R, Raynal M, Laudié M, Grellet F, Delseny M, Morris P C, Guerrier D, Giraudat J, Quigley F and Clabault G 1996 Further progress towards a catalogue of all Arabidopsis genes: analysis of a set of 5000 non-redundant ESTs. *Plant J* **9**: 101-124.

David A H, Jason P and Michael S 2007 Protein microarray technology. In: *Mechanisms of Ageing and Development* **128**: 161-167.

Feizi T 2003 Use of nitrocellulose surfaces for the immobilisation of carbohydrates and neoglycolipids. *Curr Opin Struc Biol* **13**: 637-645.

Fiehn O 2001 Combining genomics, metabolome analysis, and biochemical modelling to understand metabolic networks. *Comp Funct Genom* **2**: 155-168.

Fiehn O 2002 Metabolomics – the link between genotypes and phenotypes. *Plant Mol Biol* **48**: 155–171.

Fields S and Song O 1989 A novel genetic system to detect protein-protein interactions. *Nature* **340**: 245 – 246.

Fodor S P, Read J L, Pirrung M C, Stryer L, Lu A T, Solas D 1991. Light-directed, spatially addressable parallel chemical synthesis. *Science* **251**: 767-773.

Gupta P K, Roy J K and Prasad M 1999 DNA chips, microarrays and genomics. *Current Science* **77**: 875-884.

Henzel W J, Billeci T M, Stults J T, Wong S C, Grimley C and Watanabe C 1993 Identifying proteins from two-dimensional gels by molecular mass searching of peptide fragments in protein sequence databases. *PNAS* **90**: 5011-5015.

Hieter P and Boguski M 1997 Functional genomics: it's all how you read it. *Science* **278**: 601-602.

Hurt J, Thibodeau S, Hirsh A, Pabo C and Joung J 2003 Highly specific zinc finger proteins obtained by directed domain shuffling and cell-based selection. *PNAS* **100**: 12271–12276

Joung J K, Ramm E I and Pabo O Carl 2000 A bacterial two-hybrid selection system for studying protein–DNA and protein–protein interactions. *PNAS* **97**: 7382-7387.

Junker V, Contrino S, Fleischmann W, Hermjakob H, Lang F, Magrane M, Martin M J, Mitaritonna N, Donovan C and Apweiler R 2000 The role SWISS-PROT and TrEMBL play in the genome research environment. *J Biotechnol* **78**: 221-234.

Karas M and Hillenkamp F 1988 Laser desorption ionization of proteins with molecular mass exceeding 10000 daltons. *Anal Chem* **60**: 2299–2301.

Kersten B and Feilner T 2007 Generation of plant protein microarrays and investigation of antigen-antibody interactions. *Methods Mol Biol* **355**: 365-378.

Kim S H, Shin D H, Choi I G, Schulze-Gahmen U, Chen S and Kim R J 2003 Structure-based functional inference in structural genomics. *Struct Funct Genomics* **4**: 129-130.

Kussmann M, Affolter M, Nagy K, Holst B and Fay L B 2007 Mass spectrometry in nutrition: understanding dietary health effects at the molecular level. *Mass Spectrom Rev* **26**: 727-750.

Leung H and An G 2004. Rice functional genomics: Large scale discovery and applications to plant improvement. *Adv Agron* 82: 55-111.

Lockhart DJ and Winzeler EA 2000. Genomics, gene expression and DNA arrays. *Nature* **405**: 827-836.

Ma X and Gang D 2006 Metabolic profiling of Turmeric (*Curcuma longa* L) Plants Derived from *in vitro* Micropropagation and Conventional Greenhouse Cultivation. *J agric Food chem* **54**: 9573-9583.

MacBeath G and Schreiber S 2000 Printing proteins as microarrays for high-throughput function determination. *Science* **289**: 1760-1763.

Marshall E and Pennisi E 1998 Hubris and the human genome. *Science* **280**: 994-995.

Mashego M R, Wu L, Dam J C, Ras C, Vinke J L, Winden W A, Gulik W M and Heijnen J J 2004 MIRACLE: mass isotopomer ratio analysis of U-C-13-labeled extracts. A new method for accurate quantification of changes in concentrations of intracellular metabolites. *Biotechnol Bioeng* **85**: 620-628.

Moco S, Bino R J, Vorst O F, Verhoeven H A, Groot J C, Beek T A, Vervoort J J and Vos C H 2006 A liquid chromatography-mass spectrometry-based metabolome database for tomato. *Plant Physiol* **141**: 1205 - 1218.

Moco S, Bino R J, Vorst O, Verhoeven H A, Groot J, Beek T A, Vervoort J and Vos C H 2006 A Liquid Chromatography-Mass Spectrometry-Based Metabolome Database for Tomato. *Plant Physiol* **141**: 1205-1218.

Mrkish M and Su J 2002 A modified surface for microarrays and SPR. *Chem Biol* 41: 4175-4181.

Mueller L A, Zhang P and Rhee S Y 2003 AraCyc: a biochemical pathway database for Arabidopsis. *Plant Physiol* **132**: 453–460.

Nakagawa H, Jiang C, Sakakibara H, Kojima, Honda I, Ajisaka H, Nishijima T, Koshioka M, Homma T, Mander L and Takatsuji H 2004 Overexpression of a petunia zinc-finger gene alters cytokinin metabolism and plant forms. *Plant J* **41**: 512-523.

Newman T, Bruijn F J, Green P, Keegstra K, Kende H, McIntosh L, Ohlrogge J, Raikhel N, Somerville S and Thomashow M 1994 Genes galore: a summary of methods for accessing results from large-scale partial sequencing of anonymous Arabidopsis cDNA clones. *Plant Physiol* **106**: 1241-1255.

Oliver S G, Winson M. K, Kell D B and Baganz F 1998 Systematic functional analysis of the yeast genome. *Trends Biotechnol* **16**: 373–378.

Polyak K and Riggins G J 2001 Gene discovery using the serial analysis of gene expression technique: implication for cancer research. *J Clinic Onco* **19**: 2948-2958.

Pruess M, Kersey P, Kulikova T and Apweiler R 2006 Databases and resources for in silico proteome analysis. *Methods Biochem Anal* **49**: 397-414.

Ratner DM 2005. Advancing the burgeoning field of genomics. *Biotech* **17**: 8-11.

Roessner U, Hegemann B, Lytovchenko A, Carrari F, Bruedigam C, Granot D and Fernie A R 2003 Metabolic Profiling of Transgenic Tomato Plants Overexpressing Hexokinase Reveals That the Influence of Hexose Phosphorylation Diminishes during Fruit Development. *Plant Physiol* **133**: 84-99.

Roessner U, Willmitzer Land Fernie A R 2001 High-resolution metabolic phenotyping of genetically and environmentally diverse potato tuber systems. Identification of phenocopies. *Plant Physiol* **127**: 749–764.

Sauter H, Lauer M, Fritsch H 1991 Metabolic profiling of plants - A new diagnostic technique. In: *Synthesis and chemistry of agrochemicals II.* ACS Symposium American Chemical Society, Washington 288-299.

Schena M, Shalon D, Davis RW and Brown PO 1995 Quantitative monitoring of gene expression patterns with a complementary DNA microarray. *Science* **270**: 467-470.

Schwender J and Ohlrogge J B 2002 Probing in vivo metabolism by stable isotope labeling of storage lipids and proteins in developing *Brassica napus* embryos. *Plant Physiol* **130**: 347-361.

Seifar R M, Zhao Z, Dama J, Winden W, Gulik W and Heijnen J 2008 Quantitative analysis of metabolites in complex biological samples using ion-pair reversed-phase liquid chromatography–isotope dilution tandem mass spectrometry. *J Chrom* **1187**: 103–110.

Shin I, Cho J W and Boo D W 2004 Carbohydrate Arrays for Functional Studies of Carbohydrates. *Comb Chem Screen* **10**: 565-574.

Somerville C and Somerville S 1999. Plant functional genomics. *Science* **285**: 380-383.

Sullivan N, Sun Y, Binley J, Lee J, Barbas C F, Parren P, Burton D R and Sodroski J 1998 Determinants of human immunodeficiency virus type 1 envelope glycoprotein CD4 and monoclonal antibodies. *J Virol* **72**: 6332-6334.

Thimm O, Blasing O, Gibon Y, Nagel A, Meyer S, Kruger P, Selbig J, Muller L A, Rhee S Y and Stitt M 2004 MAPMAN: a user-driven tool to display genomics data sets onto diagrams of metabolic pathways and other biological processes. *Plant J* **37**: 914–939.

Tonge R, Shaw J, Middleton B, Rowlinson R, Rayner S, Young J, Pognan F, Hawkins E, Currie I and Davison M 2001 Validation and development of fluorescent two-dimensional gel electrophoresis proteomics technology. *Proteomics* **1**: 377-396.

Tuteja R and Tuteja N 2004. Serial analysis of gene expression: Application in human studies. *J Biomed Biotech* **2**: 113-120.

Unlu M, Morgan M E and Minden J S 1997 Difference gel electrophoresis: a single gel method for detecting changes in protein extracts. *Electrophoresis* **18**: 2071-2077.

Velculescu V E, Zhang L, Vogelstein B and Kinzler K W 1995 Serial analysis of gene expression. *Science* **270**: 484-487.

Vikram A, Prithiviraj B, Hamzehzarghani H and Kushalappa A 2004 Volatile metabolite profiling to discriminate diseases of McIntosh apple inoculated with fungal pathogens. *J Sci Food Agric* **84**: 1333–1340.

Wakatsuki T, Hada A and Ryo A 2001 Use of serial analysis of gene expression (SAGE) technology. *J Immun Methods* **250**: 45-66.

Wang D 2002 Tools of glycobiology. *Nat Biotechnol* **20**: 275-281.

Wilkins M R, Sanchez J C, Gooley A A, Appel R D, Humphery I, Hochstrasser D F and Williams K L 1996 Progress with proteome projects: why all proteins expressed by a genome should be identified and how to do it. *Biotech Genet Eng Rev* **13**: 19-50.

Yamamoto K and Sasaki T 1997 Large-scale EST sequencing in rice. *Plant Mol Biol* **35**: 135-144.

Zhu H, Bilgin M, Bangham R, Hall D, Casamayor A, Bertone P, Lan N, Jansen R, Bidlingmaier S, Houfek T, Mitchell T, Miller P, Dean R A, Gerstein M and Snyder M 2001 Global analysis of protein activities using proteome chips. *Science* **293**: 2101-2105.

7

INCREASING YIELD POTENTIAL OF GRAIN LEGUMES : PHYSIOLOGICAL DETERMINANTS IN RELATION TO CEREALS

PUSHP SHARMA AND VIRENDER SARDANA

Legumes have been recommended as a part of cropping systems since ancient times. The key benefit of legume, of course, is their ability to form symbiosis with specific bacteria that allows the fixation of atmospheric dinitrogen (N_2). The fixed N is used in legume plants to synthesize the essential amino acids, nucleic acids and other nitrogenous compounds required to sustain and grow plant tissue. The ability of legumes to fix atmospheric N_2 can make these species especially valuable assets in cropping systems.

In roughly, the past half century, however, the role of legumes in cropping systems has often diminished in favour of increasing emphasis on cereals. Rather than relying on N input to the cropping system from legumes in many parts of the world N fertilizer is applied in increasing amounts in the production of cereals. However, now, the continuous production of cereals with applications of large amounts of fertilizer is facing several constraints. For several perspectives, legumes can become critical components of alternate cropping systems, particularly if legume yields can be enhanced. In industrial countries, the abundant application of N-fertilizers to fields has raised environmental concerns and regulations have been established to control nitrogenous compounds in ground water and water ways. In developing countries, the opposite concern exists in the lack of manufactured N-fertilizer causing severe limitations in cereal yields. Nitrogen fertilizer availability in the developing countries is constrained by the cost of manufacture, distribution and credit for farmers to purchase fertilizer.

The high protein concentrations of legume tissues make them desirable as feed and food sources. The vegetative tissues of virtually all legumes are high in protein concentrations, which make them very desirable as forage for animals. Seeds of grain legumes are high in protein so that when digestive inhibitors in some species are inactivated, they are excellent foods. Human consumption of seeds of grain legumes is important world wide including common beans (*Phaseolus* sp.), Soybean (*Glycine max*), peas (*Pisum sativum*), Peanut (*Arachis hypogaea*), cowpea (*Vigna ungiculata*), chickpea (*Cicer orientinum*) and pigeonpea (*Cajanus cajan*).

To sustain legume production and stimulate expanded use of legume in cropping system, there is a need to further increase legume yield both in terms of plant mass and total plant N accumulation. The topics to be considered for possible improvement in order to increase yields include plant development, carbon assimilation, nitrogen assimilation and stress amelioration. Most of the work with legumes has been done with grain legumes pod/legumes, so much of the discussion will focus on pulse crops.

PLANT DEVELOPMENT

The basic scheme of temperature - driven plant development in legumes is not unlike that of cereal species. Rates of germination and emergence, leaf emergence and flowering are all quantitatively linked to the temperature to which the plants are subjected. Rates of development increase in an approximately linear manner until temperature becomes too high and results in inhibition of development. In addition many legumes are sensitive to the day length under which they develop. Growth related limitations are discussed hereby

Germination and seedling vigour

Grain legumes, pulses in general have slow growth rate at the time of germination and seedling stage. Most of the long duration pulses exhibit slow dry matter accumulation in the initial stages of growth. Due to slow development of leaf area, canopy, poor interception of solar radiation leads to slow growth rate, poor seedling vigour and ultimately yields. This has been well illustrated in chickpea and pigeonpea by Khanna-Chopra and Sinha (1990). Higher rates of germination and seedling vigour as reported in large-sized seeds of cowpea provides competitive advantage for sunlight in mixed cropping. Since pulses are often cultivated in mixed cropping systems, these attributes constitute a major hurdle in realizing higher yield potentials. Temperature extremities viz. both low and high decreases the rate of germination and thus, growth rates. Therefore, soaking of seeds prior to sowing increases the germination percentage and also reduces the time required for germination. Old seeds have higher germination percentage than new seeds, which is due to the presence of chemicals or hard seed coat. This leads to lower percentage of germination, seedling vigour and root/shoot ratio. Therefore, it is desirable to enhance seedling vigour.

Even the fresh hard seeds of lentil and mungbean exhibit lower germination, which can be improved by storing the seeds for some time before sowing.

Leaf area

More the leaf area, more will be the interception of solar radiation, higher will be photosynthetic activity and subsequently higher productivity. Most of the pulses have low leaf area index (LAI). Therefore, increase in the leaf area to some extent would enhance the transfer of photoassimilates to the sinks.

Seed weights at different nodes have been directly correlated with the leaf area at these nodes (Savithri *et al.* 1978), thus, clearly illustrating larger the leaf area more will be interception of light which leads to higher grain yield.

In most of the legume crops, the light interception by the canopy is low due to low leaf area. Leaf area increases gradually reaches maxima followed by a decline. So, LA and ultimately LAI varies with the age of plant. Leaf area of mungbean, averaged on genotypes was 58, 371, 683, 530 and 226 cm^2 $plant^{-1}$ at 15, 30, 45, 60 and 75 DAS (Prasad and Srivastava, 1999) and of late sown chickpea was 14, 82, 243, 379 and 288 cm^2 $plant^{-1}$ at 25 and 50 DAS, at flowering, 10 and 20 days after flowering (Ganguly *et al.* 1999). Leaf area is low at early stages of growth but also during reproductive phase LA declines drastically, leading to low photosynthesis and ultimately low yields. Increase in LA with irrigation has been reported by Sharma *et al.* (unpublished).

Therefore, selection of varieties based on their physiological pattern and manipulation of agronomic practices particularly time of sowing, plant population, planting geometry etc. can improve LAI and subsequently increase productivity per unit time. Little is known about the genetic regulation of development rate of plants. Some legumes are handicapped by especially slow production of leaves causing low crop growth during early stages of development Pigeonpea, for example, has small leaves that are produced at a slow rate so that the time to achieve full radiation interception is substantially delayed as compared to other species (Muchow, 1985). Consequently, major increase in yield potential of pigeonpea and other relatively small leaved species seem possible if the rate of leaf area development could increase.

Growth phenology

The phenology of grain legumes is strongly responsive to photoperiod (Summerfield and Roberts, 1985). Temperature not only affects the length of vegetative growth phase but also the initiation of flowering, by increasing the mean temperature during the vegetative phase, hastens flowering in mungbean (Poehlman, 1991) and chickpea (Sharma *et al.* 2005). Developing short duration varieties for multiple cropping systems and maximizing grain yield, the vegetative phase should be terminated with flowering and assimilates be channelized into pod production. It is also argued that early flowering plants are low yielders due to short stature, few nodes, smaller leaf area and shorter grain filling period as compared to the late flowering plants. So, the growth pattern of genotypes should be such that it suits best to the environment or the growing season for which it has been targeted. Longer reproductive phase results in the development of better sink for higher harvest index (Kumar *et al.* 2001). To improve and stabilize yield, efforts are needed to optimize the resources by manipulating both genotypes through breeding and environment via agronomic manipulation in order to exploit genetic potentials to the maximum and minimize environmental constraints on the realization of that potential (Lawn 1981). It is desirable, therefore, to match crop phenology to the climate/environment, which prevails during the growing season in a given location.

CARBON ASSIMILATION

Leaf photosynthetic potential of legumes seems to be as high as C_3 cereals. Measurements of leaf photosynthetic (P_n) rates of field grown soybean have been reported over 40 μmol $m^{-2}s^{-1}$ and a number of observations near or above 30 μmol $m^{-2}s^{-1}$ have been reported for field grown peanut (Bell *et al.* 1994). Photosynthetic rates (range) expressed as mg $CO_2dm^2hr^{-1}$ in some pod legumes are lentil (10-30), cicer (25-40), *Phaseolus vulgaris* (15-40) and cowpea (12-30.7), which are similar to wheat, barley and cotton in their photosynthetic rate but certainly poorer than those of maize, sorghum and millets. Several studies on the variability in photosynthesis rates in soybean (Sinclair 1980). Phaseolus vulgaris (Evans *et al.* 1973), chickpea (Shanthakumari and Sinha 1972) and mungbean (Sinha and Khanna unpublished) clearly show that varietal differences are common in these species. Further, photosynthetic rates changes considerably during the lifecycle of legumes. The fact that virtually no data has been reported on the leaf photosynthetic activity of other grain legumes under field conditions may offer an important research opportunity for characterizing growth and yield potential in other legume species.

While legumes are capable of high leaf photosynthetic rates, these rates contrast with cereals in that they are achieved only with a much greater investment in leaf N content per unit leaf area than required in cereals. For example, leaf N contents of soybean have shifted to about 0.7 g N m^2 greater levels as compared to rice to achieve comparable photosynthetic rates (Sinclair and Horie 1989). The additional leaf N in soybean seems to be needed to achieve photosynthetic rates comparable to C_3 cereals is seemingly disadvantage. The basis for this difference between cereals and legume has not been explained. There appears to be no evidence that photosynthetic biochemistry of legumes is less efficient than that of C_3 cereals. One hypothesis is that leaves of legumes simply store more protein than cereals because of the higher relative abundance on N in legume plants as a result of N_2 fixation. Consistent with this hypothesis is the presence in many legumes leaves of a layer of paraveinal mesophyll cells (Lansingh and Franceschi, 2000) that are rich in glycoproteins (Klauer *et al.* 1996). The paraveinal mesophyll cells have little photosynthetic capability and seem to function as storage and transport cells. The physiological role of the paraveinal mesophyll in legume leaves certainly deserves further attention as contributors of the carbon and N economy of these species.

Radiation use efficiency

The carbon assimilation capacity of the total crop can be effectively expressed by Radiation Use Efficiency (RUE), defined as the ratio of mass accumulated to cumulative intercepted total solar radiation. Sinclair and Horie (1989) calculated the potential RUE of soybean based on leaf photosynthetic capacity as being about 1.2 g MJ^{-1}. This estimate was lower than that anticipated for C_3 cereals (1.4 MJ^{-1}) as expected based on the difference in the higher energy content of the soybean mass. In practice, there have been only a few reports of soybean RUE as high as the 1.2 g MJ^{-1} suggested by Sinclair and Horie (1989). Most experimental observations have reported RUE values in the range of 0.65 to 1.15 g MJ^{-1} with a median of approximately 0.9 g MJ^{-1} (Sinclair 2004).

Field observations of RUE may frequently be lower than theoretical values because RUE is sensitive to many stresses which decreases leaf photosynthesis rate. Consequently, water deficit or biotic pests that result in decreased leaf photosynthetic rate will result in decreased RUE. Probably, the most important environmental factor decreasing RUE is water deficit. Soil dehydration that results in loss of leaf photosynthetic activity will result directly in a decrease in RUE. Especially, in legumes that may have shallower roots than cereals, the onset of decrease in RUE may occur early in the drying cycle.

Focused research to understand the commonly observed lower RUE than theoretically anticipated for legumes may lead to insights for increasing yields. Management or genetic options can be worked out to sustain high leaf photosynthetic rates of legume leaves and consequently sustained high RUE.

Photosynthetic efficiency

Increase in LAI may not be correlated with increased productivity because all the radiation captured may not be utilized into yield. It is the conversion efficiency which determines the efficiency of photosynthetic active radiations (PAR) which is converted into yield. Variation in photosynthetic efficiency has been observed in genotypes of many pulse crops (Table 1). Photosynthetic rate and RUBP-carboxylase activity were positively correlated with grain yield in chickpea (Pandey and Singh, 1996). Thicker leaves have higher photosynthetic capacity, which compensates for slower growth rate. Canopy structure to capture higher quantity of solar radiation could not be compensated for lower photosynthetic efficiency of pods relative to leaf in top podding pulses like mungbean, pigeonpea and cowpea as significant quantity of radiations are intercepted by flowers and pods. Due to lack of assimilate supply, yields are lowered as observed due to partial defoliation at different stages of growth (Sharma *et al.* unpublished) and shading during flowering (Saxena *et al.* 1983) Pulse crops are grown in mixed and intercropping system, therefore, shading of pulses could influence photosynthesis and productivity.

FLOWER AND FRUIT ABSCISSION

Large number of reproductive sinks viz. flowers, fruits and seeds, are produced in grain legumes. But, only a small fraction of them ends up in seed production. Due to physiological constraints, there is a large scale abscission of flowers and fruits. The major reason for low productivity of pulses is the poor realization of sink capacity due to flower and fruit shedding (Hebblethwaite *et al.* 1984). Plants are unable to meet the

TABLE 1 : Photosynthetic rates (P_n) rates and photosynthetic characteristics in pod legumes

Characters	*Species*	*P_n (mg $CO_2 dm^{-2} hr^{-1}$)*
Photosynthetic rates (P_n)	Soybean	12-24
	Lentil	10-30
	Cicer	25-40
	Phaseolus vulgaris	15-40
	Cowpea	12-30.7
Photosynthetic characteristic		
Photosynthetic rates	Pod legumes	Normally 20-30
Temperature range		Optimum between 25°-30°C
Light saturation		10,000 lux
CO_2 compensation point		About 50 ppm
Photorespiration		Present
Main carboxylating enzyme		RUBP
Kranz anatomy		Absent

carbon and nitrogen demands of developing sinks. Thus, the greater requirement of C and N and also continuous translocation from vegetative parts leads to the senescence of vegetative tissues and reduction in the seed filling period, thus, lowering the yield. Intra-plant competition for assimilates between the reproductive and vegetative parts has been attributed by Gates *et al.* (1983) to higher flower and pod abscission. Other reasons for lower fruit and seed set in legume crops are briefly enlisted.

HORMONAL/NUTRITIONAL REGULATION

Sometimes pollination and fertilization occur but still the flowers abscise. Shedding of flowers with ovules is due to hormonal or nutritional imblance.

Pollination failure

Pulses are generally self-pollinated crops. Sometimes during flower development, under adverse climatic conditions, pollination does not occur and flowers wither away.

Environmental conditions

Environmental factors particularly temperature affects pollen germination, pollen tube growth and thus, fertilization leading to flower abortion in chickpea Sharma *et al.* 2005). Water stress too has similar effect on mungbean (Singh, 1990a). Lower pod set may also result from cloudy weather, low night temperature and soil moisture. Even if under conducive temperature's for pollen germination only a small percentage of flowers set fruit.

Availability of photoassimilates

Excessive production of sinks (flowers) but poor pod set is probably due to limited availability of assimilates. The analysis of carbon and nitrogen economy of chickpea and pigeonpea by Singh (1990b) has suggested the selection criterion for cultivars *i.e.* low photorespiration, delayed leaf abscission, efficient mobilization of C and N from stem and leaves, higher photosynthetic and N-fixing rates during seed filling phase, for maximizing the efficiency of photosynthates utilized for seed weight and protein production.

Insects : In summer mungbean, Thrips cause flower shedding as a result of which number of pods formed and grain yield declines significantly (Chhabra and Kooner, 1985).

Any of these factors or a combination might be true. In determinate genotypes, there is increased competition for photosynthesis between developing fruits and nitrogen fixing nodules and fruit setting is better than indeterminate genotypes. The degree of synchrony of flowering and pod set could increase the potential sensitivity of water supply during reproductive phase of growth in grain legumes.

It is well documented that growth regulators helps to enhance the productivity of legumes. This could be the direct or indirect effect on flower retention (Sharma *et al.* 2005).

SOURCE-SINK RELATION WITH YIELD

Large number of flowers are produced in most of the grain legumes but only a few of them produce pods and set seeds. There exist high sink potential to give higher yield but it is not realized. Major limiting factor for realizing higher yield was the insufficient supply of assimilates to the young pods during later part of the reproductive phase, leading to unfilled pods in higher proportion as observed in *Kabuli* chickpea (Pandey *et al.* 1978).

Yield parameters viz. pods, pod weight node^{-1}, number of seeds and seed weight in the late formed pods decline, which indicates the limited supply of assimilates/nutrients during pod filling. Towards the apex, both fruit and seed weight declines and some of the fruit do not contain seeds Flowers are produced continuously in indeterminate plants, but these flowers do not develop into fruits, even if these flowers produce fruits, seeds are not formed. This is due to the limitation of leaves (source) at the upper flowering nodes. Moreover, the first formed fruits and seeds may exert some regulatory effect on fruit formation and seed development by regulating leaf area or floral biology.

Though the N-content of vegetative phase is more than double even then the extra N is not translocated to seeds. The site(s) that limit translocation have not been identified though they appear to be at or close to the developing seeds for genetic manipulation of N translocation. Atkins *et al* (2005) have cloned an asparaginase and two seed amino acid transporters from developing lupin seed. Constructs from these genes coupled to a seed specific promoters lead to the generation of transgenic over expression and anti-sense lines where significant increase in pod set and seed number has been achieved.

The determinant growth habit has potential advantage as the competition from vegetative sink to reproductive sink is reduced, for assimilates, resulting in improved productivity.

Low biomass production is the potential constraint for higher productivity in short duration determinate type of pod bearing legumes, which could be increased by higher plant density (Gates *et al.* 1983). Higher biomass production and reduction in the number of late formed branches may help in increasing grain yield in indeterminate type of faba bean (Silim and Saxena, 1990). Sinha (1974) has reported that sink potential was only 15% in cowpea against 75% in wheat as per theoretical calculations. It was inferred that sink potential was more than adequate to give yields comparable with cereals. However, poor sink utilization/ seed filling could be due to poor photosynthesis, limitation in nitrogen fixation, canopy structure, hormonal regulation, soil properties, water management etc.

PARTITIONING OF ASSIMILATES

The translocation of photosynthates from source to developing sinks depends upon the crop duration particularly vegetative and reproductive phases. Actually, it is the relative partitioning of current assimilates and the extent of remobilization of stored assimilates to the reproductive parts, which matters. Generally, the partitioning will be more if reproductive phase is longer and remobilization of assimilates is significant from vegetative to reproductive parts. Remobilization accounts for a very small portions of seed and major contribution is from current assimilation. Extent of remobilization depends on duration and nature of the crop. Remobilization may not take place from stem and leaves but from pods, the latter contributes about 9-15% of seed weight in rainfed chickpea (Davies *et al.* 1999).

Partitioning of assimilates depends upon the photoperiod under cloudy weather conditions/adverse photoperiod, partitioning of assimilates to reproductive sinks is reduced and there is excessive vegetative growth. Optimum photoperiod results in termination of vegetative growth and initiates reproductive growth, prolonging the reproductive phase and diverts major part of current assimilates to the desirable reproductive sinks.

Indeterminate growth habit is common in grain legumes, with no distinction of vegetative phase as in cereals. Vegetative phase continues even during pod development as a result, there is stiff competition between the vegetative growth and pod/seed development for assimilates. This is an important character limiting yield in grain legumes. Shinano *et al.* (1993) reported low productivity in grain legumes was due to greater respiratory loss from shoots rather than N translocation to the reproductive organs. The substantial increase in productivity in cereals has been possible by increasing the HI than the increase in total biomass (Jain 1986). Similarly, the greatest potential for physiological improvement in productivity of tropical grain legumes lies with increasing the proportion of biomass partitioning into seed. Genotypes with high partitioning efficiency towards reproductive sink and higher dry matter production at pod formation have been recommended for breeding in pulse crops by Manian *et al.* (1989).

Increase in harvest index has been a major component of yield increase for cereals and a few grain legumes. Soybean, for example, commonly has a harvest index of approximately 0.40 or greater. For a number of grain legumes, harvest indices can be substantially below this level. Pigeonpea, for example, may have harvest index as low as 0.20 (Lawn and Troedson, 1990). Success with other crops in increasing harvest index indicates that such selection in breeding programmes could be a practical approach to achieve rapid grain yield increase in some legumes.

NITROGEN FIXATION

The capability of legumes to undertake symbiotic N_2 fixation is a critical distinguishing trait of legumes, and causes legumes to be a highly desirable component of many cropping systems. There are, however, several aspects of the N_2 fixation process that result in unique physiological considerations. One of these is the regulation of O_2 concentration in the nodule. Nitrogen fixation is catalyzed by nitrogenase, which is highly sensitive to oxygen and can be readily denatured when exposed to O_2. Therefore, the metabolic step leading to N_2 fixation is a paradox in that the reaction requires large amount of energy supported by oxidative respiration while the basic N_2 fixing enzyme is destroyed by the O_2 that is essential for respiration.

Nodules are well designed to meet the dual requirement of high O_2 supply rate for respiration within the nodule and maintenance of low O_2 concentration. The internal volume of the nodule where fixation takes place is surrounded by a shell, 2 or 3 cells thick, which acts as an O_2 diffusion barrier. The O_2 permeability of the barrier is variable and adjusts to maintain a low O_2 concentration in the internal volume of the nodule (Serraj *et al.* 1999), oxygen readily diffuses in the internal volume of nodules through a network of gas spaces. To facilitate diffusion of O_2 from the air spaces to the site of respiration within N_2 fixing cells, nodules of many species have O_2 transporters in these cells to which O_2 binds and facilitates diffusion. The transporter in grain legumes is "leghemoglobin" which gives the interior of healthy nodules a pinkish/red colour when they are cut open and the leghemoglobin is exposed to atmospheric oxygen concentration.

Amides and/or ureides are the nitrogenous compounds synthesized in nodules from ammonia produced in the primary step of N_2 fixation. Those grain legumes that transport ureides seem to have a higher N_2 fixation rates than those that transoport amides. Ureides are efficient for N transport since their C:N ratio is 1:1. However, the solubility of ureides is low (Sprent, 1980) and this fact may contribute to the requirement by ureide-transporting species of relatively high water fluxes out of nodules via the xylem (Serraj *et al.* 1999). Since, the source of the water for xylem outflow is phloem flow into the nodule, interruption of phloem flow into nodules results in rapid decreases in N_2 fixation rates. In addition, feedback of nitrogenous compounds from the mother plant exerts a control on nodule activity (Serraj *et al.* 1999). The tight regulation of nodule activity by water and nitrogen flow results in a close balance between N supply and N requirement to support plant growth.

In cropping situation, it would seemingly be beneficial to maximize N_2 fixation to increase yield and/or protein content of the plant products and to increase the production of fixed N for the overall cropping system. Currently, there is little information about what may be the ultimate limit on N_2 fixation rate. The highest N_2 fixation rate for soybean appears to be about 0.6 g N $m^{-2}d^{-1}$, which was reported for plants grown on a gravel hydrophonic culture in field (Harper, 1971). One possibility is that N_2-fixation rate in soybean might simply be limited by the rate of formation of vegetative storage 'compartments' in which to deposit the fixed N. One research approach to investigate maximizing N_2-fixation rate would be to undertake studies with legumes with large tubers (Tropical Legumes, 1979). These tubers might provide both energy reserves and storage capacity, to investigate the limit for maximum N_2 fixation rate.

Another major limitation on N_2-fixation rate when legumes are rotated with a cereal in a cropping system is the large amounts of N-fertilizer that may be applied to the cereal. High level of residual nitrate in the soil will inhibit nodulation and N_2 fixation activity (Streeter, 1988). Delayed nodule formation and initiation of N_2-fixation can result in a period after the soil nitrate is exhausted when there is little or no N_2 fixation. Consequently, there may be several weeks during critical stages of plant development when N supply within the plant is restricted and the legume suffers N-deficiency symptoms and loss of yielding capability.

There has been considerable interest in developing cultivars with N_2 fixation that is tolerant to residual nitrate for use following heavily fertilized cereals in cropping rotations. A goal has been to identify genotypes that can develop nodules in the presence of nitrate so that N_2-fixation activity is initiated prior to or as soon as the soil nitrate is exhausted. Several soybean genotypes have been identified that have less sensitivity of N_2-fixation rates to nitrate than most soybean (Serraj *et al.* 1992).

Another approach to enhance nitrate tolerance was the identification of genotypes that form unusually large numbers of nodules in the absence of nitrate. Even though nitrate resulted in the same relative decrease in these 'super nodulating' lines, as normal lines the number of nodules on these lines under nitrate treatment was still high (Carroll *et al.* 1985; Gremaud and Harper, 1989). However, the retention of a reasonable nodule population when subjected to nitrate has failed to result in yield increase (Maloney and Oplinger, 1997).

Nitrogen fixation in the species that transport ureides from the nodules to the shoot have been found to be extremely sensitive to soil water deficits (Serraj and Sinclair 2002). Nitrogen fixation rates in the ureide-transporting species e.g. soybean and cowpea, begins to decline as result of soil drying well in advance of any other physiological process. This sensitivity may in part be a consequence of the need for high water flux to and from nodules to sustain ureide transport. In addition, there appears to be an especially strong negative feedback from the shoot to roots (Serraj *et al.* 1999) involving elevated ureide concentrations that signal decreased N_2-fixation rates. Nitrogen fixation in amide transporting species, on the other hand, does not appear to be any more sensitive to water deficits than other physiological responses.

The sensitivity of N_2-fixation to drying soil has proved to be greater in ureide-transporting legumes than any other physiological processes in these plants (Serraj *et al.* 1999). This sensitivity has been confirmed in field studies employing various water-deficit treatments on a sandy soil with no residual N (Sinclair *et al.* 1987). So, nitrogen accumulation reflected N_2-fixation and the results were expressed relative to the well-watered treatment. It was inferred that in every water deficit treatment the relative nitrogen accumulation decreased to a greater extent than mass accumulation. Further, Purcell and King (1996) showed directly a loss of soybean yield as a result of N_2-fixation sensitivity to soil drying. In a field experiment, plots subjected to drought were either left with no added nitrate or nitrate was added. Those plots to which nitrate was added as a source of N during the drought had an 18% greater yield than the plots untreated with nitrate and dependent solely on N_2-fixation. Eight lines were found to have substantial N_2-fixation tolerance by Sinclair *et al.* (2000) in soybean. In addition, these eight tolerant lines were found to exhibit little feedback inhibition of N_2-fixation rates when supplied with high levels of ureides (Sinclair *et al.* 2003).

Nitrogen assimilation and partitioning

A large quantity of nitrogen is required for growth and yield in grain legumes. A part of this requirement is via biological nitrogen fixation (BNF). Nodulation and N_2-fixation is maximum during flowering and early pod development followed by a decline. This decline coinciding with greater need of N for developing sinks *i.e.* seed (Sharma and Sharma, 2008). During pod-filling stage, nodulation and nitrogen fixation is reduced due to the greater sink capacity *i.e.* preferential mobilization of assimilates for seed development, as a result less photosynthates are available for nodule growth and maintenance. Since, the current nitrogen fixation declines, the N-requirement of developing seeds are met by translocation of N from subtending leaves. Rubisco (the major enzymatic protein of leaves) is mobilized, resulting in reduction of photosynthetic efficiency of leaves (Grover *et al.* 1985).

Leaf area declines during pod development leading to leaf senescence because a large quantity of carbon and nitrogen is left unused. Poor mobilization of N from vegetative to reproductive parts, accompanied by inadequate BNF, leads to poor productivity in grain legumes. Although, pulses (also grain legumes) have the ability to draw adequate N from soil, BNF and remobilization from leaves, yet the greater N demand remains to be a major constraint for productivity. It is important to enhance the rate of photosynthesis during seed filling so that enough of carbohydrates are available for enhancing N uptake and its utilization for the development of seed (Ganeshaiah and Uma Shanker, 1987).

Pulses are considered as valuable food crops, because of high protein content in the seeds. Therefore, their nitrogen requirement is much higher as compared to cereals.

Nitrogen content in seed is several times higher than in other plant parts viz. leaves, stems, roots etc. The proportion of total nitrogen recovered in seeds exceeds grain HI *i.e.* NHI (Nitrogen Harvest Index). The level of nitrogen in the seeds ranges from 3-7% while in plant residues, it is 0.75-1.5%. Thus, for most grain legumes, a higher HI necessitates an even higher NHI.

Hence, the comparison of grain legumes with cereals is greatly unjustified. However, amino acid composition of grain legumes is complementary to cereals since they are relatively richer in lysine.

STRESS AMELIORATION

Major abiotic stresses affecting legume crops are impact of temperature, salinity, moisture, solar radiations and micronutrient deficiency.

Moisture stress

Grain legumes are highly sensitive to water stress. Drought leads to yield reduction because of less dry matter accumulation and increased abscission. In, general under water stress, there is decrease in leaf water potential, stomatal conductance, net assimilation of CO_2 and transpiration. Water stress imposed at flowering stage, leads to reduction in dry matter accumulation in different plant parts. Yield losses were higher when stress was imposed at flowering than at pod filling stage. However, cumulative effect of water stress at both the stages has been reported by Sekhon *et al.* (2004).

Water stress at initial stages limits sources, leading to the reduction in total biomass, whereas at reproductive phase (initiation of flowering, initiation of podding and both) leads to flower and pod abscission, thereby decreasing the sink strength. Significant reduction in leaf area, leaf area duration, relative leaf growth, specific leaf weight (SLW) and yield have been observed in mungbean and urdbean by Sharma *et al.* (unpublished). Genotypes having higher root dry matter, root length and volume are tolerant to drought. Irrigation increases dry matter accumulation than yield. Excess water enhances vegetative growth, thus, creating competition for developing desirable sinks. Flooding creates anoxia (oxygen deficiency) which leads to decline in LAI, nodulation, yield and yield attributes. However, irrigation applied at initiation of flowering, podding and at both the stages enhances leaf area and leaf characteristics, chlorophyll content, photosynthetic efficiency and ultimately yield as compared to stressed plants (Sharma *et al.* unpublished).

The special limitation of soil drying on N_2 fixation of ureide-transporting legumes, has been already discussed in this chapter. Besides N_2-fixation, growth and water loss by grain legumes varies with soil water content in the same pattern as observed for all crop species (Sinclair and Ludlow, 1986). There is no influence on plant gas exchange rates until only $1/3^{rd}$ of the transpirable water remains in the soil. At this $1/3^{rd}$ point, the plant gas exchange rates decreases approximately linearly until the stomata are closed. Once, there is no transpirable water remaining in the soil plants enter a survival phase in which they are no longer growing but continue to extract soil water at a very slow rate to maintain plant water balance and avoid desiccation. Interestingly, there are very large differences among grain legumes in the ability to survive these severe stress conditions. Soybean has a very poor ability to survive severe stress while pigeonpea can survive long term stress (Sinclair and Ludlow, 1986). However, Serraj and Sinclair (2002) argued that these differences in survival of severe drought are likely not to be particularly important in high production agriculture. If the total amount of available water is so limited to cause crop survival to be a concern, then yield potential is likely to be severely restricted and little or no yield can be harvested from such a crop. A water-deficit response in soybean is the slow wilting phenotype PI-416937. This genotype was identified as to maintain turgor with soil drying such that the plants wilted several days later than the other genotypes (Solane *et al.* 1990). The basis of this unique phenotypic expression is speculated to be a result of either deeper rooting in the soil or a water conservation strategy in which plants use water at a slightly lower rate than other lines depending on the environment through the growing season, such traits could be of considerable benefit in some dry seasons and result in little or no penalty in wet seasons.

Temperature

Legume production has been extended over a wide range of climates from cold, northern latitudes, to hot, tropical regions. Much of the expansion in the range of adaptation has resulted from empirical selection of superior performance of individual genotypes for new climate zones. There have, however, been research programs targeted specifically at identifying genetic variation within legume species for adaptation to various temperature regimes.

In peanut, cold night temperatures of less than 15°C resulted in lost leaf photosynthetic activity during the following day (Sinclair *et al.* 1994; Bell *et al.* 1994). Bell *et al.* (1994) studied eight genotypes and found significant differences in the influence of cold night temperature while the threshold for the negative influence of temperature did not seem to vary among genotypes, some genotypes (OAC Ruby and OAC Garroy) had smaller decreases in photosynthetic rates at lower temperatures than the other genotypes. These results indicated that selection in legumes for photosynthetic tolerance of cold temperature is possible.

Sink strength in terms of flower and pod formation are directly linked with temperature. Significant decline in bud and flower drop has been recorded in pigeonpea and chickpea under low temperatures (Singh *et al.* 1997). Time of sowing is directly related to temperature too. For example early sown pigeonpea attains highest LAI, intercepts maximum solar radiation, produces highest total dry matter and yield (Patel *et al.* 2000). Late sowings curtails vegetative period resulting in reduction in relative growth rates, biomass, yield and yield attributes (Singh *et al.* 2003).

Low temperature had adverse effect on the reproductive physiology in chickpea germplasm (Sharma *et al.* 2005). Pollen fertility was reduced significantly when the temperature falls below 3°C. Pollen germination, pollen tube growth and fertilization was hampered by the cold spell. Low temperature reduced flower retention. However, in the genotypes which flowered and set pods during cold spell, seeds could not develop because low temperature hindered the translocation. In cowpea, high night temperatures (>30°C) resulted in flower abscission (Warrang and Hall, 1984). The high night temperature was shown to result in indehiscence of anthers and low pollen viability. Out of 58 cultivars only three cultivars had high pod-set under high night temperatures (Warrang and Hall, 1983).

Under field conditions, it is hardly possible to separate the temperature and photoperiod effects. Therefore, attempt in the form of agronomic practices or breeding objectives which would confirm to the requirements of the location. On the basis of the studies carried out by Summerfield (1975, Personal Communication) has proposed the model (Fig.1) for the physiology of yield in cowpea and soybean in the tropics.

Solar radiations

Pod legumes particularly summer grown mungbean genotypes are higher yielders because of high solar radiation, higher crop growth rates (RGR, NAR), specific leaf weight (SLW), leaf area index (LAI) than in the rainy seasons with cloudy weather (Singh *et al.* 1985). There is a direct positive correlation between PAR interception and crop duration. There are adverse effects of shading on legumes in intercropping systems. Defoliation at different stages of crop growth limits source strength, reduces light interception, photosynthetic efficiency and yield in mungbean (Sharma, unpublished). Shading during reproductive stage has more severe effects on yield than vegetative phase (Hudge *et al.* 1993).

Salinity

Salinity invariably affects germination and growth of plants through a check on the nitrogen-fixing machinery, protein synthesis, carbohydrate metabolism, enzyme synthesis etc., which in turn have a tremendous negative effect on harvest index of the crops. Salt injury influences several physiological processes and productivity, which has gained immense importance in the last couple of years. It is the toxicity of ions penetrating cells as well as augmentation in the osmotic potential of the medium, which subsequently hinder the penetration of water into the roots (Munns and Keyser, 1981). The accumulation of salts within the plant tissues caused a disturbance in the plant growth and development thus, altering the degree of hydration of cytoplasm during the period of osmotic adaptations (Sands and Clarke, 1977). The extent of injury varied with plant species and within same species, it varied with the type of predominant ions, their concentration, the physiological stage of growth at which the plant encounters the saline stress. Various physiological processes, nitrogen fixation, growth and productivity of legumes in response to abiotic stresses has been vividly discussed by Sharma and Sardana (2008).

Nutrition

Both inadequate nitrogen supply via soil or BNF reduce biomass, leaf area, leaf area duration, nitrate reductase, grain nitrogen and total nitrogen per unit area (Bhattacharya and Ali, 2002) whereas nitrogen application enhances these characteristics along with the increase in chlorophyll, photosynthetic activity and nitrate reductase in nodules (Sekhon *et al.* 2002c). Nitrogen limitation in cropping system can be minimized by using legumes but often there remains a limitation on plant growth due to inadequate phosphorus. Certain legumes, however, appear to have the capacity to make a major contribution in improving the availability of P in rotational cropping system. Several species have been identified that exhibit substantial superiority in recovering soil P that is unavailable to other species. For example, Ae *et al.* (1990) reported that pigeonpea thrived on an Alfisol soil with most of the P bound to iron while maize, sorghum, pearlmillet and soybean died from P-deficiency within one month after sowing. Ishikawa *et al.* (2002) found that roots of pigeonpea encreated large amounts of citric and piscidic acid and these contributed to the mobilization of P in the soil. They concluded, however, that this mechanism was inadequate to account for all of the P taken up by the plant. The root cell-walls of pigeonpea contains a small molecule that has P-solubilizing abilities (Ae and Shen, 2002) and likely contributes to the recovery of bound P from the soil. Peanut also exhibited the same P solubilizing abilities, which was associated with survival on low P-available soil while maize, sorghum and soybean died. Ohwaki and Hirata (1992) also reported that chickpea exuded more citric and malic acid than five other legumes indicating that this species may be another good candidate for use on soils with low P availability.

In addition to differences among species in recovering non-soluble P, there are genotypic differences within species Lynch and Beebe (1995) compared 364 common bean genotypes under P-deficient conditions and found substantial differences in growth. It was confirmed in a comparison to the six contrasting cultivars of common bean that there were substantial differences in the ability to recover P from different forms of immobile P (Yan *et al.* 1996). Ishikawa *et al.* (2002) included four cultivars of pigeonpea in their study of P recovery from various sources and found distinct differences among genotypes, which depended on the form of immobilized P. Phosphorus application increased LAI, crop growth rate, chlorophyll content, lessened abscission of flowers and increased grain yield in mungbean (Garai and Datta, 2002), cowpea (Rao and Subramanian, 1990) lentil and field pea (Singh *et al.* 2001) by enhancing leaf water potential, stomatal conductance, light interception, water use efficiency and decreasing canopy temperature (Yadav *et al.* 1994).

Nutritional efficiency has been attributed to a number of mechanisms including root processes affecting availability of nutrients in rhizosphere, root structure and geometry, translocation from roots to shoot and foremost remobilization with shoots understanding the physiology, genetics and regional distribution of efficiency for a particular nutrient should assist in developing nutrient efficient varieties for productivity enhancement.

Expansion in legume production is likely to continue for the foreseeable. Protein demand will increase as human diets improve and/or shift to greater amounts of meats, virtually necessitating an increase in the production of all grain legumes. The increasing demand will result in a number of opportunities and challenges. The opportunities are to use innovative approaches to fully integrate grain, legume production into cropping system., Approaches which take full advantage of legumes ability to fix atmospheric nitrogen symbiotically. Following approaches/thrust areas should be looked into for increasing grain legumes productivity.

- Efforts are needed to redesign the varieties with appropriate growth habit and efficient source-sink kinetics. In grain legumes wild traits like photo-thermo sensitivity, indeterminate growth habit which causes intra-plant competition for photosynthate resulting into poor partitioning. It may be presumed that determinate types under good management would partition photosynthate to yield components with greater efficiency through early seedling vigour, faster leaf development, synchronous flowering, slow senescence and increased rate of dry matter production.
- All the heritable characters viz. morphological and physiological processes responsible for linear productivity in grain legumes can be manipulated by genetic manoeuvres to irradicate undesirable factors and incorporate desirable traits which minimize the adverse effect on yield.
- Indepth studies and systematic efforts are required for identifying desirable traits from the beginning *i.e.* higher seedling vigour, faster vegetative growth, higher response to nutrients, irrigation and other inputs, thermo and photoperiod insensitivity, improved root system, efficient partitioning, lower flower and fruit drop, synchronous maturity, more biomass production are the needs of all the pulse/legume crops.
- Understanding the physiological mechanisms that enable plant to adopt to various stresses via tolerance or avoidance mechanism, thus, maintaining growth by imparting stability to the membrane will help in selection of stress tolerant cultivars.
- In short duration, pod legumes increase in the reproductive phase, pod filling period, higher tolerance to water stress during reproductive phase in long duration pulses under rainfed conditions and under residual moisture in rabi pulses. Likewise resistance against waterlogging, high temperature (Kharif season), both high and low temperature (Rabi season), improved BNF capacity, resistance to insect-pest and pathogen should be attempted through selection, breeding and biotechnological tools.
- Inorder to genetically manipulate N translocation, constructs from the cloned genes coupled to a seed specific promoter should be used to generate transgenic over-expression and anti-sense lines. Atkins *et al.* (2005) have generated transformants using a flower specific promoter and significant increase in pod set and seed number has been recorded.
- The possible underlying mechanism particularly endogenous level of hormones responsible for the hormonal imbalance related to flower and fruit shedding requires outmost attention. Further role of plant growth substances in enhancing

sink strength, efficient assimilate partitioning and increasing yield is required to be studied in detail.

- Most productive genotypes could be identified through selection for yield in high productivity environments and for HI in low productivity conditions. This will impart the much desired respite to pulses against unfavourable environment via increase in inherent capacity for higher productivity.
- Development of dual-purpose grain legume cultivars with high N_2-fixation capability to produce a modest grain crop in the growth year for immediate economic benefit to the grower and to generate a reasonable amount of high N-content plant residue to be incorporated into the soil for use in the subsequent cereal crops.
- Full exploitation of some grain legumes to recover P from the soil could be a very important asset in many cropping situations. The challenges of fully exploiting legumes are likely to be focused on their sensitivity to a number of environmental stresses.

Continued success in increasing the productivity of grain legumes and their use in cropping systems will likely be well rewarded with improved economic and environmental viability of crop production over many areas of the globe.

REFERENCES

Ae N Arihara J Okada K Yoshihara T Johansen C 1990 Phosphorus uptake by pigeonpea and its role in cropping systems of the Indian subcontinent *Sci* **248** : 477-480

Ae N Shen RF 2002 Root cell wall properties are proposed to contribute to phosphorus (P) mobilization by groundnut and pigeonpea *Plant Soil* **245** : 95-103

Atkins CA Ferguson M Penny M Smith C 2005 Source and sink relationship in food legumes Abst. A-28. In 4th International Food Legumes Research Conference October 18-22, 2005 New Delhi, India

Bell MJ Gillespie TJ Ray RC Michaels TE Tollenaar M 1994 Peanut leaf photosynthetic activity in cool field environments *Crop Sci* **34** : 1023-1029

Bell MJ Gillespie TJ Ray RC Michaels TE Tollennr M 1994 Peanut leaf photosynthetic activity in cool field environments *Crop Sci* **34** : 1023-1029

Bhattacharya A Ali M 2002 Physiological traits and seed yield in chickpea under high nitrogen input *Legume Res* **25** : 9-14

Carroll BJ McNeil DL Gresshoff PM 1985 A supernodulation and nitrate tolerant symbiotic (NTS) soybean mutant *Plant Physiol* **78** : 34-40

Chhabra KS Kooner BS 1985 Problem of flower shedding caused by thrips, *Megalurothrips distalis* (Karnyl) on summer mung bean, *Vigna radiata* (L.) Wilczek and its control *Tropical Pest Management* **31** : 186-188

Davies SL Turner NC Siddique KHM Plumer JA and Leport L 1999 Seed growth of *desi* and *kabuli* chickpea (*Cicer arietinum* L.) in a short-season Mediterranean type environment. *Aust J of Exp Agri* **39** : 181-188

Evans AN 1973 Genetic improvement of *Phaseolus vulgaris* In Nutritional improvement of food legumes by breeding Protein Advisory Group, United Nations, New York

Ganeshaiah KN Uma Shankar R 1981 Auto-parasitization of leaves in pulses – N limiting or C limiting *Curr Sci* **50** : 942-944

Ganguly SB Bhattacharya A Singh DN 1999 Physiological studies in chickpea varieties : Physiological determinants and their relationship to productivity under late seeding *Ind J of Pulses Res* **12** : 65-74

Garai AK Datta JK 2002 Sources of phosphate fertilization and cycocel spray on some physiological attributes in seed yield in mung *Crop Res* (Hisar) **23** : 300-304

Gates P Smith ML Boulter D 1983 Reproductive physiology of *Vicia faba* L. In : The *faba* bean *Vicia faba* L. (ed PD Hebblewaite) pp. 133-142 London : Buntherworks

Gremaud MF Harper JE 1989 Selection and initial characterization of partially nitrate tolerant nodulation mutants of soybean *Plant Physiol* **89** : 169-173

Grover A Koundal KR Sinha SK 1985 Senescence of attached leaves regulated by developing pods *Physiol Plantarum* **63** : 87-92.

Harper JE 1971 Seasonal nutrient uptake and accumulation patterns in soybeans. *Crop Sci* **11** : 347-350

Hebblethwaite PD Scott RK Kogbe JOS 1984 The effect of irrigation and bees on the yield and yield components of *Vicia faba* L. In *Vicia faba* : Agronomy. Physiology and Breeding (eds PD Hebblethwaite TEK Dewkins MC Heath Ghock Wood) pp 71-93 The Hague : Martinus Nijhoft/Dr W Junk

Hudge VS Sardar MB Khan GM Chavan MH Shete DM 1993 Effect of low light on growth and yield in black gram *Annals of Plant Physiol* **7** : 182-186

Ishikawa S Adu Gyamfi JJ Nakamura T Yoshihara T Watanabe T Wagastsuma T 2002 Genotypic variability in phosphorus solubilizing activity of root exudates by pigeonpea grown in low-nutrient environments *Plant Soil* **245** : 71-81

Jain HK 1986 Eighty years of post Mendelian breeding for crop yield : Nature of selection process and further potential. *Ind J of Genet and Plant Breeding* **46** (Supplement) **30** : 30-53

Khanna Chopra R Sinha SK 1990 What limits the yield of pulses? Plant Processes or plant type . In : Proceedings of the International Congress of Plant Physiology Vol.1 (eds SK Sinha PV Sane SC Bhargava PK Agrawal) pp 268-278 New Delhi : Society for Plant Physiology and Biotechnology

Klauer SF Franceschi VR Ku MSB Zhang D 1996 Identification and localization of vegetative storage proteins in legume leaves *Am J Bot* **83** : 1-10

Kumar J Pannu RK Rao BV 2001 Development of a short duration chickpea for the subtropics *Int Chickpea and Pigeonpea Newsletter* **8** : 7-8

Lansingh AJ Franceschi VR 2000 The paraveinal mesophyll : a specialized path for intermediary transfer of assimilates in legume leaves *Aust J Plant Physiol* **27** : 757-767

Lawn RJ 1981 The potential contribution of physiological research to pigeonpea improvement In : Proc of the International Workshop on Pigeonpea Vol 1 (ed Nene YL) pp 151-164 Hyderabad : ICRISAT

Lawn RJ and Troedson RJ 1990 Pigeonpea : Physiology and yield formation : In The Pigeonpea (Eds. YL Nene SD Hall VK Sheila) pp 179-208 Wallingford : CAB International

Lynch JP Beebe SE 1995 Adaptation of beans (*Phaseolus vulgaris* L.) to low phosphorus availability *Hort Sci* **30** : 1165-1171

Maloney TS Oplinger ES 1997 Yield and nitrogen recovery from field grown supernodulating soybean *J Prod* Agric **10** : 418-424

Manian K Natarajaratnam M Ramaswamy P Balasubramanian M Sreerangaswamy SR 1989 Significance of dry matter production and its partitioning in horsegram genotypes. *Madras Agri J* **76** : 512-516

Muchow RC 1985 Canopy development in grain legumes grown under different soil water regimes in a semi-arid tropical environment Field crops *Res* **11** : 99-109

Munns DW Keyser HH 1981 Tolerance of rhizobia to acidity aluminium and phosphate *Soil Sci Soc Am J* **34** : 519-523

Ohwaki Y Hirata H 1992 Difference in carboxylic acid exudation among P-starved leguminous crops in relation to carboxylic acid contents in plant tissues and phospholipid level in roots. *Soil Sci Plant Nutr* **38** : 235-243

Pandey CS Singh M 1996 Genotypic variation in photosynthetic efficiency in relation to productivity in gram (*Cicer arietinum*) genotypes. *Ind J of Agri Sci* **66** : 466-469

Pandey RK Saxena MC Singh VB 1978 Effect of some growth regulators on yield and yield components of *Kabuli gram* (*Cicer arietinum* L.) *Pantnagar J of Res* **3** : 175-177

Patel NR Mehta AN Shekh AM 2000 Radiation absorption growth and yield of pigeonpea cultivars as influenced by sowing dates *Expt Agri* **36** : 291-307

Poehlman JM 1991 The Mungbean New Delhi Oxford & IBH Publishing Co. Pvt. Ltd.

Prasad S Srivastava JP 1999 Physiological analysis of yielding ability in mungbean *Ind J Pulses Res* **12** : 49-56

Purcell LC King CA 1996 Drought and nitrogen source effects on nitrogen nutrition, seed growth and yield in soybean *J Plant Nutrition* **19** : 969-993

Rao DG Subramanian VB 1990 Effect of P on nodulation N harvest index and growth : Correlation between nodulation and other plant parts in cowpea under well watered and dryland conditions *Ind J of Plant Physiol* **33** : 275-281

Sands R Clarke ARP 1977 Response of radiata pine to salt stress I water relations osmotic adjustment and salt uptake *Aust J Plant Physiol* **4** : 637-646

Savithri KS Ganapathy PS Sinha SK 1978 Fruit and seed development in mungbean (*Phaseolus aureus* Roxb) *J Agri Sci Cambridge* **90** : 551-556

Saxena NP Natarajan M Reddy MS 1983 Growth and yield of chickpea, pigeonpea and groundnut. In : Proc of the Symposium on Potential Productivity of Field Crops under different Environments (ed Y Yoshida) pp 281-305 Manila IRRI

Sekhon HS Sharma P Singh G 2004 Effect of irrigation on growth and yield on kharif mungbean (*Vigna radiata* (L.) Wilczek) genotypes *J Pl Sci Res* **20** : 102-105

Sekhon HS Singh G Sharma P Sharma AK 2002c Effect of *Rhizobium* inoculation and nitrogen on biological nitrogen fixation growth and yield of mungbean (*Vigna radiata* (L.) Wilczek) *Envi & Ecol* **20** : 282-290

Serraj R Drevon JJ Obaton M Vidal A 1992 Variation in nitrate tolerance of nitrogen fixation in soybean (*Glycine max*) Bradyrhizobium symbiosis *J Plant Physiol* **140** : 366-371

Serraj R Sinclair TR 2002 Osmolyte accumulation : Can it really help increase crop yield under drought conditions *Plant Cell Envir* **25** : 333-341

Serraj R Sinclair TR Purcell LC 1999 Symbiotic N_2 fixation response to drought *J Exp Bot* **50** : 143-155

Shanthakumari P Singh SK 1972 Variation in chlorophylls and photosynthetic rate in cultivars of Bengal gram (*Cicer arietinum* L.) *Photosynthetica* 189-194

Sharma P Sardana V 2008 Physiological processes nitrogen fixation growth and productivity of legumes in response to abiotic stresses. In Advanced Topics in Biotechnology and Plant Biology (eds. Malik CP Bhavneet K Chitra W) pp 323-398 MD Publishers Pvt. Ltd., New Delhi

Sharma P Sekhon HS Sandhu JS 2005 Cold tolerance studies in chickpea (*Cicer arietinum* L.) In : Proceedings of the 4th International Food Legumes Research Conference (IFLRC IV) p 86 October 18-22, 2005, IARI, New Delhi

Sharma P Sharma Poonam 2008 Biological Nitrogen fixation : Legume-Rhizobium physiology, phytohormones mediated response to nodulation and nitrogen fixation : In Advanced Topics in Biotechnology and Plant Biology (eds. Malik CP Bhavneet K Chitra W) pp 399-445 MD Publishers Pvt Ltd., New Delhi.

Shinano T Osaki M Tadano T Barrow NJ 1993 Comparison of production efficiency among field crops related to nitrogen nutrition and application. *Plant and Soil* **155-156** : 207-210.

Silim SN Saxena MC 1990 Composition of dry matter accumulation, partitioning and yield of determinate and indeterminate faba bean (*Vicia faba* L.). In Proc. of the *Int. Congress of Plant Physol* Vol. 1 (Eds. SK Sinha PV Sans SC Bhargava PK Agrawal) pp 289-298 New Delhi Society for Plant Physiol and Bichem.

Sinclair TR 1980 Leaf CER from post-flowering to senescence of field grown soybean cultivars *Crop Sci* **20** : 196-200

Sinclair TR 2004 Improved carbon and nitrogen assimilation for increased yield in "Soybeans" : Improvement, Production and Uses (ed J Specht R Boerma) (Am Soc Agronomy, Madison, WI)

Sinclair TR Bennet JM Drake GM 1994 Cool night temperature and peanut leaf photosynthetic activity Proc Soil Crop Sci Soc Florida **53** : 74-76

Sinclair TR Horie T 1989 Leaf nitrogen, photosynthesis and crop radiation use efficiency : A review *Crop Sci* **29** : 90-98

Sinclair TR Ludlow MM 1986 Influence of soil water supply on the plant water balance of four tropical grain legumes *Aust J Plant Physiol* **13** : 329-341

Sinclair TR Muchow RC Bennet JM Hammond LC 1987 Relative sensitivity of nitrogen and biomass accumulation of drought in field grown soybean *Agron J* **79** : 986-991

Sinclair TR Purcell LC Vadez V Serraj R King CA Nelson R 2000 Identification of soybean genotypes with N_2 fixation tolerance to water deficits *Crop Sci* **40** : 1803-1809

Sinclair TR Vadez V Chenu K 2003 Ureide accumulations in response to Mn nutrition by eight soybean genotypes with N_2-fixation tolerance on soil drying *Crop Sci* **43** : 592-597

Singh BB Singh DP Singh NP 1997 Genetic variability in pigeonpea germplasm for cold tolerance *Ind J of Genet and Pl Breeding* **57** : 425-430

Singh DN 1990a Influence of water on pollen germination and fruit set in mungbean. In : Proceedings of the International Congress of Plant Physiology Vol.1 (eds SK Sinha PV Sane SC Bhargava PK Agrawal) pp 320-322 New Delhi : Society for Plant Physiology and Biotechnology

Singh G Sekhon HS Sandhu JS Singh SJ Gumber RK Randhawa AS 2003 Effect of location and seed rate on three genotypes of mungbean *Tropical Sci* **43** : 116-120

Singh G Sekhon HS Sharma P 2001 Effect of *Rhizobium* vesicular arbuscular mycorrhiza and phosphorus on the growth and yield of lentil (*Lens culinaris*) and field pea (*Pisum sativum*) *Envir & Ecol* **19** : 40-42

Singh R 1990b Assimilation and partitioning of carbon and nitrogen during growth and development of grain legumes. In : Proceedings of the International Congress of Plant Physiology Vol.1 (eds SK Sinha PV Sane SC Bhargava PK Agrawal) pp 261-267 New Delhi : Society for Plant Physiology and Biotechnology

Sinha SK 1974 Yield of grain legumes. Problems and prospects. *Indian J. of Genet and Pl Breeding* **34** : 988-994

Solane RL Patterson RP Carter TE Jr 1990 Field drought tolerance of a soybean plant introduction *Crop Sci* **30** : 118-123

Sprent JL 1980 Root nodule anatomy, type of export product and evolutionary origin in some Leguminosae. *Plant Cell and Envir* **3** : 35-43

Streeter J 1988 Inhibition of legume nodule formation and N_2 fixation by nitrate *Crit Rev Plant Sci* **7** : 1-23

Summerfield RJ Roberts EH 1985 Grain legumes species of significant importance in world agriculture In A Handbook of Flowering Plants (ed Halery AH) pp 61-184 Boca Raton Florida CRC Press

Tropical Legumes : Resources for the Future 1979 National Academy of Sciences, Washington DC

Warrang MoA Hall AE 1984 Reproductive response of cowpea (*Vigna unguiculata* (L.) Walp) to heat stress II Response to night air temperature *Field Crops Res* **8** : 17-33

Yadav SK Chander K Kumar A 1994 Response of late sown gram (*Cicer arietinum*) to irrigation and phosphorus *Ind J of Agri Sci* **64** : 24-28

Yan X Lynch JB Beebe SE 1996 Utilization of phosphorus substrates by contrasting common bean genotypes *Crop Sci* **36** : 936-941

8

ENDOPHYTIC FUNGI : A BIOCONTROL AGENT

KARTIKEYA TIWARI, MUKESH K. UPADHYAY AND PAYAL LODHA

The control of pests and diseases affecting cultivable plants is made mainly by the use of chemical compounds. The use of agrochemicals to control insects and plant pathogens represents, however, a high risk to field workers, consumers and environment. They are responsible, for eliminating important species of insects that control other pests and microorganisms that are performing a critical role in the environment, inhibiting growth and multiplication of useful microbes. The natural and biological control is an alternative way that may contribute to reduce or even eliminate the use of chemical products in agriculture. The natural and biological control of agricultural pests gained more attention during the past decades especially through endophyte microorganisms. Therefore, this way to control pests are important to reduce the costs and protect the environment.

Petrini (1986) defined all organisms that colonize internal plant tissues are referred as endophytes and include everything from virulent foliar pathogens to mycorrhizals root symbionts. Fungi that cause visible symptoms of plant diseases have been separately classified as pathogens, regardless of the degree of penetration into the host plant's tissue. By default, fungi that cause asymptomatic infections entirely within the tissues of the host plants called endophytic fungi (Carroll, 1986). However, some mycologists also include fungi that inhabit plant organs at some point in their life cycle without causing apparent harm to their host, as well as latent internal pathogens (Petrini, 1992).

In the continual search by agricultural industries for new products, natural selection has been found to be superior to combinatorial chemistry for discovering novel substances from endophytic fungi that have the potential to be developed into new industrial products especially mycoherbjudes and pharmaceutical compounds. Since natural products are adapted to a specific function in nature, the search for novel secondary metabolites should concentrate on organisms that inhabit novel biotopes and endophytic fungi inhabit such a biotope. From the last 12 years, 6500 endophytic fungi from herbaceous plants and trees, screened for biological activities, and have isolated and determined the structures of the biologically active compounds. The potential role of the endophyte and its biologically active metabolites in its association with its host has been investigated by large number of workers (Barker *et al.* 1984; Prestidge *et al.* 1984; Ahmad *et al.* 1985). It may be hypothesized that the interaction of fungal endophyte–plant host is characterized by a finely tuned equilibrium between fungal virulence and plant defense. If this balance is disturbed by either a decrease in plant defense or an increase in fungal virulence, disease develops. Not only the endophytes synthesize metabolites to compete first with epiphytes and then with pathogens in order to colonize the host, but presumably also to regulate metabolism of the host in their delicately balanced association. The utilization of a biotope such as that of the fungal endophyte is one aspect of intelligent screening, another very important one is the taxonomy of the fungus in order to avoid redundant structural isolations.

The endophyte communities in angiosperms are frequently dominated by species of Diaporthales and those in gymnosperms by species of Helotiales. Divergence of angiosperms and gymnosperms coincides with the divergence of the Diaporthales and the Helotiales in the late Carboniferous about 300 million years ago, indicating that the Diaporthalean and Helotialean ancestors of tree endophytes had been associated, respectively, with angiosperms and gymnosperms; some are, however, opportunists and can cause disease after the host has been weakened by some other factor. Mutualism of tree endophytes is often assumed, but evidence is mostly circumstantial. The sheer impossibility of producing endophyte-free control trees impedes proof of mutualism. Some tree endophytes exhibit either a pathogenic or a putatively mutualistic behaviour depending on the situation. A wide diversity of microorganisms are present in most plant parts of both gymnosperms and angiosperms, especially the leaves where the tissue is apparently healthy, these may be endophytes, epiphytes or latent pathogens. The distinction among endophytes, epiphytes (those that live on the surface of plants) and phytopathogens (those that cause diseases to plants) is of pure didactic meanings. There is a gradient separating them and thus, it comes to a situation where it is a very difficult task to draw limits to discriminate each category. Endophytic microorganisms only received considerable attention in the last 20 years, when their capacity to protect their hosts against insects-pests and other pathogens. As endophytic fungi and bacteria started to be better analysed, it became clear that they could confer other important characteristics to plants, such as greater resistance to stress conditions (*i.e.* water), alteration in physiological properties, production of phytohormones and other compounds of biotechnological interest. In addition to the economical aspects, the study of endophytic microorganisms has strong academic interests, concerning the discovery of new microbial species, mainly when tropical hosts are investigated. It was demonstrated that the presence of these microorganisms in their respective hosts could result in the reduction of insect attacks. Webber (1981) was probably the first researcher to report an example of plant protection giving by an endophytic fungus, in which the endophyte *Phomopsis oblonga* protected elm trees against the beetle *Physocnemum brevilineum*. It was suggested that the endophytic fungus *P. oblonga* was responsible for reducing the spread of the elm Dutch disease causal agent *Ceratocystis ulmi* by controlling its vector, the beetle *P. brevilineum*. Gaynor and Hunt (1983) observed in several ryegrasses that high fungi infection is correlated with a decrease in the attack frequency of the Argentine steem weevil, *Listronotus bonariensis*. These authors also showed that these associations may be complex and factors like the level of nitrogen fertilizers may as well affect attack frequency. Lasota *et al.* (1983) had also correlated in the white spruce *Picea glauca*, death rate of the Homoptera *Adelges abietis* when galls are infected with the endophytic fungus *Cladosporium sphaerosperum*. In this review the major themes related to the control of insects-pests by endophytes will be addressed, together with the perspective of applying this knowledge to the development of a biological control by means of modern methods.

Endophytic microorganisms and their effect on insects

It was confirmed that the plant protection process promoted by endophytes is more complex than one might have expected. Although, in most of the published reports on the subject they show a mutualistic interaction, that is, positive in respect to the host and negative to the insects-pests and there are also reports of neutral or even antagonistic interactions. In addition to the reports mentioned above (Webber, 1981; Funk *et al.* 1983; Gaynor and Hunt, 1983; Lasota *et al.* 1983; Barker *et al.* 1984; Prestidge *et al.* 1984; Ahmad *et al.* 1985), that correlate endophytic fungi with negative effects towards insects, In fact, Clay (1989) stressed that, although many endophytic fungi have already been isolated from around 80 grass species of temperate countries, only six genera are the most frequent. The situation, since Clay's review (1989) about ten years ago, has gained minor changes. Ahmad *et al.* (1986) showed that the blue grass billbug *Sphenophorus parvulus* is less frequent on plants infected with endophytic fungi and verified the same results in grasses attacked by the Southern armyworm *Spodoptera eridania*. Saha *et al.* (1987), studying endophytes in fine fescue, verified that endophytic fungi from the genus *Acremonium* were associated with host resistance to *Blissus leucopterus hirtus* and the same observation was made by Mathias *et al.* (1990), in *L. perenne-A. lolii* interaction. Kindler *et al.* (1990) verified that *Acremonium* reduced the attacks of the aphid *Diuraphis noxia* in *Lollium* as well as in *Festuca* and the same observations were made by Clement *et al.* (1990; 1992). Kanda *et al.* (1994) reported the preference of larvae from the bluegrass webworm *Parapediasia teterrella* for diets with endophyte-free plants, to a point that the larvae would starve to death if only plants infected with *Acremonium* were available. In the field, endophyte-free species were severely attacked by insects, whereas those infected with *Acremonium* stayed almost free of insect larvae. As it will be seen in more details, endophytic fungi (Clavicipitaceae, Ascomycetes) synthesize alkaloids during plant infection, reducing the survival and the development of *S. frugiperda* in Graminae and Cyperaceae. The authors concluded that this is a general characteristic of infected grasses, that is, insect resistance and consequential improved performance of plants in nature. Considering other grasses, the studies are sparse. However, the host protection process displayed by endophytic fungi against insects seems to be general. For example, alcoholic extracts from *Poa ampla* infected or not with *Neotyphodium typhnium* were assayed against mosquito larvae. Only the extracts from the infected plants were effective against the insect, whereas the extract obtained directly from the fungi was not (Ju *et al.* 1998). Although plants other than grasses can be protected by endophytes, the data on these cases are not so conclusive nor numerous. Sherwood-Pike *et al.* (1986) described in oak (*Quercus* sp.), the role of the endophytic fungus *Rhabdocline parkeri* against the insect *Contarinia* sp. Petrini *et al.* (1989) found that the endophytic fungi *Phyllosticta* sp. and *Hormonema dematioides* were dominant in the balsam fir *Abies balsamea*. The first fungus colonized preferentially galled needles attacked by *Paradiplosis tumifex*, being a candidate for biological control. The examples given so far include researches performed in field and lab conditions, as well as comparisons between field and lab experimental data. Choice tests on diets composed of endophytes-free and infected plants are often used along with several choice combinations. Although, the majority of the works cited above are quite conclusive, the results do not always finish up in the control of insects by endophytes. As it will be shown, there are cases of neutrality or even improvement of insect performance in plants colonized by endophytic microorganisms.

Insect control displayed by endophytic fungi

Since the pioneer works in the field, the capacity of endophytic fungus to repel insects, induce weight loss, reduction of growth and development reduction and even to increase pest death rate, was correlated with toxin production. In several cases, it was shown that the mode of action of certain fungi was based on the capability to render the plant unpalatable to several types of pests like aphids, grasshoppers, beetles, etc (Carroll, 1988; Clay, 1988a; 1988b). Bacon *et al.* (1977) established for the first time a correlation between an endophytic fungi, *Epichloë typhina* and the toxicity of its host; It is now a fact that several toxins are produced by endophytic fungi and that these substances confer host protection against different herbivors. We suggest for more details on toxins the review of Bacon and Hills (1996) in which the most important toxins found in *L. perenne,* like ergot alkaloids of two types, ergopeptine and clavine and neurotoxines called lolitrems are described. It is not known if the precursor is converted in lolitrems by the plant or if the fungus is not capable of synthesizing it in pure culture but is able to do so while inside the plant. Examples of toxin production by endophytic fungi, notably those colonizing grasses are considerably abundant in the literature. In 1988, Prestidge and Gallagher established a relationship between the presence of the fungus *A. lolii* in *Lolium perenne* and the growth, survival and feeding behaviour of *Listronotus bonariensis* larvae. In this case, the reduction in insect attacks towards infected plants was due to a strong toxin, lolitrem B, also toxic to mammals. This toxin, once added to insect diets, reduced insect growth and survival. Its assimilation occurs by ingestion but not by absorption through the insect integument. In certain cases, the production of toxin by the endophytic fungi was a plausible explanation for interactions resulting in natural insect control. Thus, Clark *et al.* (1989) showed in *Abies balsamea* and red spruce *Picea rubens,* that from 900 samples of fungal isolates, five produced toxic substances and three of those produced powerful toxins that, once extracted and given to insects, caused death and decreased development rate of *C. fumiferana*. Few fungi generally produce alkaloids, mainly peramine and ergovaline. Peramine, lolitrem B and ergovaline were found in *Lolium* and *Festuca* infected with *Acremonium coenophialum* and *A. lolii* and in *Festuca longiflora* infected with *E. typhina*. Individuals of *Rhopalosiphum padi* and *Schizaphis graminum* did not survive in grasses containing the alkaloid loline. On the other hand, ergovaline did not affect both insect species. The methanolic extracts of *F. arundinacea* infected with *A. coenophialum*, contain lolines of fungal origin, which are able to alter feed behaviour and weight of insects-pests. Diets amended with extracts containing loline derivatives reduced weight and altered certain behaviours of *S. frugiperda* and *O. nubilalis*, notably in the first insect (Riedell *et al.* 1991). In certain cases, the production of toxin by the endophytic fungi was a plausible explanation for interactions resulting in natural insect control. Thus, Clark *et al.* (1989) showed in *Abies balsamea* and red spruce *Picea rubens,* that from 900 samples of fungal isolates, five produced toxic substances and three of those produced powerful toxins that, once extracted and given to insects, caused death and decreased development rate of *C. fumiferana*. Many other cases of insect control involving toxins produced by endophytic microorganisms can be found in the literature. Siegel *et al.* (1990) verified the presence of the alkaloids N-formil, N-acetil loline, peramine, lolitrem B and, ergovaline during plant attack by aphids. Several grasses infected with *Acremonium* spp. and *E. typhina* were analysed. Patterson *et al.* (1992) observed in *Lolium* and *Festuca* that the production of alkaloids by *Acremonium* reduced attacks of the Japanese beetle *Popilla japonica*. Other works find that ambient variations may affect toxin production by endophytic fungi. Breen (1992) verified that changes in temperature and seasons modify toxin levels in the plant. This is the case of the peramine toxin, produced in *L. perenne* by *A. lolii*. The antixenosis towards the aphid *S. graminum* is dependent on peramine and the endophyte concentration so, a natural consequence is that the effect on the insect-pest will also be modified. Two new active toxins against the Spruce budworm *C. fumiferana* were found in an unidentified endophytic fungus infecting the wintergreen *Gaultheria procumben* (Findlay *et al.* 1997). Alkaloids from *N. lolii* and *L. perenne* are capable of altering insect behaviour. Several of these alkaloids were added to the diet of adult individuals of the Coleoptera *Heteronychus aratur*. Peramine, lolitrem B, lysergol-type alkaloids, festuclavine and lisergic acid showed no effects on the insect. Ergovine showed moderate effects whereas ergotamine, ergovaline from the ergot-type alkaloid family seem to be responsible for the plant resistance (Ball *et al.* 1997b). Recently, Miles *et al.* (1998) showed that endophytic isolates of *Neotyphodium* sp. produce N-formilonine and a paxiline analogous in the host *Echinopogum ovatus*. These compounds show insecticidal activity against *L. bonariensis* and other insects. The majority of works related to toxin production was performed in grasses. However, Calhoun *et al.* (1992), for the first time, identified toxic products synthesized by endophytic fungi in woody plants and that were able to modify growth and death rates in larvae of the spruce budworm C.

fumiferana feeding on balsam fir. The endophytes in this case were identified as *Phyllosticta* and *Hormonema dematioides* and the toxic compounds were mainly heptelidic acid and rugulosine.

Modifications of endophytes

Environmental factors like temperature, hydric stress, soil pH, insect-pest resistance and other ecological parameters may also affect the endophyte concentration and toxin production. Breen (1994), idealized a model with all of these interactions affecting the increase or decrease in plant resistance to insects-pests in the presence of endophytes. Arechavaleta *et al.* (1989) had already verified in the *F. arundinacea* endophyte *A. coenophialum* that the production of ergo-peptide alkaloids increased in soil amended with nitrogen fertilizers and during a moderate hydric stress, which consecutively altered insect-pest control. A study about the interaction between nitrogen fertilization, pests and the endophytic fungus *A. coenophialum* showed that in a general manner, insects like *S. frugiperda* developed better in nitrogen fertilized plants not infected by the endophyte. However, considering this and other insects, the occurrence of *A. coenophialum* blocks insect development. The results, therefore, are quite variable and do not allow us to draw any general correlation between nitrogen fertilizer and endophyte-mediated pest control in *F. arundinacea* (Davidson and Potter, 1995). In the same manner, Bultman and Conard (1998) verified interactions among several factors like nutrient levels and plant damage during endophytic fungi control of *S. frugiperda* in the host *F. arundinacea*. The insect pupal mass was negatively influenced by the endophyte in low nutrient levels but the effect was negligible in plants highly fertilized. This study showed again that these interactions are complex and influenced by several factors, in this case by fertilizer application. The recent studies, in soil less cultivation conditions, conduced by Richardson *et al.* (1999) verified that the Chewings fescue *F. rubra* subspecies *commutata* showed differences in alkaloid contents in respect of endophyte colonization and amounts of several nitrogen sources. Fluctuation in the contents of the alkaloids ergovaline and peramine in colonized plants was verified, which will consecutively alter insect resistance. Clay *et al.* (1993), analized changes resulting from these plant-plant interactions in insect control and presence of endophytic fungi. They also analysed the effects of a member of the genus *Acremonium* in relation to attacks of *S. frugiperda* larvae to the tall fescue *F. arundinacea* and the red fescue *F. rubra* grown separately or mixed with other grasses. In general, insect attacks were reduced in hosts colonized by endophytes, because the plants were less chosen by the pests. However, different degrees of insect attacks were observed according to the plant associations analysed. These studies demonstrated that plant interactions wouldchange the effects of endophytic fungi in insect control.

Effect of endophytic entomopathogenic fungi on plants for pest control

It is widely known that Fungal species like *Metarhizium anisopliae, Beauveria bassiana* and others are often used in the biological control for agriculture insects-pest management. If these fungi inhabit plant tissues as endophytes, they will obviously be able to act in insect control. Even in situations where they are not present in the plant, artificial inoculation techniques may circumvent this obstacle. As discussed in this review that endophytic fungi that do not colonize insects and cause disease, may be used as a natural insect control as well as an effective insect control if they are artificially inoculated into plants and display characteristics against the pests. Unfortunately, works concerning this promising line of research are scarce and, when were conduced, involved a temperate climate system, as usual, where the pest control problem is not so urging as in plants. Clay (1989), in a review on the potential of insect control by endophytic fungi, stressed that insect biocontrol may be improved by the development of artificial inoculation techniques, as those developed by Latch and Christensen (1988). As endophytic fungi, notably members of the genus *Acremonium*, are able to control some insects-pests, it is important to know if they can be inoculated in endophytefree plants and, if they can transmit to the host the capacity to resist to certain pests. That was done with success by Koga *et al.* (1997), by infecting *F. arundinacea* and *L. perenne* seeds with the endophytic fungus *Acremonium* and, as a result, obtained plants resistant to the bluegrass webworm *P. teterrella*. It was observed that not only the inoculation worked well, but also the mutant fungus was kept in competition with wild type ones. These results demonstrated that inoculation of valuable strains for biological control will not only be used with success but, modified microorganisms, once inoculated into a host, will be able to coexist and even overcome wild type strains of the same species. An important point, at least in grasses, resides in the fact that several fungi species may be transmitted by seeds,where the biocontrol becomes able to be passed to the next.

Effects on the host plant physiology with impact on insect control

In certain plants, toxin production is not solely a fungal function. There are fungus-plant interaction factors acting on this production and toxins may be produced and dissipated throughout the host tissues. Evidently, it may result in insect control even if the endophyte is not present in the attacked sites. The development and nutrition of the moth *Plutella xylostella* larvae were studied in brassicas. Larvae were affected by the ingestion of leaves, before the endophytic fungus *Acremonium alternatum* reached these leaves. There were also differences in feed efficiency of males and females. Contrasting with their utility in insect control, toxin and mycotoxin produced by endophytic fungi may cause problems to cattle feeding on colonized plants. In *L. perenne* infected by *N. lolii*, peramine production plays a major role in controlling the Argentine stem weevil *L. bonariensis*. It was verified that with plant ageing, the amount of peramine decreases in leaves and reaches the lower levels during the inflorescence phase in eight fungus plant interactions (Ball *et al.* 1997a). Other seemingly simple interactions may occur like the increase of stroma induced by the endophytic fungus *E. typhina* in host plants, resulting in an increase in stroma-mediated transpiration and consecutively reduction in herbivorous insect attacks (White *et al.* 1993).

Molecular biology improved pathogenic properties of endophytes

The recombinant DNA technology has been applied as a tool to improve endophytic microorganisms, aiming to the introduction of new characteristics of agronomic interests as biological control of pests. Probably, Fahey (1991) described the first work directed at the introduction of a heterologous gene in an endophytic microorganism with the purpose of insect control. As a member of the biotechnological company Crop Genetics International, described the major steps in the construction of an endophytic bacterium with the purpose of insect control. This was achieved through the of an insecticidal toxin in the host plant. The genetically modified microorganism is able to secrete toxin inside the plant, protecting it against attacks of target insects. The author described several advantages of inoculating this bacterium into corn plants as a way of protecting it against insects: the product requires only one application in seeds or aspersion over young plants; there is no need for a large amount of the product because the bacterium multiplies after inoculation; the genetically modified bacterium stays restricted to the inner parts of the plant and thus, there is no dissemination to other plants; the process does not yield toxic compounds; this and other similar products require a shorter time to be technologically developed because it is easier to modify a microorganism than a plant; the modified microorganism is not transmitted to progeny seed of inoculated plants and therefore, from a commercial point of view, it must be acquired constantly; the process has wide applicability because it can be used for other plant species; bacteria multiplication is high inside the plant, the endophytic bacteria may be found inside hosts at high densities. Consequently, false positive results may be observed in plant transformation protocols involving, for instance, enzymatic activity detection. This was the case of Tor *et al.* (1992), while transforming plants of the genus *Discorea* with a vector expressing *Gus* (b - glucuronidase). Although, *Gus* activity was observed in putative plant transformants, this activity was due to endophytic bacteria inhabiting the plant material. The engineered bacterium exhibited insecticidal activity in artificial diets but not *in planta*. Lampel *et al.* (1994) used an improved integrative vector that, although showed some instability, resulted in toxin production *in planta* after all.These transformed bacteria were then inoculated in corn crops and showed to reduce damages caused by the European corn borer *Ostrinia nubilalis* (Tomasino *et al.* 1995). In a review on endophytic fungi from the genus *Acremonium*, Van-Heeswijck and McDonald (1992) launched the idea of using engineered endophytic fungi in the control of insects and diseases affecting *L. perenne*. Murray *et al.* (1992) introduced into a lolitrem B-deficient strain of *Acremonium* (designated 187B) a b –glucuronidase (*Gus*) gene. The transformation protocol yielded 700-800 transformants/mg DNA and transformants were successively introduced into *Lolium* where *Gus* activity was observed. In a similar manner, Tsai *et al.* (1992) developed a transformation system for *A. coenophialum*. Several genes from endophytic fungi and related to toxin production are now being cloned and studied in depth. Wang *et al.* (1999) cloned a gene coding for a DMAT Synthase (dimethyl tryptophane synthase) that is responsible for the first step in the synthesis of ergot by the fungus *Claviceps purpurea*. Genes related to the latter were identified in the endophytic fungi *Balansia obtecta* and *Neotyphodium* spp., known to produce, respectively, ergobalansine and ergovaline. Tudzynski *et al.* (1999) also cloned a gene (DMATS) of *C. purpurea* involved in the ergot synthetic pathway *in planta*. Young *et al.* (1999) found in *Epichloë* a cluster containing genes for the synthesis of mycotoxins like paxiline and lolitrem B. In our laboratory in the Escola Superior de Agricultura " Luiz de Queiroz", Universidade de São Paulo at Piracicaba, for some years we have been developing studies with endophytic microorganisms from banana, corn, citric plants, among others. Efficient transformation systems have been established for some of these endophytic microorganisms as well as protocols for reintroducing these microorganisms into plants.

CONCLUSION

Endophytic microorganisms were studied in more details only from the 80's onwards. Soon after that, they started to be recognized as being of great importance for the hosts, protecting the plants against pests, including among others, insects, nematodes and plant pathogenic fungi and bacteria. They also cause physiological modifications in their hosts, such as making them more resistant to hydric stress. The role of endophyes protecting plants against insects and diseases has been quite well studied. However, the research is based mainly on endophytes from hosts and, besides, the work has been concentrated in some genera of grasses. Apart from isolated studies, only more recently, efforts are being directed to endophytes from tropical plant hosts. This is a broad field of investigation that is almost entirely open to new findings. The results presented in this review show that, as expected, great diversity has been found among endophytes isolated from tropical hosts. They must, of course, play important roles as, for instance, protecting plants against pests and environmental stresses, as already found in temperate isolates of endophytes. Thus, it is expected that new ways of interactions between endophytes and their hosts will be certainly found. It is also expected that new biological agents produced by endophytes will be discovered with the increase of studies focusing on these fungi.

REFERENCES

Ahmad S Govindarajan S Funk CR and Johnson- Cicalese JM 1985 Fatality of house crickets on perennial ryegrass infected with a fungal endophyte *Entomologia Experimentalis et Applicata* **39**:183-190

Ahmad S Johnson-Cicalese JM Dickson WK and Funk CR 1986 Endophyte-enhanced resistance in perennial ryegrass to the bluegrass billbug *Sphenophorus parvulus Entomologia Experimentalis et Applicata* **41**:3-10

Arechavaleta M Bacon CW Hoveland CS and Radclife DE 1989 Effects of tall fescue endophyte on plant response to environmental stress *Agronomy Journal* **81**:83-90

Bacon CW and Hills NS 1996 Symptomless grass endophytes: products of coevolucionary symbioses and their role in the ecological adaptation of grasses In: Endophytic fungi in grasses and woody plants Redlin SC and Carris LM (edts) American Phytopathologycal Society Press St Paul pp 155-178

Ball OJP Barker GM Prestidge RA and Lauren DR 1997a Distribution and accumulation of the alkaloid peramine in *Neotyphodium lolli*-infected perennial ryegrass *Journal of Chemical Ecology* **23**:1419-1434

Ball OJP Miles CO and Prestidge RA 1997b Ergopeptide alkaloids and *Neotyphodium lolli*-mediated resistance in perennial ryegrass agains adult *Heteronynchus arator* (Coleoptera: Scarabaeidae) *Journal of Economic Entomology* **90**:1382-1391

Breen JP 1992 Temperature and seasonal effects on expression of *Acremonium* endophyte-enhanced resistance to *Schizaphis graminum* (Homoptera: Aphididae) *Environmental Entomology* **21**:68-74

Breen JP 1994 *Acremonium*-endophyte interactions with enhanced plant resistance to insects *Annual Review of Entomology* **39**:401-423

Bultman TL and Conard NJ 1998 Effects of endophytic fungus nutrient level and plant damage on performance of fall armyworm (Lepidoptera: Noctuidae) *Environmental Entomology* **27**:631-635

Bultman TL Borowicz KL Scneble RM Couldron TA and Bush LP 1997 Effect of a fungal endophyte on the growth and survival of two *Euplectrus* parasitoids *Oikos* **78**:170-176

Calhoun LA Findrlay JA Miller JD and Whitney NJ 1992 Metabolites toxic to spruce budworm from balsam fir needle endophytes *Mycological Research* **96**:281-286

Carroll G (1986) The biology of endophytism in plants with particular reference to woody perennials In: Microbiology of the Phylloplane Fokkema NJ and Van der Heavel J (edts) Cambridge University Press London UK pp 205-222

Carroll G 1988 Fungal endophytes in stems and leaves: from latent pathogens to mutualistic symbionts *Ecology* **69**:2-9

Clark CL Miller JD and Whitney NL 1989 Toxicity of conifer needle endophytes to spruce budworm *Mycological Research* **93**:508-512

Clay K 1988 Fungal endophytes of grasses A defensive mutualism between plants and fungi *Ecology* **69**:10-16

Clay K 1989a Fungal endophytes of grasses Their potential as biocontrol agents *Mycological Research* **92**:1- 12

Clay K 1989b Clavicipitaceous endophytes of grasses: their potential as biocontrol agents *Mycological Research* **92**:1-12

Clay K Marks S and Cheplick GP 1993 Effects of insect herbivory and fungal endophyte infection on competitive interactions among grasses *Ecology* **74**:1767- 1777

Clement SL Lester DG Wilson AD and Pike KS 1992 Behaviour and performance of *Diuraphis noxia* (Homoptera: Aphididae) on fungal endophyte-infected and uninfected perennial ryegrass *Journal of Economic Entomology* **85**:583-588

Clement SL Pike KS Kaiser WJ and Wilson AD 1990 Resistance of endophyte-infected plants of tall fescue and perennial ryegrass to the Russian wheat aphid (Homoptera: Aphidiae) *Journal of the Kansas Entomological Society* **63**:646-648

Davidson A W and Potter DA 1995 Response of plant-feeding predatory and soil-inhabiting invertebrates to *Acremonium* endophyte and nitrogen fertilization in tall fescue turf *Journal of Economic Entomology* **88**:367-379

Fahey JW Dimock MB Tomasino SF Taylor JM and Carlson PS 1991 Genetically engineered endophytes as biocontrol agents: A case study in industry In: Microbial Ecology of Leaves Springe Verlag New York pp 402-411

Findlay JA Buthelezi S Li GQ Seveck M and Miller JD 1997 Insect toxins from an endophytic fungus from wintergreen *Journal of Natural Products* **60**:1214-1215

Funk CR Halisky PM Johnson MC Siegel MR Stewart AV Ahmad S Hurley RH and Harvey IC 1983 An endophytic fungus and resistance to sod webworms: association in *Lolium perenne BioTechnology* **1**:189-191

Gaynor DL and Hunt WF 1983 The relationship between nitrogen supply endophytic fungus and Argentine stem weevil resistance in ryegrass Proceedings of the New Zealand Grassland Association **44**:257-263

Ju Y Sacalis JN and Still CC 1988 Bioactive flavonoids from endophyte-infected blue grass (*Poaampla*) *Journal of Agricultural and Food Chemistry* **46**:3785-3788

Kanda K Hirai Y Koga H and Hasegawa K 1994 Endophyte-enhanced resistance in perennial ryegrass and tall fescue to bluegrass webworm *Japanese Journal of Applied Entomology and Zoology* **38**:141-145

Kindler D Breen JP and Springer T 1990 Russian wheat aphid resistance in cool-season grasses *Journal of Economic Entomology* **84**:685-692

Lampell JS Canter GL Dimock MB Kelly JL Anderson JJ Uratani BB Foulke Jr JS and Turner JT 1994 Integrative cloning expression and stability of the *cryIA*(c) gene from *Bacillus thuringiensis* subsp *kurstaki* in a recombinant strain of *Clavibacter xyli* subsp *cynodontis Applied and Environmental Microbiology* **60**:501-508

Lasota JA Waldvogel MG and Shetlar DJ 1983 Fungus found in galls of *Adelges abietis* (L) (Homoptera: Adelgidae): identification within tree distribution and possible impact on insect survival *Environmental Entomology* **12**:245-246

Murray FR Latch GCM and Scott DB 1992 Surrogate transformation of perennial ryegrass *Lolium perenne* using genetically modified *Acremonium* endophyte *Molecular and General Genetics* **233**:1-9

Patterson CG Potter DA and Fannin FF 1992 Feeding deterrency of alkaloids from endophyte-infected grasses to Japanese beetle grubs *Entomology Experimental and Applied* **61**:285

Prestidge RA Zijpp S and Badan D 1984 Effects of Argentine stem weevil on pastures in Central Volcanic Plateau *New Zealand Journal of Experimental Agriculture* **12**:323-331

Richardson MD Cabrera RI Murphy JA and Zaurov DE 1999 Nitrogen form and endophyte-infection effects on growth nitrogen uptake and alkaloid content of Chewings fescue turfgrass *Journal of Plant Nutrition* **22**:67-79

Riedell WE Kieckhefer RE Petroski RJ and Powell RG 1991 Naturally-occurring and synthetic loline alkaloid derivatives: Insect feeding behaviour modification and toxicity *Journal of Entomological Science* **26**:122-129

Saha DC Johnson-Cicalese JM Halisky PM Heemstra MI and Funk CR 1987 Occurrence and significance of endophytic fungi in the fine fescues *Plant Disease* **71**:1021-1024

Sherwood-Pike M Stone JK and Carroll GC 1986 *Rhabdocline parkeri* a ubiquitous foliar endophyte of Douglas fir *Canadian Journal of Botany* **64**:1849-1855

Siegel MR Latch GCM Bush LP Fannin FF Rowan DD Tapper BA Bacon CW and Johnson MC 1990 Fungal endophyte-infected grasses: Alkaloid accumulation and aphid response *Journal of Chemical Ecology* **16**:3301-3316

Tomasino SF Leister RT Dimock MB Beach RM and Kelly JL 1995 Field performance of *Clavibacter xyli* subsp *cynodontis* expressing the insecticidal protein gene *cryIA* (c) of *Bacillus thuringiensis* against European corn borer in field corn *Biological Control* **5**:442-448

Tor M Mantell SH and Ainsworth C 1992 Endophytic bacteria expressing beta-glucuronidase cause false positives in transformation of *Dioscorea* species *Plant Cell Reports* **11**:452-456

Tsai HF Siegel MR and Schardl CL 1992 Transformation of *Acremonium coenophialum* a protective fungal symbiont of the grass *Festuca arundinacea Current Genetics* **22**:399-406

Tudzynski P Holter K Correia T Arntz C Grammell N and Keller U 1999 Genetics of alkaloid biosynthesis in *Claviceps purpurea*: evidence for a gene cluster *Fungal Genetics Newsletter* **46** (Supplement):140

Van-Heeswijck R and McDonald G 1992 *Acremonium* endophyte in perennial ryegrass and other pasture grasses in Australia and New Zealand *Australian Journal of Agriculture Research* **43**:1683-1709

Wang J Machado C Panaccione D and Schardl C 1999 Ergot alkaloid biosynthesis genes cloned from *Claviceps* and *Balansia Fungal Genetics Newsletter* **46** (Supplement):120

Webber J 1981 A natural control of Dutch elm disease *Nature London* **292**:449-451

White Jr JF Glenn A E and Chandler K F (1993) Endophyte-host associations in grasses XVIII-Moisture relations and insect herbivory of the emergent stromal leaf of *Epichloe Mycologia* **85**:195-202

Young C McMillan L and Scott B 1999 Molecular cloning of an indole-diterpenoid gene cluster from *Penicillium paxilii Fungal Genetics Newsletter* **46** (Supplement):121

9

DNA-DELIVERY METHODS TO RAISE TRANSGENICS

C P MALIK, SARITA BHURADIA AND AMAN VERMA

Genetic transformation occurs frequently in nature in prokaryotes. The transfer of genes from one organism to another is termed horizontal gene transfer (Richardson and Palmer, 2007). For example, bacteria can acquire virulence factors, as well as antibiotic resistance genes, which may lead to the breakdown of the efficacy of antibiotics. Horizontal gene transfers are rare in higher eukaryotes, but years ago scientists found that a pathogenic bacterium, *Agrobacterium*, could transfer genes from its genome to its plant hosts, where expression of the transferred genes caused crown gall disease (Chilton *et al.* 1977).

The development of *Agrobacterium*-mediated genetic transformation, and direct transformation by biolistics, *i.e.*, the high velocity delivery of DNA attached to metal particles, led to the first generation of transgenic plants and the rapid application of this technology to crop improvement.

GENETIC ENGINEERING AS A DRIVING FORCE FOR MODERN AGRICULTURE

Genetic engineering is a powerful tool for improving crop quality and productivity, and reducing information systems for biotechnology Plant Research labor and resource utilization of farming. For example, farmers saved up to an estimated 60% of costs for pest control by growing Bt (insect resistance) cotton in certain regions of the US in 1997, according to a USDA report (Price *et al.* 2003). The reduction of pesticide spray has many other advantages for both the environment and humans. Because of these and other benefits, the adoption of three primary GE crops (corn, cotton and soybean) for two traits, HT (herbicide tolerance) and Bt, has dramatically increased since the first commercial introduction in 1996. According to recent USDA survey the acreage of GE is shown in the Fig. 1 below.

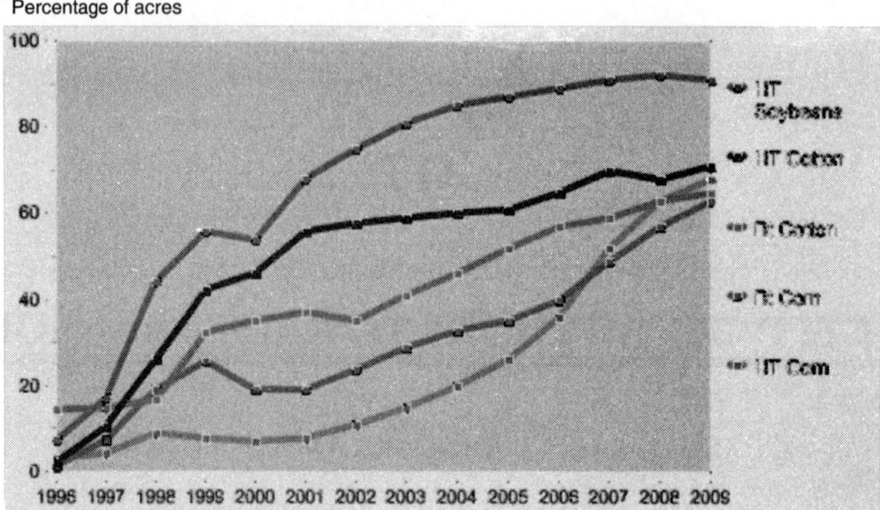

Data for each crop category include varieties with both HT and Bt (stacked) traits.
Sources: 1996-1999 data are from Fernandez-Cornajo and McBr'de (2002).

Fig. 1. Rapid growth in adoption of genetically engineered crops continues in the U.S.

Plant transformation is widely adopted as a technique to improve crop plant traits and to gain deep understanding how plants work. Successful plant transformation depends on the stable introduction of desirable transgene (s) into the genome of the plant. Several methods have been developed to accomplish this and many divergent crop plant species have been successfully transformed.

Two processes are involved to accomplish the genetic modification of plants and these comprise: the potential to transfer and integrate DNA into the genome of a host cell. The second process is the ability to regenerate fertile plants from the genetically transformed cell.

The first production of recombinant DNA molecule was accomplished in the early 1970's using restriction enzymes (Smith and Wilcox, 1970; Meselson and Yuan, 1968). Subsequently, the first stably transformed plants were reported in the early 1980's (Herrera-Estraella, 1983). Currently transgenic crops are increasingly becoming a dominant feature of the agricultural landscape of USA and other countries. In due course of time US Department of Agriculture animal and Plant Health Inspection Services permitted field trials of these engineered herbicide-tolerant varieties in the USA. With continued research transgenic plants have a pivotal role in crop improvement and are also being exploited in pharming industry (see Fig.1).

The first fertile transgenic monocot was developed through regeneration of transformed protoplast with naked DNA using electroporation or PEG techniques, including rice (Datt *et al.* 1990; Zhang *et al.* 1988; Zhang and Wu, 1988) and maize (Rhodes *et al.* 1988). For details refer to the reviews by Jones (2006) and Darbani *et al.* (2008).

Compared with traditional systems, plants have several advantages for the molecular pharming of pharmaceutuical proteins (Teli and Timko, 2004) and these are : low cost of production, rapid scalability, the absence of human pathogens and the ability to fold and assemble complex proteins accurately. Transgenic plants are widely used for large scale production of recombinant proteins like plantibodies and plantigenes.

As mentioned earlier, the genetic modification of plants depends on two main processes; first, the stable introduction of new genetic material into plant cells (transformation) and second, is utilization of tissue culture methods to regenerate transgenic plants from single transformed cell.

There are two types of gene delivery methods:

- Biological and
- Non-biological systems

Transgenic plant production process is a very complex and multistage procedure which involves three discrete phases:

- Delivery of exogenous DNA
- Stable integration of exogenous DNA into the genome of the host cells and
- Recovery of a viable transgenic plant

In the following we shall discuss the two systems:

BIOLOGICAL GENE TRANSFER

Agrobacterium-mediated transformation:

The first record of successful dicots transformation utilized oncogenic and disarmed strains of *Agrobacterium tumefaciens* and a range of explants types (Gasser and Fraley, 1989; Klee *et al.* 1987; Wordragen, and Dons, 1992). In the 1970's the rational gene transfer of exogenous DNA into crops using *A. tumefaciens* was revolutionary. However, regeneration for protoplast proved more difficult for many crops species and alternative more easily regenerable explants were developed. For instance, different explants were used in various species *e.g.* in tomato leaf discs (McCormick *et al.* 1986), in soybean cotyledon nodes (Hinchee *et al.* 1988) and immature cotyledons (Parrott *et al.* 1989) .

Agrobacterium was the logical and natural transformation candidate to consider since it naturally transfers DNA (T-DNA) located on the tumor inducing (Ti) plasmid into the nucleus of host plant cells and stably incorporated DNA into the plant genome (Chilton *et al.* 1977). In transgenic plant production methods the *Agrobacterium*-mediated transformation technique is the most widely used and powerful technique. However, many aspects remain to be unraveled *e.g.* genotype independent transformation of many economically important crop and forest species (Stanton, 2003; DeLa Riva *et al.* 1998). Through this method it is feasible to introduce large segments of DNA with minimal rearrangement and with fewer copies of inserted transgenes at high efficiency and with low cost (Gelvin, 2003; Gheysen *et al.* 1998; Hansen and Wright, 1999; Jones, 2006).

The development of *Agrobacterium* for plant transformation required the removal of oncogenic functions to generate disarmed strains and modification to permit the insertion of foreign DNA between the border sequences and to reduce the large size of the native Ti plasmid (which in *Agrobacterium* strain C-59 is 214233 bp (Goodner, *et al.* 2001). Two approaches have been used: one involves cis-integration of transgenes via homologous recombination or co-integration and in the other involves plant genetic engineering.

Agrobacterium-mediated transformations remain popular and the most effective among the most dicotyledons plants where *Agrobacterium* is naturally infectious.

Tissue culture/*in vitro* and in plants transformations: Transformation targets can be classified into two categories:

- Tissue culture and
- *in planta* methods

Tissue culture method utilizes the totipotency and plasticity of plant cells. This method depends on two principle routes by which plants can be recovered. These are somatic embryogenesis which is an asexual propagation process. Here somatic cells differentiate into embryos-like structures with shoot and root meristems. The other route comprises organogenesis in which the ability of some plant tissues *e.g.* hypocotyls, cotyledons, leaf bases or callus derived from them get reorganized into shoot meristems which can subsequently be rooted to generate complete plant (s). *In planta* method or germ line transformation methods depend on the ability to introduce naked DNA or T-DNA into gametes at the time of fertilization. In this way tissue culture is avoided. *Arabidopsis thaliana* can be transformed by several in plants methods including vacuum infiltration (Alough and Bent, 1998) transformation of germinating seeds (Feldmann and Marks, 1987) and floral dipping (Alough and Bent, 1998).

However, recently *Agrobacterium* inoculation of germinating seeds of rice has brought about transformation efficiency higher than 40%(Supertova *et al.* 2005) but ranged from 4.7 to 70% for the flower infiltration method and from 2.9 to 27.6% for the seedling infiltration method. For inoculation of manually impossible infection by viruses a method was developed called agro inoculations (Grimsley, 1990). The agro inoculation method invariably enhanced infection efficiency even when manual inoculation is possible (Hayes, *et al.* 1988; Biddens *et al.* 1989; Lesser *et al.* 1992). The optimization of agro infection –plant interaction is crucial for efficient transformation. Besides many other factors (the type of explants) are also important and they must be suitable to permit the recovery of whole transgenic plant (De la Riva *et al.* 1998; Frame *et al.* 2002; Zhang *et al.* 2003). The fundamental limitations of *Agrobacterium* based system led to the reduced ability of them to transfer DNA into monocots cells. Moreover, this system relies completely on the functioning of *vir*-proteins which target only the nucleic genome and not the organelle genome.

Viral-based transformation

Viral based transformation systems provide very high expression temporarily in plant cells. Potato virus X (Ruiz *et al.* 1998), barley stripe mosaic virus (Holzberg *et al.* 2002) and gemni virus vectors (Kjemtrup *et al.* 1998; Turnage et al,., 2002) are some of plant viral mediated transformants. They could be viral RNA or cDNA which can be used via dicot inoculation. This has proven to be the most efficient way of introducing cDNA-derived viral RNA into plants (see Jia *et al.* 2003 and references there in).Viral based transformation provide rapid and high level expression and this is the chief benefit of this technique. The main disadvantage of this technique is that they have less infectivity using average or large length transgene (Porta and Lomonossoff, 2002 Marillonnet *et al.* 2005).Other disadvantages are- large number of integrated transgene copies, chromosomal rearrangement, appearance of disease symptoms and low frequency of stable transformations.

Non-Biological Transformation Methods

Though *Agrobacterium* based transformations are preferred for most plant species there are some non-biological (NBM) based or direct gene transfer methods used chemically or physically for DNA uptake by the protoplast, intact cells or even whole plant. The NBM are potentially suitable for all species and all plants materials. They also do not require vector systems dependence. In the following we shall discuss some of the widely used NBM:

Particle bombardment/biolistic Method

The bombardment method was first described in 1987 for the production of transgenic (Sanford *et al.* 1987). Subsequently, the method has been found to be more useful for the transformation of recalcitrant cereals. A variation in the micro projectile transformation is an electric discharge Particle Acceleration Device (DPAD). Gene constructs for ballistic can be in the form of circular or linear plasmids or a linear expression cassette. Embryogenic cell cultures are likely the best explant to use for ballistic transformation because they can be spread out as uniform targets of cells and also they have high recovery potential (Kikkert *et al.* 2004). .Mannitol or sorbitol can be also effective as osmoticum in bombardment medium as they cause higher rates of stable transform ants for all suspension cultured cells (Kikkert *et al.* 2004). Particle bombardment has emerged as a reproducible method for rice transformation, in commercially important conifers species (*Picea glauca*), which was achieved via embryogenesis callus tissue as explants (Ellis *et al.* 1993), larch (Klimaszewska *et al.*1997) as well as *Pinus radiata* (Walters *et al.* 1998) and loblolly pines (Connett *et al.* 2002).

There are several **advantages** of this methodology compared with alternate propulsion technology and these have been discussed by McCabe and Christou, 1993, Darbani *et al.* 2008.

- The range of species and genotypes which can be engineered.

- High transformation frequencies.
- Elimination of cell death following hypersensitive responses to agro bacterium (Perl *et al.* 1996).
- No need to kill Agro bacterium after transformation.
- The use of linear expression cassettes called clean gene technology eliminates the possibility of plasmid backbone integration into the target (Fu *et al.* 2000).

Disadvantages

- The transformation efficiency might be lower than that with *Agrobacterium*
- It is more costly
- Intracellular targets are random and DNA is not protected.
- High frequency for complex integration pattern and multiple copy insertions that cause gene silencing and variation of transgene expression (Dai *et al.* 2001; Darbani *et al.* 2008). Efforts to reduce the incidence of recombination and reduction in the amount of DNA bombarded in each cell have led to the development of 'agrolistics'. This was developed to enhance efficiency of ballistics and yield simple integration patterns (Taylor and Fauquest, 2002). Agrolistics is a refinement to micro particle bombardment technology in which *vir* gene are co- bombarded.

Electrotransfection

Electroporation is the application of strong electric field pulses to cells and tissues and is known to cause some type of structural rearrangement of the cell membrane resulting in a temporary increase in porosity and providing a local driving force for ionic and molecular transport through the pores. In vitro introduction of DNA into cells is now the most common application of electroporation. There are several physical factors which are shown to affect the electrotranfection efficiency and these are created transmembrane potential by imposing pulse electric field, extent of membrane permeation, duration of the permeated state, mode and duration of molecular flow. Global and local concentrations of DNA forms of DNA and tolerance of cells to membrane permeation and the heterogeneity of the cells population (Hui,1995; Wearer, 1995).

Electroporation has several advantages over ballistic: it does not require the expensive particle gun apparatus associated consumable supplies and licensing and has worked well for stable transformations experiments (Shewry *et al.* 1994.). The efficiency of electroporation is relatively high: 6.25% for embryos and 54.6% for callus clusters, which is fully comparable with the eat results obtained for this species after micro bombardment (Deshayes *et al.* 1985). Despite apparent advantages of this technique, it is not used very often because of irreproducible results (sometimes!) and due to non-mendelian segregation of progeny. On an average this technique gives about 55% survival rate which up to 57% of the survivors expressed a marker gene (Songstad *et al.* 1995).

Silicon carbide fibers: This method is one of the recently developed methods to deliver DNA to plants. Physical and chemical characteristics 0f silicon carbide make them capable of penetrating the cells without killing them. There are several examples of SCMT success including maize (Kappler *et al.* 1992; Wang *et al.* 1995; Petalino *et al.* 2000), rice (Komatsu *et al.* 2006), wheat (Brisibe *et al.* 2000), tobacco (Kappler *et al.* 1990), and Agrostic stolonifera (Dalton *et al.* 1998). This system has several advantages and some are: rapid, inexpensive, easy to set up and effective on wide variety of cell types.

Among the disadvantages the main are: low rate of transformations,low efficiency, cell damage, inhalation of fibers as health hazards (Komalsu *et al.* 2006); Wang *et al.* 1995; Songstad *et al.* 1995). Liposome mediated transfection:

A derivation of PEG-mediated transfection is the liposome mediated transformations technique. Cationic liposome are positively charged lipids and are increasingly used for DNA uptake due to their favorable interaction with negatively charged DNA and cell membranes. In this approach foreign DNA must be encapsulated in a spherical lipid bilayer termed a liposome to prepare lipoplexes (Gad *et al.* 1990). After endocytosis (Fukunaga *et al.* 1983) the DNA is free to recombine and integrate into the host genome. Successful transfection based on the system was reported for tobacco (Dekeyser *et al.* 1990), wheat (Zhu *et al.* 1993 and potato Sawahel, 2002). The advantages of this method are that it is relatively non-toxic (Anonelli and Stadler, 1989), and is very simple to perform with readily available chemical reagents. This method is highly reproducible and efficient and requires no sophisticated apparatus (Antonelli and Stadler, 1990; Felgner *et al.* 1987). Wave and beam mediated transfection: Ultrasound for transfection has been widely used in animals cells and tissues. In this technique explants are suspended in a few milliliters of sonication medium in a microcentrifuge tube. Plamid DNA (and possibly carrier DNA) is then added and after rapid mixing the samples are ready for sonication. The cells are finally transferred to fresh growth medium, In this step, sound frequency and exposure time are determined for maximal uptake and efficiency (Lin *et al.* 2005). Ultrasound treatment increases membrane permeability (Bommannan *et al.* 1992) thereby felicitating the entrance of macromolecules into cells for transformation (Wyber *et al.* 1997). Stable transformations of tobacco by leaf pieces sonication was achieved via 1500 to 2000 fold longer time ultrasound treatment compared with protoplast sonication (Zhang *et al.* 1991).Another instance for intact tissue sonication based transformation is reported in potato tube discs subjected to ultrasonication (Scwahel, 1996). Sonication-assisted *Agrobacterium* mediated transformation (SAAT) is a new techno logy and the method involves subjecting the plant tissue to short periods of ultrasound in the presence of *Agrobacterium*. Trick and Finer (1993) reported that, in all tissues tested, the SAAT treatment greatly enhanced the level of transient expression. SAAT could be efficiently used for transformation of woody trees especially eucalyptus sp. (Monica *et al.* 2004). The method was newly developed and required further assessing for both of the different experimental conditions and plant species.

Injection based method: Microinjection is one of the most effective techniques of transforming animal cells and has been used for the transformation of plant cell as well (Banks and Evans, 1976). A pivotal feature of this technique is the use of low melting point (LMP) agarose both for holding protoplasts during microinjection and for their subsequent culture (Harwood and Davies, 1990). Several plant species such as tobacco (Schnorf *et al.* 1991), Petunia (Griesbach, 1987), oilseed rape (Neuhaus *et al.* 1987)and barley (Holm *et al.* 2000) have been transformed using microinjections. The injection of inheritable materials using a hypodermic syringe is called macroinjection. This is an approach to the delivery of exogenous DNA that could circumvent the expenses, time and genome stress associated with the transformation of cells in culture and subsequent plant regeneration. The technique has many advantages *e.g.* it is very precise from the delivery view point, is capable for the transformation of any protoplast or single cell system and has extremely high transformation efficiency. The chief disadvantage of this technique is the likelihood for the production of chimera plants with only a part of the plant having transformed. However, from this chimeric plant, transformed plants of single cell can be subsequently obtained. To overcome these problems pollen tube pathway (PTP) utilizing the normal fertilization cycle was developed to eliminate the regeneration phase and in this approach the stigma ia cut off and a drop of DNA solution is applied to the cut end of the style of recently pollinated plant florets. This procedure was used by Luo and Wa (1988) for the first time in rice. For more details a reference may be made to a review by Malik and Thind (1993). The procedure is extremely slow and requires expensive equipment and moreover it is very tedious. Transgene Integration and Gene targeting:

Molecular analyses of integration sites resulting from integrated DGT suggests that transgenes insert via double stranded illegitimate recombination, utilizing the plant's own DNA repair mechanism at a single cor sometimes more than one locus. The locus may contain one or more/multiple co[pies of the trangenes which may have undergone rearrangements and /or many have generated short length of "filler" DNA homologous to flanking plant genome DNA (Kohli *et al.* 1998; Suitashev *et al.* 2000; 2002).The processing transfer and integration of T-DNA into plant chromosomal DNA have been studied, both biochemically and in *Agrobacterium* strains containing mutant forms of the vir, chv, psc and Att alleles (for detailed reviews pl see Christie, 1997; Ward *et al.* 2002; Zupan *et al.* 1998; 2000). While the precise mechanism of T-DNA integration remains unclear, it is likely that DNA in the T-complex is made double-stranded just prior to integration *via* non-homologous end joining. The conventional way of overcoming the problem of transgene rearrangements high copy number or unintentional disruption of native genes is to generate a large number of lines and screen for single-copy events. However this unpredictability is a major concern for opponents of GM technologies, and gene targeting, by homologous recombination an alternative mofre precise method of genetic manipulation. It is commonly used in m ice in prokaryotes in the nuclear genomes of simple plants such as moss and the chloroplast genome (Capecchi 2001;

Evans *et al.* 2001; Smithies, 2001) but appears to operate at low frequency in higher plants. In over four million *Arabidopsis* plants transformed to induce two simultaneous mutations, only one was identified as a true gene targeting event without ectopic T-DNA insertions (Hanin *et al.* 2001). A more encouraging result was reported by Terada *et al.* (2002) who obtained sex rice plants in which the waxy gene had been disrupted by true homologous recombination with a frequency of 6.5 x 10^{-4}. With further development gene targeting technologies could provide a means through which any gene could be mutated and reintroduced into its original chromosomal location.

BENIGN METHODS OF SELECTION AND REMOVAL OF MARKER GENES:

Selectable marker genes are used routinely for research purposes, enabling highly efficient DNA cloning procedures in *E.coli,* yeast and plant transformation. However, perceived risks of horizontal and vertical gene transfer have made the use of such selectable genes undesirable especially where field trials are proposed and efforts have been made to develop benign selection systems or to reduce the dependence on selectable markers genes in the transformation process. The most widely adopted benign selection system utilizes the manA gene encoding phosphomannose isomerase (PM1) (Syngenta) which converts the predominant carbon source available to the plant, mannose-6-phosphate to fructose-6-phosphate for respiration (Joersbo, 2001; Joersbo *et al.* 1999). Not only are the untransformed cells deprived of a carbon source but also the unutilized mannose-6-phosphate accumulates and has additional negative effects including inhibition of glycolysis possibly due to phosphate starvation. PM1 selection method has been shown to be an effective selection system for crop species including wheat and maize (Negotto *et al.* 2000; Reed *et al.* 2001; Wang *et al.* 2001) and pearl millet (O'Kennedy *et al.* 2004).

In a small proportion of lines the co-bombardment of trait and marker genes on separate plasmids generates integrations into unlinked loci and the potential to identify progeny individuals that are null for the marker gene through segregation away from the trait gene. However, even in lines that have lost the plant selectable plasmids by segregation the bacterial selectable marker, origin of replication and plasmid by segregation, the bacterial selectable marker, origin of replication and plasmid backbone linked to the trait gene remain in the transgenic plant. In an attempt to overcome this, Fu *et al.* (2000) used DNA fragments comprising only the transgene expression cassette to transform rice biolistically. They found that, not only was the vector backbone eliminated from the plant line but also the proportion of low copy number structurally simple transgenic loci were enhanced. Purifying clean DNA fragment containing only the sequences necessary for transgene expression is not a trivial task and *Agrobacterium* T-DNAs should be inherently cleaned because in an ideal model, only the sequences delineated by the left and right borders are transferred.

TABLE 1 : Selected reports detailing the explant and A. tumefaciens strain/plasmid used in A. tumefaciens transformation of rice, maize, wheat and barley, PCE, pre-cultured immature embryos, EC, embryogenic callus cultures, E, fresh immature embryos.

Variety (S—Spring) (W—winter)	*Explant*	*Agrobacterium strain (vector)*	*Transformation Frequency (%)*	*Plant Selection*	*No. of plants reported*	*Ref.*
Wheat						
Bobwhite (S)	PCIE EC	C58 pMON18365	1.4-4.3	G418	>100	(Cheng *et al.* 1997)
Fielder (S)	PCIE	AGL0 pBGX1	1.8	GFP, Timentin, Bialaphos	4	(Weir *et al.* 2001)
Florida (W)	IE	AGL1 pAL154/156	0.3-3.3	PPT (c-phosphinothricin)	39	(Wu *et al.* 2003)
Cadenza (S)	IE	AGL1 pAL154/156	2.5	PPT (c-phosphinothricin)	5	(Wu *et al.* 2003)
Ceery-5 (S)	EC	LBA4404 pHK21	1.2-3.9	Glufosinate ammonium	17	(Khanna and Daggard, 2003)
Bobwhite (S)	PCIE, EC	C58 pMON18365	4.4	Glyphosate	3354	(Hu *et al.* 2003)
Bobwhite (S)	PCIE, EC	C58 pMON18365	4.8-19	Paromomcyin Glyphosate	154	(Cheng *et al.* 2003)
Bobwhite (S)	PCIE	C58 pPTN155	0.5-1.5	G418	13	(Haliloglu and Baenziger, 2003)
Barley						
Golden Promise (S)	IE	AGL1 pDM805	4.2	Bialaphos	5.4	(Tingay *et al.* 1997)
Golden Promise (S)	IE	AGL1 pBGXYNR: pBHXYNR	6.0	Bialaphos	24	(Patel *et al.* 2000)
Golden Promise (S)	IE	AGL1; AGL0 pWBVec82b	2-12	Hygromycin	63	(Mathews *et al.* 2001)
Schooner (S)	EC	AGL0; AGL1; LBA4404	0.5-2.9	Hygromycin; Biolaphos	19	(Wang *et al.* 2001)

		pTO134; pHPAVpol; pB2Pol				
Golden Promise (S)	PCIE	AGLO pBU-B3.5S.IG	1.7-6.3	Bialaphos	42	(Trifonova, Madsen and Olesen, 2001)
Golden Promise (S)	IE	AGLO pYF133	3.5	Hygromycin GFP	11	(Fang. Akula and Altpeter, 2002)
Golden Promise (S)	IE	H228 pNRG040	N/A	Bialaphos	42	(Horvath *et al.* 2003)
Golden Promise; Schooner; Chebec and Sloop (all S)	IE	AGL1 pWBVecB; pVecB-GFP; pVecB-Gusl	0.6-12.0	Hygromycin; GFP; GUS	34	(Murray *et al.* 2004)
Maize						
Funk's G90	Shoot apices	EHA1 TiBO542; pGUS3	N/D	Kanamycin; Carbenicillin	6	(Gould *et al.* 1991)
A118 and others	BC, IE	LBA4404 pSB131; pTOK233	5-30	PPT; Bialaphos	>100	(Ishida *et al.* 1996)
A118	IE	LBA4404 pNOV117	0.7-32	Mannose	>50	(Negrotto *et al.* 2000)
Hi II	IE	LBA4404 pSB11 Ubi bar/35SGUS	40	Bialaphos	>500	(Zhao *et al.* 2002)
Hi II	IE	EHA101 pTF102	5.5	Bialaphos	50	(Frame *et al.* 2002)
Hi II	IE	LBA4404 p108-p111, based on pSB11	0.50-25	Bialaphos	>100	(Gordon-Kamm *et al.* 2002)
A118, Hi II and hybrids	IE	LBA4404 pWCO38; pWCO39	N/D	Butafenecil; mannose	>2500	(Li *et al.* 2003)

Delivery method		*Transformation efficiency*	*Rang of transformable plant species*	*Tissue culture*	*Type of explant*	*Eqiupments*	*Others*
Agrobacterium-based method		High and stable	Many species, specially dicotyledonous	With and without tissue culture phase	Different intact cells, tissues or whole plant	–	Possibility of agroinfection, combination with sonication and biolistic methods and transgene size up to 150 Kb
Viral-based method		High and transient	Virus-host specific limitation	Without tissue culture phase	In-planta innoculation	–	Rapid, efficient PTGS, inducible expression and with mosaic status
Bombardment Biolistics based methods		High	Unrestricted	With and without tissue culture phase	Intact tissues or microspores	Very expensive apparatus	Efficient for viral infection complex integration patterns, without specialized vectors and backbone-free integration and transformation of YAC with insert up to 150 kb
	Agrolistic	High	Unrestricted	With and without tissue culture phase	Different intact explants or whole plant	Very expensive apparatus	Clean and precisely integration with reduced degraded transgene events
Electro-transfection	Electro-portation	Low to high	Unrestricted	With and without tissue culture phase	Protoplasts, meristens or pollen grains	+	Fast, simple and inexpensive in contrast with biolistics
	Electro-phoresis	Moderate	Unrestricted	Without tissue culture phase	Intact tissue on whole plant	–	Rapid and simple but yields shimeric
Lipofection		Low	Recoverable	With tissue	Protoplasts	–	High efficiency with combination of

			species from protoplast	culture phase			PEG based method, simple, non-toxic and positive correlation between efficiency and lipoplex size and highly reproducible
Silicon carbide mediated		Low to high	Unrestricted	With tissue culture phase	Variety of cell types	–	Rapid, inexpensive and easy to set-up
Injection based methods	Micro injection	High	Recoverable species from protoplast	With tissue culture phase	Protoplasts	Expensive micro-manipulator	Very slow, precise, single cell targeting possibility, requires high skill, the chimeric nature of transgenic plants and ability of whole chromoseome transformation
	Micro-injection PTP*	Low	Plants with sexual reproduction	Without tissue culture phase	Cametes	–	Rapid and simple
	Others	Low	Limited (with few species)	Without tissue culture phase	Immature embryos or meristems	–	Production of chimeric plants and damaging of cells
Wave or beam mediated transformation	Sonoporation	Low	Unrestricted	With and without tissue culture phase	Protoplast, cells, tissues and seedings	+	Effective to transfect by virus particles and able to increase the *Agrobacterium* based transformation efficiency
	Laser mediated	Low	Unrestricted	With tissue culture phase	Variety of cell types	+	Rapid and simple
Desiccation based method		Low	Unrestricted	With tissue culture phase	Rehydrated embryos	–	Simple

* Pollen-tube pathway, +; Necessary, –; Unnecessary

Other approaches to marker gene elimination involve the use of site specific recombinases or transpoases to excise or relocate the marker gene for elimination by segregation. These methods have proved successful in model dicot species (see review by Ebinuma *et al.* 2001; Hohn *et al.* 2001) and have been successfully applied in rice (Hoa *et al.* 2002) and maize (Zhang *et al.* 2003).

In a radically new approach to the removal of foreign DNA it has been demonstrated that analogs of the *Agrobacteriumn* T-DNA border sequences exist in plants (so called P-DNA) and that they can function to delimit T-strand synthesis and DNA transfer during the transformation process (Rommens *et al.* 2004).

MINICHROMOSOMES: THE NEXT GENERATION TECHNOLOGY FOR PLANT

The next generation – engineered minichromosomes

Despite many successes, genetic engineering has many limitations

- The technology used to create stacked gene var is not efficient. When stacking genes in inbred lines, the two genes located on different chromosomes will segregate at a ratio of 1:2:1 following first a genetic cross, and then self-pollinate to make each homozygous. This implies that one-sixteenth of the progeny will maintain the two homozygous stacked transgenes in the next generation. Gene stacking or pyramiding is also widely used in conventional crop breeding programs to combine multiple characteristic traits.
- Another difficulty is linkage drag, *i.e.* reduction in fitness of a cvr due to deleterious genes introduced along with the beneficial gene during backcrossing. This may occur because genes closely linked to the desired genes or transgenes are usually difficult to recombine during the transfer from one line to another. Hence several generations of backcrossing is usually required to introduce a good character gene to an elite cvr without linkage drag. This problem has created a bottleneck that limits the advancement of plant genetic engineering technology from the first generation of GE crops, characterized as 'input traits', to the next generation, characterized as 'output traits' (Halpin, 2005).

Recently Yu and co workers (Yu W *et al.* 2007.) have described an engineered minichromosome system that could be used as a genetic engineering platform for the expression of an almost unlimited number of genes intended to be inherited together as an independent unit. This engineered minichromosome technology should remove the bottlenecks mentioned above, and possibly lead to the second generation of genetic engineered crops.

Chromosome engineering and minichromosomes

Minichromosomes have been produced in yeast and mammalian cells through either *de novo* construction using the minimum constituent parts of chromosomes or telomere-mediated chromosomal truncation of existing Chromosome (Farr. 1991; Harrington *et al.* 1997) (Fig. 2).

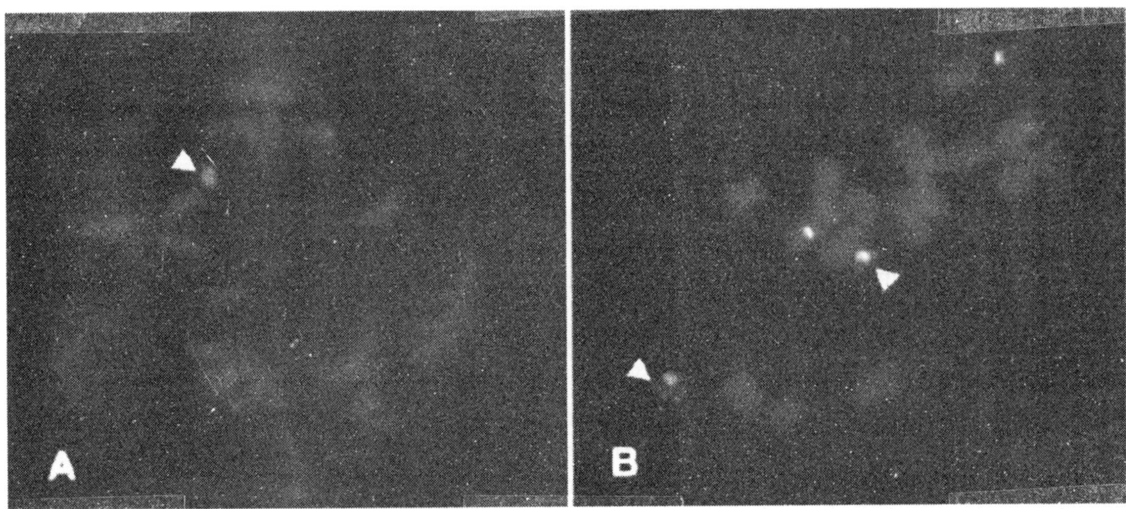

Fig. 2 Minichromosomes produced from maize B chromosome truncation. Arrows denote minichromosomes. B chromosome centromeres are labeled green with a B specific repeat probe. Transgenes carrying the truncating sequences and site specific recombination cassettes, as detected with the transgene probe, are red. Chromosomes are stained blue with DAPI. A) A minichromosome without the transgene (a copy of a stably integrated transgene is present on an A chromosome). B) Two minichromosomes with the truncating terminal transgenes (red) occurring in the same cell are shown, as well as two normal B chromosomes

However, these technologies for minichromosome construction were not successful in plants until recently when telomere-mediated chromosomal truncation in maize was reported, which causes chromosome fracture at the telomere sequence insertion site. An array consisting of 2.6 kb telomere repeats efficiently produced chromosomal truncations in maize. However, due to deficiencies in truncated chromosomes, recovery of chromosomes with large truncations is difficult to attain through the genetic transformation process. Moreover, genetic transmission of truncated chromosomes is unlikely; they are non-viable in the haploid gametophyte generation because they lack essential genes. However, the lethality of truncated chromosomes can be overcome either by using tetraploid materials in which the gametes will be diploid and the deficiency of truncated chromosomes would be compensated by additional copies of the homologs, or by using a non-essential chromosome, such as the maize B chromosome 5.

Yu *et al.* (2007) recovered a minichromosome formed by the truncation of chromosome 7 in a tetraploid plant. This minichromosome was transferred to a diploid background by repeated backcrossing and as stably maintained. By using the same set of constructs, they targeted the maize B chromosome with biolistic-mediated gene transformation. Truncated B chromosomes were recovered with much greater efficiency. The sizes of the mini B chromosomes ranged from very small to the size of the large B chromosome. These chromosomes from A and B chromosomes of maize were studied for their utility as utility as artificial chromosome for future genetic engineering in plants.

Attaching genes to minichromosomes

Minichromosomes generated by telomere truncation can have different configurations: those with transgenes and those without. This difference depends on the orientation of construct integration relative to the Centromere of the recipient chromosome, which is subsequently broken at the site of telomere integration. The transgene can segregate with either fragment of the broken chromosome. However, if the transgene is retained on the chromosomal fragment without a functional Centromere, which is essential for normal chromosome segregation during cell division, it will be lost after cell division. In this circumstance, minichromosomes lacking a transgene will be formed and could be selected if there are other transgenes in the same cell able to produce resistance to the selection agent in the transformation process (Fig. 2A). If the transgene is present on the fragment containing a centromere, it is then present on a minichromosome (Fig. 2B). In fact, both types of minichromosomes were recovered. Minichromosomes with attached transgenes were characterized in detail for mitotic and meiotic behaviors, transmission frequencies, gene expression, and the ability to use site-specific recombination.

All minichromosomes in the experiment were transmissible from generation to generation. They never paired with other chromosomes during meiosis, which minimized the chance of recombination with other normal chromosomes in the genome; thus all transgenes on the minichromosome can transmit as a unit. In addition, the research demonstrated that foreign genes could be expressed from maize B chromosomes, which otherwise are inert chromosomes with no essential genes (Fig. 3).

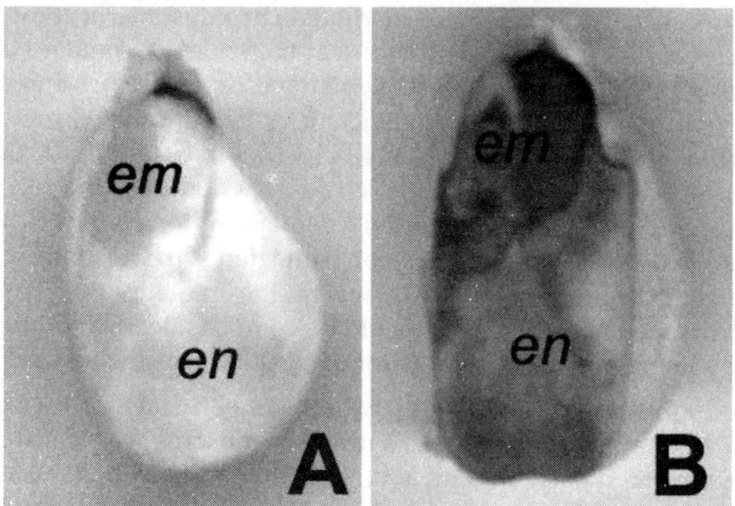

Fig. 3. GUS gene expression from a targeted gene on a miniB chromosome.

A) shows no GUS expression in the absence of the transgene. B) shows GUS gene expression in both embryo (em) and endosperm (en) of a mature kernel with a copy of the transgene on a miniB chromosome.

Minichromosomes from maize B chromosomes have other properties that make them exceptionally useful as artificial chromosome platforms. For example, mini-B chromosomes can accumulate to high copy numbers in the presence of normal B chromosomes, but remain stable in the absence of normal B chromosomes. This property might allow one to manage the copy number of the chromosome in order to increase gene expression from transgenes on the chromosome under appropriate circumstances.

Manipulation of genes on minichromosomes

Yu *et al.* (2007) also demonstrated that additional transgenes could be targeted to the minichromosome by site-specific recombination, which allowed the addition of subsequent genes. Yu *et al.*(2007) crossed a minichromosome containing a promoter less lox-DsRed gene, which encodes a red fiuorescent protein, with another transgenic plant that expresses Cre, the recombinase for Cre/lox recombination, in a 35S promoter-lox-Cre cassette. Recombination occurred between the two chromosomes at the lox site, thus activating DsRed gene expression, and at the same time placing all genetic elements distal to the 35Slox- Cre gene onto the minichromosome5. The demonstration of the Cre/lox recombination system in the minichromosome facilitates further manipulation of foreign genes onto minichromosomes. By using other site-specific recombination systems (Yu W *et al.* 2006. Ow *et al.* 2007; Wright *et al.* 2005), Yu *et al.* predict they will be able to add, delete, or replace genes on minichromosomes in order to redesign existing engineered chromosome platforms. This new technology has numerous advantages over the current genetic engineering technology.

As noted above, an immediate application of minichromosome technology is to enable stacking of genes involved with herbicide tolerance and pest resistances. Future developments could facilitate any application that requires the inheritance of multiple foreign genes as a unit. One can now contemplate adding entire biochemical pathways to plants to confer new properties or to synthesize novel metabolites in mass quantities. With mini-B chromosomes, the quantity can be increased to maximize.

CONCLUSION

Plant genetic transformation applications and the desire for higher transformation efficiency have stimulated work on inventing and improving various methods. These applications are of specific benefit to molecular genetic studies, crop improvement and production of pharmaceuticals materials. *Agrobacterium* based methods are usually superior for many species including dicots and monocots. For some angiosperms and non-angiosperms biolistic is usually used.

Biolistic is by far the most widely used direct transformation procedure both experimentally in research and commercially. In addition, it has proven to be the most efficient technique.

In the past two decades several other methods have emerged with the view point that: more efficient method might emerge; the methods may turn out to be less expensive. Moreover, biolistic and *Agrobacterium* methods and patented, and securing a commercial license is trivial.

Chloroplast targeting is one of the advantages for direct transformation since high level recombinant protein expression for chloroplast targeting. This technique has also some drawbacks like low transformation frequency, prolonged selection pressure and green part dependence for transgene products.

In plants transformation methods are immensely attractive: however, this has proven to be effective and efficient in *Arabidopsis.*

Even though plant transformation can no longer be described as a new technology, it continues to evolve and leading to more facile and efficient transformation.

REFERENCES

Antonelli N and J Stadler 1989 Chemical methods for efficient direct gene transfer to maize cells: Treatment with polyethylene glycol or polybrene J Genet Breed **43**: 113-122

Briddon RW J Watts PG Markham and J Stanley 1989 The coat protection of beet curly top virus is essential for infectivity Virology **172** (2): 628-633

Chilton MD MH Drummond DJ Merio D Sciaky AL Montoya MP Gordon and EW Nester 1977 Stable incorporation of plasmid DNA into higher plant cells: The molecular basis of crown gall tumorigenesis Cell **11** (2): 263-271

Dai S P Zheng P Marmey S Zhang W Tian S Cher RN Beachy and C Fauquet 2001 Comparative analysis of transgeneic rice plants obtained Agrobacterium-mediated transformation and particle bombardment Mol Breed **7** (1): 25-33

Darbani B A Eimanifar CN Stewart and WN Camargo 2007 Methods to produce market-free transgenic plants Biotechnol J **2** (1): 83-90

Dekeyser RA B Claes RMU De Rycke ME Habets MC Van Montagu and AB Caplan 1990 Transient gene expressions in intact and organized rice tissues Plant Cell **2** (7): 591-601

De la Riva GA J González-Cabrera R Vázquez-Padrón and C Ayra-Pardo 1998 *Agrobacterium tumefaciens:* A natural tool for plant transformation Elect J Biotechnol **1** (3): 118-113

Deshayes A L Herrera-Estrella and M Caboche 1985 Liposome-mediated transformation of tobacco mesophyll protoplasts by an *Escherichia coli* plasmids EMBO J **4** (11): 2731-2737

Farr C et al 1991 Functional reintroduction of human telomeres into mammalian cells *Proc Nati Acad Sci USA* **88** 7006-10 *Genet* 15 345-55

Feldmann KA and MD Marks 1987 *Agrobacterium* mediated transformation of germinating seeds of *Arabidopsis thaliana*: A non-tissue culture approach Mol Gen Genet **208** (1): 1-9

Felgner P T Gadek M Holm R Roman HW Chan M Wenz JP Northrop GM Ringold and M Danielsen 1987 Lipofection: A highly efficient lipid-mediated-DNA transfection procedure Proc Natl Acad Sci **84** (21): 7413-7417

Frame BR H Shou RK Chikwamba ZI Zhang CI Xiang TM Fonger SE Pegg B Li DS Nettleton D Pei and K Wang 2002 *Agrobacterium tumefaciens*-mediated transformation of maize embryos using a standard binary vector system Plant Physiol **129** (1): 13-22

Griesbach RJ 1987 Chromosome-mediated transformation via microinjection Plant Sci **50** (1): 69-77

Grimsley NH 1990 Agroinfection Physiol Plant **79** (1): 147-153

Halpin C 2005 Gene stacking in transgenic plants - the challenge for 21st century plant biotechnology *Plant Biotechnol J* **3**: 141-55

Harwood WN and DR Davies 1990 Protoplast Microinjection Using Agarose Microdrops In: Method in Molecular Biology Pollard JW and JM Walker (Eds) Vol 6 The Human Press pp: 323-333

Herrera-Estrella L 1983 Transfer and expression of foreign genes in plants PhD Thesis Laboratory of Genetics Genetics University Belgium

Holm PB O Olsen M Schnorf H Brinch-Pederse and S Knudsen 2000 Transformation of barley by microinjection into isolated zygote protoplasts Transgenic Res **9** (1): 21-32

Holzberg S P Brosio C Gross and GP Pogue 2002 Barley stripe mosaic virus-induced gene silencing in a monocot plant J **30** (3): 315-327

Hui SW 1995 Effects of Pulse Length and Strength on Electroporation Efficiency In: Methods in Molecular Biology Plant Cell Electroporation and Electrofusion Protocols Nickoloff JA Vol 55 Humana Press Inc: Totowa NJ pp: 29-40

Jefferson RA Assaying chimeric genes in plants: The genes fusion system Plant Mol Biol Rep **5** (4): 387-405

Jones HD 2006 Advances in transformation technology in plant Biotechnology: Current and future applications of genetically modified crops John Wiley and Sons London (ed) NG Halford Pp 71-90

Kaeppler H DA Somers HW Rines and AF Cockburn 1992 Silicon carbide fiber-mediated stable transformation of plant cells Theor Applied Genet **84** (5-6): 560-566

Kikkert JR 1993 The Biolistic® PDS-100/He device Plant Cell Tissue Organ Culture 33: 221-226

Kikkert JR JR Vidal and BI Reish 2004 Stable Transformation of Plant Cells by Particle Bombardment/Biolistics In: Methods in Molecular Biology Transgenic Plants: Methods and Protocols Peña L (Ed) Vol 286 Human Press Inc: Totowa NJ pp: 61-78

Kjemtrup S KS Sampson CG Peele LV Nguyen MA Conkling WF Thompson and D Robertson 1998 Gene silencing from plant DNA carried by a geminivirus Plant J **14** (1): 91-100

Lesser RM V Ziegler-Graff A Reutenauer E Herrbach O Lemaire H Guilley K Richards and G Jonard 1992 Agroinfection as an alternative to insects for infecting plants with *Beet western yellows luteovirus* Proc Natl Acad Sci **89** (19): 9136-9140

Liu Y H Yang and H Sakanishi 2005 Ultrasound: Mechanical gene transfer into plant cells by sonoporation Biotechnol Adv wwwsciencedirectcom

Luo ZX and R Wa 1988 A simple method for the transformation of rice via pollen-tube pathway Plant Mol Biol Rep **6** (3):165-174

Marillonnet S C Thoeringer R Kandzia V Klimyuk and y Gleba 2005 Systemic *Agrobacterium tumefaciens*-mediated transfection of viral replicons for efficient transient expression in plants Natl Biotechnol **23** (6): 718-723

Meselson M and R Yuan 1968 DNA restriction enzyme from *E coli* Nature 217 (5134): 1110-114

Monica TV A Carlos and ER Gonzales 2004 Method for genetic transformation of woody trees European Patent EP1448777 2004 http://wwwfreepatents onlinecom/ EP1448777html

Neuhaus G G Spangeberg O Mittelsten-Scheid and HG Schweizer 1987 Transgenic rape seed plants obtained by microinjection of DNA into microscope derived embyoids Theor Applied Genet **75** (1): 30-36

Ow DW 2007 GM maize from site-specific recombination technology what next? *Curr Opin Biotechnol* **18** :115-20

Paszkowski J RD Shillito M Saul V Mandák T Hchn B Hohn and I Potrykus 1984 Direct gene transfer to plants EMBO J **3**: 2717-2722

Porta C and GP Lomonossoff 2002 Viruses as vectors for the expression of foreign sequences in plants Biotechnol Genet Eng Rev **19**: 245-291

Price GK et al 2003The size and distribution of marker benefits from adopting agricultural biotechnology *US Department of Agriculture*

Richardson AO & Palmer JD 2007 Horizontal gene transfer in plants *J Exp Bot* 58

Ruiz MT O Voinnet and DC Baulcombe 1998 Initiation and maintenance of virus-induced gene silencing Plant Cell **10** (6): 937-946

Sawahel W 1996 Ultrasound-mediated stable transformation of potato tuber discs Biotechnol Tech **10**: 812-824

Sawahel WA 2002 The production of transgenic potato plants expressing human U-interferon using lipofection-mediated transformation Cell Mol Biol Lett **7** (1): 19-29

Schaefer DJ 2002 A new moss genetic: Targeted mutagenesis in *Physcomitrella patens* Annu Rev Plant Biol **53**: 477-501

Shewry PR A tatham NG Halford JHA Barker U Hamnapel P Gallois M Thomas and M Kreis 1994 Opportunities for maniuplating the seed protein composition of wheat and bareley in order to improve qulity Transgenic Res **3** (1): 3-12

Smith HO and KW Wilcox 1970 A restriction enzyme from Hemophilus influenzae I Purification and general properties J Mol Biol **51** (2): 379-391

Songstad DD DA Somers and RJ Griesbach 1995 Advances in alterntive DNA delivery techniques Plant Cell Tissue Organ Culture **40** (1): 1-15

Stanton BG 2003 Agrobacterium-mediated plant transformation: The biology behind the gene-jockeying too Microbiol Mol Biol Rev **67** (1): 16-37

Taylor NJ and CM Fauquet 2002 Microparticle bombardment as a tool in plant science and agricultural biotechnology DNA Cell Biol **21** (12): 963-977

Teli NP and MP Timko 2004 Recent developments in the use of transgenic plants for the production of human therapeutics and biopharmaceuticals Plant Cell Tissue Organ Culture **79**: 125-145

Trick HN and JJ Finer 1997 SAAT: Sonication-assisted Agrobacterium-mediated transformation Transgenic Res **6** (5): 329-336

Turnage MA N Muangsan CG Peele and D Robertson 2002 Geminivirus-based vector for gene silencing in Arabidopsis Plant J **30** (1): 107-114

Yu W et al 2007 Construction and behavior of engineered minichromosomes in maize *Proc Natl Acad Sci U S A* **104** 8924-9

Wang K P Drayton B Frame J Dunwell and JA Thompson 1995 Whisker mediated plant transformation: an alternative technology *In Vitro* Cell Dev Biol Plant **31** (2): 101-104

Weaver J C 1995 Electroporation Theory In: Methods in Molecular Biology Plant Cell Electroporation and Electrofusion Protocols Nickoloff JA (Ed) Vol 55 Humana Press Inc: Totowa NJ pp: 3-28

Wright DA *et al* 2005 High-frequency homologous recombination in plants

Yu W et al 2006 Telomere-mediated chromosomal truncation in maize *Proc Natl Acad Sci USA* **103**: 17331-6

Zhang W S Subbarao P Addae A Shen C Armstrong V Peschke and L Gilbertson 2003 Cre/lox mediated gene excision in transgenic maize (*Zea mays* L) plants Theor Applied Genet **107** (7): 1157-1168

Zhu Z B Sun C Liu G Xiao and X Li 1993 Transformation of wheat protoplasts mediated by cationic liposome and regeneration of transgenic plantlets Chin J Biotechnol **9**: 257-261

10

NITRIC OXIDE (NO): A VERSATILE GASEOUS HORMONE

AMAN VERMA, C. P. MALIK, VARSHA CHATURVEDI AND V. K. GUPTA

INTRODUCTION

Umpteen research papers continue to flood the scientific journals with insights into the biological activity and potential clinical uses of nitric oxide (NO). The organization of redox signaling and the use of NO to transmit information, modulate biological processes or create cellular damage are highly complex. Recent reports provide an exceptional picture of NO production, regulation of NO bioactivity through detoxification reactions and of biochemical events through which NO transduces signals into cellular responses, especially during disease resistance. Furthermore, other exciting reports on NO function are in seed germination, seedling growth and reproduction support the view that NO is a 'do it all' molecule that plays a crucial role during the entire lifespan of the plant.

NO is seemingly emerging as an important endogenous signalling molecule in plants, mediates many developmental and physiological processes including xylogenesis, programmed cell death, pathogen defence, flowering, stomatal closure, and gravitropism (Lamattina *et al.* 2003; Neill *et al.* 2003; Delledonne, 2005; Lamotte *et al.* 2005). Experimental evidence in support of such signalling roles for NO has typically been obtained *via* the application of either NO or NO donors (NO itself is a reactive gas with a short half-life in air), *via* the measurement of endogenous NO and through the manipulation of endogenous NO content by chemical and genetic means. There are potential complications with using NO donors (Floryszak-Wieczorek *et al.* 2006) and undoubtedly technical problems associated with assaying the NO content of and release from plants (Planchet and Kaiser, 2006*a*, *b*). Moreover, in some situations, NO can be released in far higher amounts than would probably be required to effect biological responses which raises the question of how it can actually function as a biological signal. NO also have paradoxical effects. For example, it is growth promoting at low concentrations, but quite inhibitory or toxic at high concentrations (Beligni and Lamattina, 1999) and being reactive is perhaps unlikely to travel far between or even within cells. It may be that in the rush of enthusiasm to ascribe biological roles to NO some problems have been overlooked and with hindsight some of the experimental data may require re-evaluation.

This review focuses on NO evolution and its role in plants and perhaps, raises more questions than answers. The subsequent review (Verma *et al.* in press) shall discuss NO metabolism and its divergent roles in plants *e.g.* seed germination and seedling growth, cell growth and differentiation; organogenesis; flowering; pigment synthesis; Cell metabolism (photosynthesis, respiration, function of peroxisomes and senescence); We shall also discuss NO and its cross-talk with different plant hormones (IAA, GA, Cytokinin, ABA, Ethylene salicylic acid) as well as calcium. In recent years role of NO in overcoming different stresses has also been demonstrated and these include drought stress, salinity and temperature stress, heavy metal stress and UV-radiation and ozone stresses. The role of NO in biotic stresses shall also be included.

WHAT IS NO?

Until recently NO was known principally as a hazardous molecule that caused pollution and acid rain, and that destroyed the ozone layer and damaged ecological environments. After receiving negligible biological research interest for decades, the unique biological attributes of NO were finally becoming evident, and NO was chosen as "Molecule of the Year" by Science magazine in 1992. In 1998, the researchers Ferid Murad, Louis J. Ignarro, and Robert Furchgott were awarded the Nobel Prize recognizing their pioneering work with NO (The Nobel Prize in Physiology or Medicine 1998). While examining the effect NO generation and removal in plants of various chemicals on the artery wall, Furchgott observed the secretion of a substance in endothelium with a shorter activity half-life than acetylcholine (Furchgott RF, Zawadzki JV, 1980). Meanwhile Murad *et al.* (1979) reported that cGMP levels increased via activation of the enzyme guanylate cyclase (GC) during neurodilator- induced relaxation of smooth muscle cells. Based upon the convergence of findings by Furchgott (1988), the substance which caused relaxation of artery muscles was named "endothelium- derived relaxing factor" (EDRF). Four years after the adoption of this term, it became evident that EDRF was in fact NO when Murad, Ignarro and colleagues (1987, 1979), while working with nitrodilators and NO, discovered that these substances caused cGMP-related vasodilatation in smooth muscle tissue. The Nobel Prize recognizing NO research brought the NO molecule great attention in basic and clinical biomedical sciences. Today it is known that many cell types synthesize NO and that it mediates many physiological processes, such as neurotransmission, immunity, resistance, and regulation of the cell-cell communication.

Following this sudden surge of NO into biological research, its divergent effects in both animals and plants were examined. Previously the NO-related studies in plants were focused on its involvement in atmospheric pollution. NO release in plants and its effects on plant growth were defined for the first time in the 1970s [Anderson and Mansfield (1979); Klepper (1979)]. Subsequently, NO-related studies in plants were limited to those by a few pioneering researchers such as Leshem and Haramaty (1996) and Laxalt *et al.* (1997), until a remarkable rise in NO and plant biology studies [Beligni and Lamattina (2001a; b); Del Rio *et al.* (2004); Durner and Klessig (1999); Graziano and Lamattina (2005); Neill, *et al.* (2003); Neill *et al.* (2002b)] occurred when NO was demonstrated to play a role in plant defense mechanisms.

CHEMISTRY OF NO

NO is a colorless and gaseous molecule, which can easily diffuse through cell membranes due to its low molecular density and lipophilic quality. Furthermore, NO can act as a physiological messenger that interact strongly with other molecules and has a half life of just a few seconds (Durner and Klessig 1999; Neill *et al.* 2002a). As its electron structure is suitable for electron transfer, there are three inter-convertible forms of NO: the highly reactive, uncharged free radical ($NO^{\cdot}$) with an unpaired electron, the nitrosonium cation (NO^+) and nitroxyl anion (NO^-). $NO^{\cdot}$ rapidly reacts with O_2 to form NO_2, and rapidly breaks down into nitrite and nitrate in an aqueous solution. The highly reactive and destructive ion peroxynitride ($OONO^-$) is produced when $NO^{\cdot}$ reacts with superoxide (O_2^-) or H_2O_2. $NO^{\cdot}$ also reacts with iron and iron-containing proteins, thus forming iron-nitrosyl complexes. The highly promiscuous reactivity of NO and its ability to easily cross cell membranes make it a good signaling molecule in plants. NO^+ mediates electrophoretic attack on reactive sulfur, oxygen, nitrogen and aromatic carbon centers, with thiols being the most reactive group. Nitrosation of SH (*i.e.* S-nitrosation) centers of many enzymes or proteins have been described and such modification are reversible and protein-S-nitrosation↔denitrosation represents an important mechanism for regulation of signal transduction.

NO BIOSYNTHESIS

It is pivotal to elucidate the mechanisms by which NO is biosynthesized in plant cells. However, despite all the research efforts over the last 10 years or so, there remains much to be known. Most work has focused on two potential enzymatic sources of NO in plants, nitric oxide synthase (NOS) and nitrate reductase (NR), but recent research has also alluded to other potential sources of NO in different compartments of plant cells (Fig. 1). Prior to the complete sequencing of the *Arabidopsis* genome, early work on NO signalling in plants used pharmacological inhibitors of NO-generating enzymes to indicate the potential source of NO. In addition to non-specific inhibitors of NR, these inhibitors included compounds such as NG-nitro-L-arginine methyl ester (L-NAME) and NG-monomethyl-L-arginine acetate (L-NMMA), analogues of arginine expected to function as competitive inhibitors of NOS. Tungstate, that probably replaces molybdenum in the NR enzyme, has also been used as a potential NR inhibitor. The inhibition of physiological responses such as programmed cell death, stomatal closure or root growth by these compounds suggested that either NOS or NR were likely to be sources of endogenous NO (Neill *et al.* 2003). Early work by del Rio and colleagues (2002) used immunogold labelling to indicate that NOS-like enzymes were present in pea peroxisomes, but the cloning of a pea homologue of NOS was not reported.

Further, the peroxisomal NOS activity was biochemically characterized and arginine-dependent NO accumulation measured by chemiluminescence and electron paramagnetic resonance (EPR) spectroscopy (Corpas *et al.* 2004a). In addition, constitutive NOS activity, which appeared to be developmentally regulated, was detected in the leaves, stems, and roots of pea seedlings (Corpas *et al.* 2006). More recently, arginine-dependent, salinity-induced increases in NOS activity have also been demonstrated in olive (Valderrama *et al.* 2007). What still remains to be achieved, however, is the identification of genes encoding the enzymes responsible for these activities.

The first genetic evidence for a NOS-like enzyme in plants came from the work of Crawford and colleagues who identified an *Arabidopsis* orthologue of mammalian NOS, named AtNOS1 (Guo *et al.* 2003). The encoded protein, AtNOS1, had similarity to one from a snail that was possibly involved in NO synthesis. Importantly, AtNOS1 was shown to possess the biochemical characteristics of NOS in that it reduced arginine to citrulline when assayed with a commercial NOS-assay kit. More importantly still, a T-DNA insertion mutant, AtNOS1, was identified that produced much reduced levels of NO in guard cells and roots in response to ABA. Collectively, these data strongly suggested that AtNOS1 was truly a source of NO in *Arabidopsis.* Indeed, other studies also confirmed that the mutant, AtNOS1, was deficient in NO synthesis and action (He *et al.* 2004; Zeidler *et al.* 2004 and Bright *et al.* 2006).

However, the function of AtNOS1 as NOS has now been seriously questioned, with three parallel reports in 2006 discussing the problems associated with this concept. It now appears that AtNOS1 may not actually be a NOS at all. Certainly, it has been difficult to demonstrate typical NOS activity. Various researchers have been unable to reproduce the results of the earlier work and detect citrulline when using the arginine-to-citrulline conversion kit and working with either AtNOS1 or related enzymes from other species (Zemojtel *et al.* 2006). Other tests to detect NO arising from the activity of this enzyme have also failed (Crawford *et al.* 2006). The current view is that, although AtNOS1 may not be a NOS *per se*, it is somehow involved in NO synthesis or accumulation. Hence, a name change to *Arabidopsis thaliana* Nitric Oxide Associated 1 (AtNOA1) has been suggested (Crawford *et al.* 2006). AtNOS1, or as know called as, AtNOA1, has a conserved GTPase domain and because it is probably targeted to the mitochondria (Guo and Crawford, 2005), it has been speculated that it may be a GTPase involved in mitochondrial ribosome biogenesis. Presumably impaired AtNOA1 activity would then be expected to result in impaired mitochondrial function and thus, altered NO levels (Zemojtel *et al.* 2006). However, GTPase activity is yet to be shown for AtNOA1. Moreover, Guo (2006) argues that the lack of detection of NOS activity from

AtNOA1 and its homologues could be due to the fact that their NAD (P) H-dependent activity may be very low compared with that of their mammalian counterparts. It was also suggested that AtNOA1 may use the stable NO synthesis intermediate, N-w-hydroxyarginine (NOHA), rather than arginine to produce NO. Such an intermediate would not be detectable by the traditional NOS assay. In addition, it has been suggested that there may be other co-factors yet to be identified in plants which could be required to regulate AtNOA1 activity (Guo, 2006).

Realistically, these recent developments lead us back to square one: no plant *NOS* gene has yet been identified. They also highlight a major caveat in the field of NO research in that the traditional NOS assay kits that have been used for mammalian enzymes may not be appropriate for plants. Additional methods such as EPR spectroscopy and chemiluminesence to detect arginine-dependent NOS activity would be suitable alternatives (Corpas *et al.* 2004*a*). Nevertheless, the substantial pharmacological data resulting from the use of NOS inhibitors to inhibit biological responses and NO production do indicate that there must be enzymes in plants that are thus affected. By definition, these enzymes are NOS enzymes, but how they use arginine to make NO remains to be seen. Possibly a plant NOS will have little sequence similarity with its mammalian counterpart, but will still contain domains which allow its redox functions to occur. Such domains could even be located on different polypeptides which could be brought together following a signalling event. If such a scenario were to be correct, the identification of the NO-producing enzyme in plants could be more difficult than previously thought, but one may assume that the enzyme would have a redox function and contain binding sites for redox prosthetic groups such as flavin or haem.

Another enzymatic source of NO is NR. The primary function of the NR family of enzymes in plants is one of nitrogen assimilation by converting nitrate to nitrite. However, NR can also convert nitrite to NO *via* a NAD (P) H-dependent reaction. This was shown originally *in vivo* using mutant NR soybean plants (Dean and Harper, 1986), but has also been shown *in vitro* using purified NR and plant extracts (Rockel *et al.* 2002; Neill *et al.* 2003). The first genetic evidence of a physiological role for the generation of NO by NR was in ABA-induced stomatal closure in *Arabidopsis* (Desikan *et al.* 2002). NR uses nitrite as a substrate to generate NO. The inhibition of NR activity with tungstate inhibited both ABA and nitrite-induced stomatal closure and prevented NO generation (Desikan *et al.* 2002; Bright *et al.* 2006), thus implying that a NR-like enzyme did play a role in generating the NO normally required to produce these responses. Importantly, it has also been possible to show the inhibition of purified NR activity *in vitro* using tungstate (Bright *et al.* 2006). *Arabidopsis* contains two NR genes, *NIA1* and *NIA2*, which have a high degree of coding sequence similarity and result in two isoforms which are 83.5% identical at the amino acid level, but which show some localized areas of sequence divergence in the first 90 N-terminal amino acids and in various other regions within the two proteins. Use of the *Arabidopsis nia1nia2* NR double mutant confirmed a role for one or both of the encoded enzymes in guard cell responses to ABA (Desikan *et al.* 2002). The observation that nitrite did not induce NO generation in *nia1nia2* guard cells suggests that this requirement for NR reflected its *in vivo* capacity to produce NO from nitrite.

However, as pointed out by Crawford (2006), a lack of NR may have several effects on plant nitrogen metabolism and indeed Modolo *et al.* (2006) have reported that the arginine content of *nia1nia2* leaves is substantially reduced. Recent work using single NR mutants indicates that NIA1, which is usually present at a much lower abundance than NIA2, is the source of NO during ABA signalling (Bright *et al.* 2006). These data also suggest that the aberrant NO biology in *nia1nia2* is due specifically to the lack of NIA1 as opposed to generally aberrant nitrogen metabolism. Interestingly, Yu *et al.* (1998) similarly concluded that NIA1 and NIA2 have distinct signal transduction and nitrogen assimilatory roles. NO generation by NR is stimulated by hypoxic conditions and in spinach and maize NR-mediated NO generation can be modulated by the phosphorylation status of the NR (Rockel *et al.* 2002). Thus, a potential regulatory mechanism may exist *in vivo*. Increasing endogenous nitrite concentrations, either by dark treatment or by antisense-inhibition of endogenous nitrite reductase activity (Morot-Gaudry *et al.* 2002 and Rockel *et al.* 2002), increases NO emission. NR-mediated NO generation has also been demonstrated in roots with a potential physiological role that of mediating aerenchyma formation, having been suggested (Dordas *et al.* 2003).

As shown in tobacco, mitochondrial reduction of nitrite to NO can also be a major source of NO with tissue nitrite concentrations being a major limiting factor and NR function obligatory (Planchet *et al.* 2005). However, it is not clear whether or not this occurs in both leaves and roots (Gupta *et al.* 2005 and Modolo *et al.* 2005). In addition, soybean chloroplasts have recently been identified as a source of NO *via* arginine or nitrite (Jasid *et al.* 2006). However, in this case, the enzymes regulating both arginine and nitrite-dependent NO formation are not yet known. Apoplastic, non-enzymatic conversion of nitrite to NO at low pH has also been demonstrated in the barley aleurone layer (Bethke *et al.* 2004).

A plasma membrane-bound, root-specific enzyme, nitrite-NO oxidoreductase (Ni-NOR), may also function as a further source of NO. This enzyme was identified biochemically *via* its NO-generating activity. However, unlike NR, it does not use NAD (P) H as a cofactor, but uses cytochrome *c* as an electron donor *in vitro* and has a comparatively reduced pH optimum. However, neither its physiological role nor its genetic identity is yet known (Stohr and Stremlau, 2006).

Other enzymes may also be involved in NO production (Corpas *et al.* 2004b). For example, in animals, xanthine oxidoreductase (XOR), under hypoxic conditions, can produce NO in preference to H_2O_2 (Millar *et al.* 1998). However, Planchet and Kaiser (2006b) were unable to observe any NO production from recombinant xanthine oxidase. Interestingly, Arnaud *et al.* (2006) demonstrated a plastid-located, iron-induced NO burst in *Arabidopsis* that, although susceptible to inhibition by L-NAME, required neither AtNOS1 nor NR. Such novel NO sources await characterization.

HOW TO GET RID OF NITRIC OXIDE

It is likely that biologically active molecules such as NO are rapidly removed or metabolized following initial signalling events. It is also possible that increased rates of

NO accumulation or emission actually reflect reduced rates of removal rather than increased generation. Thus, the importance of determining how NO levels are controlled is of prime importance (Fig. 1B). Simple chemical reactions are often responsible for the removal of NO from solution. Nitric oxide is inherently unstable and will readily react with oxygen to form nitrite and nitrate. As described above, nitrite can act as a precursor to NO and may have some biological activity (Gladwin *et al.* 2005).

The free radical nature of NO implies that it will readily react with other radicals that might also be present. In both animals and plants, NO is often produced at the same time and in the same place as Reactive Oxygen Species (ROS) such as superoxide anions. Superoxide and NO will react in a stoichiometric manner to produce peroxynitrite ($ONOO^-$). Although it has been noted that plant cells, unlike animal cells, appear resistant to peroxynitrite (Delledonne *et al.* 2001), it may have intrinsic signalling properties. Whether or not this turns out to be the case, the level of NO can be instrumental in controlling ROS levels in cells and *vice versa*. It has been noted that the basal rates of NO production in leaves are often under-estimated because the NO reacts rapidly with superoxide anions (Vanin *et al.* 2004).

NO reacts readily and reversibly with either thiol groups in the cysteine residues of proteins or with the tripeptide glutathione (GSH) and protein S-nitrosylation may be a key facet of NO signalling. Glutathione concentrations are typically 2–3 mM in plant cells (Ball *et al.* 2004) and thus, formation of S-nitrosylated glutathione (GSNO) could have a large impact on the concentration of free NO. GSNO is metabolized by the enzyme GSNO reductase (Diaz *et al.* 2003; Fig. 1B) and this enzyme may be instrumental in controlling the bioavailability of NO and the formation of protein S-NO groups, thereby regulating such NO-regulated processes as, for example, plant pathogen defence responses (Feechan *et al.* 2005).

In addition to reacting with thiol groups, NO can also interact with transition metals, particularly iron which is often associated with haem as in guanylyl cyclase and it has long been recognized that NO can react with haemoglobins. Non-symbiotic haemoglobins (nsHbs) from barley, alfalfa, and *Arabidopsis* are known to react with NO resulting in its removal from solution. Nitrate is formed in a NAD(P)H-dependent reaction with the oxidized haem intermediate being rereduced by either the NAD(P)H or FADH or, as in the case of the barley haemoglobin, by methaemoglobin reductase (Perazzolli *et al.* 2006). *Arabidopsis* AHb1 is also S-nitrosylated (Perazzolli *et al.* 2004). Interestingly, nsHbs are induced by certain treatments where NO generation might be enhanced, for example, by low partial pressures of O_2 (Trevaskis *et al.* 1997) or by nitrate or nitrite and by NO itself (Wang *et al.* 2000; Ohwaki *et al.* 2005; Shimoda *et al.* 2005; Sasakura *et al.* 2006). Transgenic manipulation of AHb1 affects NO evolution, which, correlated with the ability to survive hypoxic stress (Perazzolli *et al.* 2004), indicates a physiological role for AHb1 in modulating NO levels. Further evidence for the endogenous NO-detoxifying action of Hb comes from work involving plant–microbe interactions. Boccara *et al.* (2005) showed that HmpX, an *Erwinia chrysanthemi* flavohaemoglobin and virulence determinant, removed NO. Infection with an HmpX-deficient mutant of *E. chrysanthemi* triggered high levels of NO coupled to the hypersensitive response in the host plant. Sasakura *et al.* (2006) showed that an nsHb in the actinorhizal plant *Alnus firma* is highly expressed in nodules and may serve to detoxify NO.

NO PERCEPTION

Undoubtedly plants perceive and respond to NO, the mechanisms by which such perception occurs still seek clarification. There is now considerable research interest concerning this question, but as no specific plant NO receptor has been identified, work in this area has taken its lead from mammalian research. The reactive nature of NO and its ability to interact with and modify many proteins suggests that there may turn out to be many 'NO perceptors' (Fig. 2). In animal cells, soluble guanylyl cyclase (sGC) has a key role in NO signalling. NO activates sGC by binding to its haem domain stimulating a transient rise in cGMP levels which, in turn, activates a number of targets. In plants, pharmacological studies using inhibitors of NO sensitive guanylyl cyclase have implicated cGMP downstream of NO and ABA signalling in guard cells (Neill *et al.* 2003). NO induces an increase in cGMP (Durner *et al.* 1998) and laboratory findings has shown that application of ABA or the NO donor SNP to guard cell-enriched preparations from *Arabidopsis* induces a small and transient increase in cGMP that can be prevented by the application of the GC inhibitor 1H-[1, 2, 4]-oxadiazolo-[4, 3-a] quinoxalin-1-one (ODQ) or the NO scavenger PTIO. Thus, a similar mechanism of NO stimulated cGMP synthesis may also operate in plants. A key signalling molecule downstream of cGMP is cyclic ADP-ribose (cADPR) (Wendehenne *et al.* 2001). In animal cells cADPR stimulates Ca^{2+} release *via* intracellular ryanodine receptor calcium channels (RYR) and it is possible that a similar signalling mechanism operates in plants. In tobacco, cADPR elevates the expression of the genes encoding phenylalanine ammonia lysase (PAL) and the pathogenesis-related protein 1 (PR-1) in a manner that is sensitive to RYR inhibitors (Durner *et al.* 1998). These genes are also NO-regulated and cADPR antagonists reduce the expression of PR-1 (Klessig *et al.* 2000). NO is known to cause increases in the level of free Ca^{2+} (Durner *et al.* 1998; Garcia-Mata *et al.* 2003). Thus, NO may signal through cGMP, cADPR, and Ca^{2+} to promote its effects. NO, cGMP, and cADPR have all also been shown to mediate ABA-induced stomatal closure (Neill *et al.* 2003; Garcia-Mata and Lamattina, 2002). The cPTIO inhibition during this process of the ABA-induced inactivation of the Ca^{2+} dependent inward rectifying K^+ channel and activation of the outward rectifying Cl^- channel (Garcia-Mata *et al.* 2003) strongly implicates NO and Ca^{2+} in the signalling cascade that may operate.

Mammals, vertebrates, insects, and many lower eukaryotes possess sGCs with a haem domain capable of binding NO. Bacterial sGCs contain a similar NO binding domain termed the H-NOX domain (Karow *et al.* 2004; Boon *et al.* 2006). However, plant homologues of the animal NO sensitive sGC have yet to be identified. The *Arabidopsis* guanylyl cyclase, AtGC1, is apparently not activated by NO (Ludidi and Gehring, 2003). Thus, the question remains as to how and by what signalling process NO induces a rise in the level of cGMP in plants and it may be that plant enzymes that generate cGMP in response to this gas are quite different from their mammalian counterparts. Indeed, a recent report has demonstrated that the *Arabidopsis* brassinosteroid receptor BRI1 contains a domain with guanylyl cyclase activity (Kwezi *et al.* 2007) indicating that there may well be more novel plant guanylyl cyclases awaiting discovery.

The redox chemistry of NO facilitates its reaction with iron–sulphur and haem groups which are present in a number of different proteins. In addition, NO may also signal its presence through other mechanisms such as either direct S-nitrosylation or indirect *trans*-nitrosylation of either protein cysteine residues or low molecular weight compounds such as glutathione or *via* peroxynitrite nitration of tyrosine residues (Fig. 2; Mur *et al.* 2006). In animals, S-nitrosylation has been shown to regulate a number of signalling processes, stuctural proteins, and metabolic pathways and has become established as the prototype redox-based, post-translational protein modification in the animal kingdom (Wang *et al.* 2006).

In plants, evidence is now beginning to emerge that S-nitrosylation may also play an important role in NO signalling. A number of proteins became S-nitrosylated when extracts of *Arabidopsis* cell cultures were treated with GSNO and SNO-containing proteins isolated by the biotin switch method (Lindermayr *et al.* 2005). The proteins identified were involved in a wide range of cellular processes. However, their *in vivo* S-nitrosylation and its biological significance remain to be seen. The *in vitro* activity of one of three recombinant methionine adenosyl transferase isoforms has been shown to be altered by S-nitrosylation in a manner dependent on the presence or absence of Cys-114

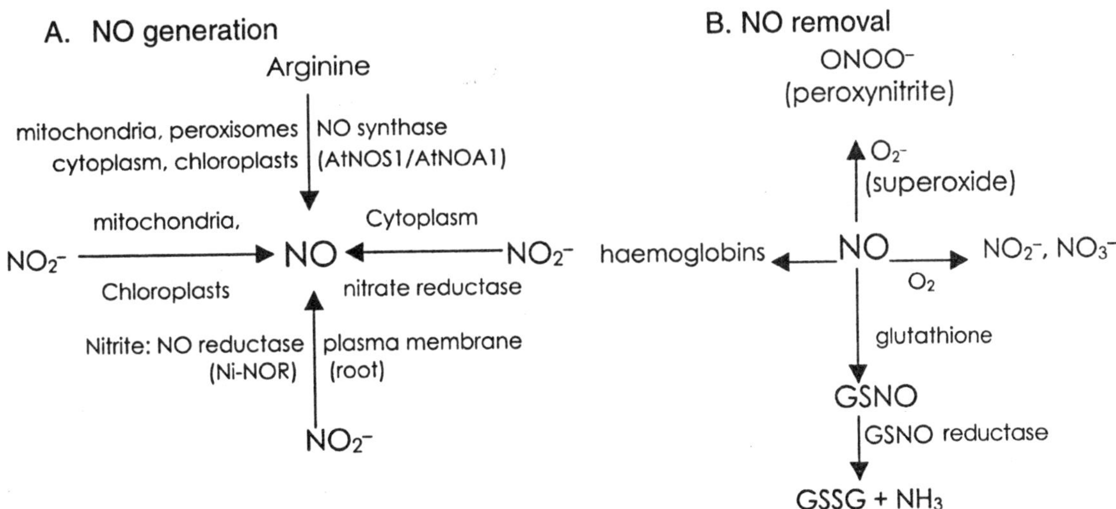

Fig. 1. NO generation (A) and removal (B) in plant cells. NO can be generated from arginine by NO synthase enzymes; AtNOA1 is required for this process in *Arabidopsis*. NO can also be produced from nitrite (NO_2^-) by nitrate reductase (NR), by nitrite:NO reductase (Ni-NOR) and *via* mechanisms yet to be fully defined in chloroplasts and mitochondria. NO reacts readily with oxygen and superoxide. NO is metabolized by reaction with haemoglobins and reacts with glutathione to form *S*-nitrosoglutathione (GSNO). GSNO is catabolized by GSNO reductase.

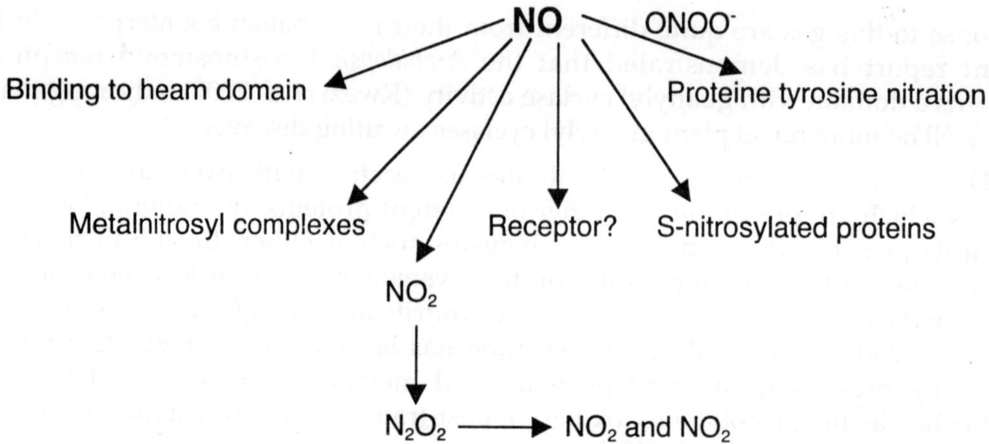

Fig. 2. Potential mechanisms of NO perception. NO may be perceived in plants by a number of mechanisms that differ depending on the cell type, intracellular location, biochemical microenvironment, and environmental stimuli. NO can bind to the haem domain in proteins such as guanylate cyclase and with metals to form metal-nitrosyl complexes. It can also react with the SH group of low molecular weight thiols such as glutathione to form *S*-nitrosoglutathione (GSNO) and, either directly or via GSNO, nitrosylate proteins to form S-nitrosylated proteins. S-nitrosylation induces conformational changes and is reversible. NO reacts with superoxide to form peroxynitrite which can then nitrate proteins on tyrosine residues. It is not yet known whether this reaction has signalling consequences.

(Lindermayr *et al.* 2006). Similarly, the activity of an *Arabidopsis* metacaspase appears to be dependent on the nitrosylation of a critical cysteine residue (Belenghi *et al.* 2007). Under lowered partial pressures of O_2 the mammalian RyR1 calcium channel also becomes S-nitrosylated on a specific Cys residue at physiologically relevant NO levels and in a manner dependent on the presence of calmodulin (Eu *et al.* 2000).

Should this occur in plants, it would have obvious relevance in terms of NO signalling and responses and there is some work suggesting that this may be the case in stomatal guard cells (Sokolovski and Blatt, 2004).

As protein S-nitrosylation can be mediated by GSNO, formed by the S-nitrosylation of GSH (Wang *et al.* 2006), the degree of protein S-nitrosylation and thus, 'NO activity', will be reflected in the availability of reactive GSNO. An *Arabidopsis* GSNO reductase, AtGSNOR1, has now been identified and its biological importance highlighted (Sakamoto *et al.* 2002; Diaz *et al.* 2003; Feechan *et al.* 2005). Loss-of-function mutations of this gene increased S-nitrosylation levels and disabled R (Resistance)-gene related defence responses against microbial pathogens (Feechan *et al.* 2005). Conversely, gain-of-function mutants were enhanced in their defensive ability. It was demonstrated that AtGSNOR1 positively regulated the signalling network controlled by the plant immune system activator, salicylic acid. In pea, both GSNO reductase activity and gene expression are decreased by cadmium stress (Barroso *et al.* 2006). Thus, there is definitely a case for S-nitrosylation being involved in signalling pathways which may include that for NO. Obviously the study of S-nitrosylation in plants is in its infancy and much work is required to determine on which specific proteins it occurs *in vivo* during the different physiological processes regulated by NO. Various proteins S-nitrosylation motifs have also been suggested based on the appropriate regions of animal proteins that are known to be affected. Wang *et al.* (2006) suggested the motif [HKR]-C-[VILMFWC]-x-[DE] as that targeted by NO and the motif [GSTCYNQ]-[KRHDE]-C-[DE] has been suggested as that targeted by GSNO. Scanning the *Arabidopsis* protein databases with these motifs yields 231 and 241 hits, respectively. While the proteins identified include a number of MAP kinases and other signalling proteins, none of those identified as being S-nitrosylated by Lindermayer *et al.* (2005) but are present in the lists of proteins generated. Thus, there is probably no substitute for laboratory-based investigations in this case. However, a number of potential, bioinformatically-generated targets could be examined in transgenic mutant complementation experiments where the highlighted Cys is either present or absent for nitrosylation in the complementing protein.

It is also possible that NO signals *via* the nitration of tyrosine residues. Tyrosine nitration is mediated by reactive nitrogen species such as the peroxynitrite anion ($ONOO^-$) and nitrogen dioxide (NO_2) which are formed during the metabolism of NO in the presence of oxidants such as superoxide radicals ($O_2^{\cdot-}$), hydrogen peroxide (H_2O_2), and transition metal centres (Radi, 2004). Although the peroxynitrite anion can cause tyrosine nitration *in vitro*, its role in this process has been questioned and alternative mechanisms have come to the fore that depend on the formation of NO_2 by the action of haem peroxidases on nitrite (Brennan *et al.* 2002). A number of recombinant *Arabidopsis* haemoglobins that exhibit peroxidase-like activity and differentially mediate nitrite-dependent protein nitration *in vitro* have been identified (Sakamoto *et al.* 2004). Endogenous protein tyrosine nitration has also been demonstrated, in mutant tobacco plants with greatly increased amounts of NO (Morot-Gaudry *et al.* 2002) and more recently in olive leaves where the amount of tyrosine nitrated proteins increased under salt stress (Valderrama *et al.* 2007). Thus, the extent and biological significance of protein nitration and whether or not what appears to be a non-reversible reaction can act as a signalling process, presumably in tandem with protein turnover, remains to be determined.

NO MOVEMENT

It is possible that NO can diffuse within a cell from a specific site of generation, say in the mitochondria, to other regions of the cell where it might induce an effect by interaction with specific target proteins. It is also possible that NO can diffuse out of the cell across the plasma membrane into adjacent cells and thereby create a small region of cells responding to NO. However, whether or not NO does diffuse within and between cells and if it does, how far it moves remains unknown. Given that cells clearly contain many proteins and other molecules that react with NO, it might be that such diffusion is limited. This could of course be the case unless the NO concentration were to be sufficiently high, not necessarily across the whole cell, but perhaps in a microlocale within the cell, so as to saturate, transiently at least, such 'NO-binding molecules' in its immediate vicinity. This would leave non-reacted NO free to diffuse across and out of cells. It is likely that cellular regions do have higher local NO concentrations either because they contain the biochemical machinery required for NO synthesis or because NO accumulates preferentially in such regions. For example, NO is more soluble in lipid than water and so may accumulate preferentially in membranes where its rates of reaction with any interacting molecules may be consequently higher (Liu *et al.* 1998). Different stimuli may activate NO synthesis either by different mechanisms and/or in different subcellular compartments and there may also be a directional focus. For example, a bacterium or fungal hypha may about only one region of a plant cell and the resultant signalling might activate NO generation in only a proximal and discrete region of the cell. If NO response proteins such as ion channels or second messenger-generating enzymes are also co-located then one could envisage local 'NO hot-spots' and NO signalling micro-domains. Although it may well be technically difficult to monitor NO transport, it may be informative to apply NO *via* a point source to the exterior or interior of a tissue and then to monitor real-time NO movement by, for instance, fluorescent imaging using a NO-sensing dye such as DAF-2DA.

An alternative, but not exclusive scenario for NO transport might be that its generation is elevated in discrete regions owing to localized stimulation resulting from the long-distance transport and site-specific accumulation of compounds such as the hormones ABA or IAA that can stimulate its production. Directional transport of IAA is well-known, particularly with respect to its role in mediating tropic responses to gravity and light. NO has been implicated in gravity signalling with localized NO accumulation being induced either by gravistimulation or asymmetric IAA application and prevented by the inhibition of IAA transport (Hu *et al.* 2005). The systemic transport of 'defence-signals' is activated by pathogen challenge and it may be that these signals also stimulate NO generation at sites distant from those of the initial pathogen perception.

Another possibility awaiting clarification is that NO precursors or 'NO storage compounds' may be transported with either NO generation or release occurring at distant sites in a manner analogous to the transport of the ethylene precursor ACC. GSNO has been suggested as one transportable form of NO and although it has not yet been unequivocally identified in plants, a recent report demonstrated cross-reactivity with an anti-GSNO antibody in pea collenchyma cells and immunofluorescence microscopy indicated that the GSNO content decreased dramatically under cadmium stress (Barroso *et al.* 2006). GSNO has recently been demonstrated in leaf vascular tissue and shown to increase under salt stress using confocal laser microscopy (Valderrama *et al.* 2007). Glutathione is also present at high (*e.g.* millimolar levels in wheat) concentrations in phloem cells (Bourgis *et al.* 1999). Arginine and nitrite could also serve as transported NO precursors. Nitrite concentrations in the phloem and xylem are unknown. However, whole tissue nitrite concentrations, which are typically 10–20 µM, can be transiently raised above this level (Rockel *et al.* 2002) which may be indicative of the movement of NO precursors. Arginine concentrations can be quite high (*e.g.* 250 µM in *Arabidopsis* leaves (Modolo *et al.* 2006) and 300–800 µM in melon phloem) and, interestingly, can be increased by ABA treatment (Mitchell and Madore, 1992).

NO GENERATION FROM PLANTS

There can be no doubt that NO is evolved from plants. Such evolution, measured as NO_x (a mixture of NO and NO_2), was first reported in the 1980s and shown to be increased by treatment with salicylic acid and various other compounds (Harper, 1981; Dean and Harper, 1986; Klepper, 1990). Wildt *et al.* (1997) measured NO emissions from several species and several other reports have demonstrated that NO evolution from plants can increase or decrease in response to treatments such as pathogen challenge, water stress, exposure to UV-B, the application of fungicides, and anoxic conditions (Lesham and Haramaty, 1996; Clarke *et al.* 2000; Magalhaes *et al.* 2000; Rockel *et al.* 2002; Hari *et al.* 2003; Conrath *et al.* 2004; Perazzolli *et al.* 2004; Mur *et al.* 2005, 2006). However, there are a number of technical and biological uncertainities with these measurements and quite varied rates of NO evolution have been estimated using a variety of different measuring techniques. Of course, it is also difficult to gauge the biological significance of the NO evolved. Typically, the rates of evolution are in the nmol g^{-1} h^{-1} range (Table 1). A key question is whether or not this NO evolution reflects increased concentrations of biologically active NO *in planta*. In some cases, this seems likely to be the case. For example, during responses to pathogen challenge there are good correlations between NO evolution and the biological responses that occur. Such correlations have been shown using either NO scavengers and NO synthesis inhibitors or virulent and avirulent pathogens. However, as pointed out by Planchet and Kaiser (2006*a*, *b*), estimates of intracellular NO content and rates of NO evolution do not always agree. For example, during anoxia NO is generated at a much higher level than is probably required for its cell signalling function. Such paradoxes remain to be resolved, but the concept of localized NO generation and action may partly explain them.

TABLE 1 : Some estimates of the rates of NO evolution from plants

Examples	*Level of NO ($nM\ g^{-1}FW\ h^{-1}$)*	*Treatments*	*Assay method*	*Reference*
Arabidopsis cultures	30–120	Control; *P. syringae*-challenged	Haemoglobin assay	Clarke *et al.* (2000)
Arabidopsis plants	0.1–1	Air; N_2	Chemiluminescence assay	Perazzolli *et al.* (2004)
Arabidopsis plants	3–17	Control	Gas chromatography	Magalhaes *et al.* (2000)
Tobacco leaves	5–20	Control; *P. syringae*-challenged	Laser photoacoustic detection	Mur *et al.* (2005)
Arabidopsis rosettes	5–30	Control; *P. syringae*-challenged	Laser photoacoustic detection	Mur *et al.* (2006)
Soybean cultures	12–600	Control; fungicide-challenged	Mass spectrometry	Conrath *et al.* (2004)
Spinach leaves	0.5–20	Air; N_2	Chemiluminescence assay	Rockel *et al.* (2002)

It is not known if NO released from one part of a plant can induce effects on either other parts of the same plant or on adjacent plants. Agricultural soils can also release substantial amounts of NO (Davidson and Kingerlee, 1997). Certainly, NO gas does have effects on plant biology. Indeed, the early work on NO concerned its effects as an air pollutant (Mansfield, 2002). NO is still an air pollutant today and plants in urban areas or close to traffic are likely to receive higher chronic and more acute exposures than those in rural areas (rural locations 7–70 nl l–1, urban 20–900 nl l–1; Environment Agency, 2006). The early work showed that NO at 50–500 nl l–1 (urban smog [NO] ≤5000 nl l–1) could retard growth and inhibit photosynthesis (Mansfield, 2002) and other studies have shown that trace amounts of NO in smoke can stimulate seed germination (Keely and Fotheringham, 1997).

NO INDUCED STOMATAL CLOSURE

Exogenous NO, supplied in the form of NO donors, was first shown to induce stomatal closure and reduce transpiration in species such as *Vicia faba*, *Salpichroa organifolia*, and *Tradescantia spp.* by Garcia-Mata and Lamattina (2001). Such findings were confirmed in epidermal peels or leaves of pea (Neill *et al.* 2002), *Arabidopsis thaliana* (Desikan *et al.* 2002), *etc*. Many different NO donors induce stomatal closure in a dose- and time-dependent manner, and their effects can be reversed by simultaneous co-incubation with the NO scavengers 2-phenyl-4, 4, 5, 5-tetramethylimidazoline-1-oxyl-3-oxide (PTIO) or 2-(4-carboxyphenyl)-4, 4, 5, 5-tetramethylimidazoline-1-oxyl-3-oxide (cPTIO) (Bright, 2006). Many such plant studies have used the NO donor sodium nitroprusside (SNP) that also generates cyanide. However, neither cyanide itself nor light-inactivated SNP induce stomatal closure (Bright, 2006), and cPTIO does not react directly with cyanide (Bethke *et al.* 2006). Moreover, the reaction product of cPTIO and NO, cPTI, which has some biological activity per se (Planchet and Kaiser, 2006b), does not inhibit ABA-induced guard cell movements. Thus, the effects mediated by NO donors such as SNP must indeed be due to the release and biological activity of NO.

The requirement for endogenous NO during ABA-induced stomatal closure has been demonstrated by both pharmacological and genetic approaches that either remove NO or prevent its generation and correlate stomatal responses with estimates of its accumulation. It has been shown that pea and *Arabidopsis* guard cells generate NO in response to ABA and that their treatment with PTIO both removes NO and inhibits ABA-induced stomatal closure. Similar guard cell responses have been shown in *V. faba* (Garcia-Mata and Lamattina, 2002). The treatment of *Arabidopsis* leaves with either L-NAME or tungstate, compounds that potentially inhibit NO generation from L-arginine or nitrite, inhibits guard cell NO production and stomatal closure (Bright *et al.*2006). Moreover, in the AtNOS1 and *nia1nia2* mutants, both ABA-induced stomatal closure and ABA-induced NO generation are impaired (Desikan *et al.* 2002).

Much of the aforementioned work has used the NO-sensitive fluorescent dye 4, 5-diaminofluorescein diacetate (DAF-2DA) and related compounds to detect the presence of NO. However, as highlighted recently by Kaiser and colleagues (Planchet and Kaiser, 2006*a*, *b*), DAF-2DA can generate art factual fluorescence arising from its reaction with

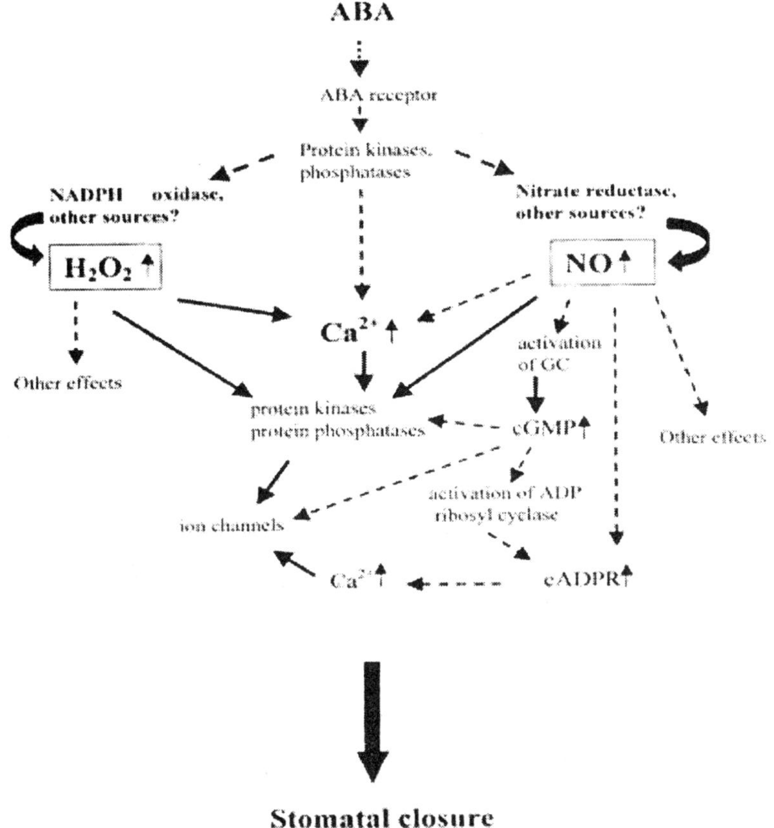

Fig. 3. The pattern and process of the ABA-induced signal transduction in plant cells

compounds other than NO. Nonetheless, the use of appropriate control dyes such as 4-aminofluorescein diacetate (4AF-DA), that do not react with NO, and the NO scavenger PTIO allows the conclusion that NO is truly being detected in guard cells. The use of other NO assays such as electron paramagnetic resonance spectroscopy or photoacoustic laser detection is difficult within the guard cell system. However, it has been possible to demonstrate ABA-enhanced NO generation in pea epidermal peels using the haemoglobin assay (Neill *et al.* 2002). Thus, information from various sources does demonstrate that ABA-induced stomatal closure requires NO generation and action (Fig. no. 3). Recent reports indicate that guard cells also generate NO in response to elicitor challenge (Melotto *et al.* 2006) and methyl jasmonate (MeJA) (Munemasa *et al.* 2007), treatments that both induce stomatal closure. As with the effect of ABA, in these cases the prevention of NO accumulation also inhibited the associated stomatal closure. Clearly there are several potential mechanisms by which NO could be generated in guard cells in response to ABA and other stimuli. ABA-induced stomatal closure and guard cell NO generation are both inhibited by potential NOS inhibitors such as L-NAME (Neill *et al.* 2002; Guo *et al.* 2003; She *et al.* 2004; Bright *et al.* 2006), and both stomatal closure and ABA-induced guard cell NO generation are impaired in the

AtNOS1 mutant (Guo *et al.* 2003 and Bright *et al.* 2006). Thus, whatever its exact function may turn out to be, the requirement for AtNOS1 (AtNOA1) is clear, as is a requirement for at least one mechanism that is inhibited by L-arginine analogues.

However, in *Arabidopsis* at least, there are also some reports indicating that ABA-induced NO synthesis in guard cells is mediated by NR. The NR substrate, nitrite, induces stomatal closure and NO production in a dose-dependent manner (Desikan *et al.* 2002; Bright, 2006; Bright *et al.* 2006). NO generation by NR is inhibited *in vitro* by tungstate, probably as a result of its competitive binding to the molybdenum-binding site in the enzyme. Tungstate also inhibits either ABA- or nitrite-induced NO generation and stomatal closure *in vivo* (Bright, 2006; Bright *et al.* 2006). The NR-deficient *nia1nia2* mutant, which contains only 0.5% of wild-type NR activity (Wilkinson and Crawford, 1993), exhibits neither guard cell NO synthesis nor stomatal closure in response to either ABA or nitrite, indicating that one or both of the two *Arabidopsis* NR isoforms are required. The observation that nitrite fails to induce NO synthesis in *nia1nia2* guard cells suggests that this requirement for NR reflects its *in vivo* ability to produce NO from nitrite. However, as raised by Crawford (2006), the lack of NR may have several effects on plant Nitrogen metabolism that could potentially limit NOS-like activity and, as mentioned previously, Modolo *et al.* (2006) have reported that levels of the NOS substrate, L-arginine, are substantially reduced in *nia1nia2* leaves. In resolving this issue, the *Arabidopsis* single mutants, *nia1* and *nia2*, have been used to study further the involvement of NR in guard cell NO synthesis during ABA-induced stomatal closure (Bright, 2006). Of the two isoforms, *nia2* accounts for 90% of the total nitrate assimilatory activity of NR in seedlings, while *nia1* accounts for the remaining 10% (Wilkinson and Crawford, 1991). The sequences of the *Arabidopsis* mRNAs encoding the two isoforms have ~70% identity and give rise to proteins which are 83.5% similar at the amino acid level (Wilkinson and Crawford, 1993). Both genes are probably expressed in guard cells (Bright, 2006). Guard cell NO synthesis and stomatal responses to ABA were examined in the loss-of-function T-DNA insertion mutant, *nia1::Ds* (Parinov *et al.* 1999), which has no *nia1* activity (Wang *et al.* 2004), and in the loss-of-function deletion mutant, *nia2-5* (Wilkinson and Crawford, 1991). Nitrite-induced stomatal closure was abolished in the *nia1::Ds* plants and ABA-induced stomatal closure was substantially reduced. NO accumulation in the guard cells of this mutant in response to these stimuli was effectively absent in comparison with that of the same cells of wild-type leaves. *nia1::Ds* stomatal closure in response to SNP was as for the wild type, indicating that these aberrant stomatal responses were not related to a general impairment in guard cell movement. In contrast to the *nia1::Ds* mutant, stomata of the *nia2-5* mutant behaved as those of wild type, closing in response to either ABA or nitrite, and their guard cells showed wild-type levels of NO accumulation after such stimulation. Thus, it would seem highly likely that the activity of *nia1*, but not *nia2*, is required for effective ABA signal transduction in guard cells during stomatal closure (Bright, 2006). This also suggests that the aberrant NO biology in the *nia1nia2* mutant is due specifically to a lack of *nia1* as opposed to generally aberrant N metabolism. Interestingly, some years ago Yu *et al.* (1998) also concluded that *nia1* and *nia2* have distinct signal transduction and nitrogen assimilatory roles. *nia1* and *nia2* have similar amino acid sequences, but there are distinct regions that differ between the two proteins. Thus, key questions must concern their subcellular localization, interacting protein partners, and activation characteristics.

NO SIGNAL TRANSDUCTION IN GUARD CELLS

Given that exogenous and endogenous NO induces stomatal closure, there must be some mechanism(s) by which NO is perceived in guard cells and the signal subsequently transduced into the alterations in membrane transport that effect ion and water movement across the tonoplast and plasma membrane. The small size, ready diffusibility, and reactive nature of NO perhaps make the ligand-specific, proteinaceous receptor interaction paradigm unlikely. On the other hand, the reactive nature of NO does make it likely that it reacts, directly or indirectly, with either several or perhaps many proteins. Thus, there may be a number of different 'NO perceptors' that exist in different cell types and during different physiological states. On the basis of mammalian work and emerging studies in plant systems, it is probable that both the transient activation of soluble guanylyl cyclase (sGC), which generates the second messenger cGMP, and the reversible modulation of protein activity *via S*-nitrosylation-induced conformational changes (Delledonne, 2005; Wang *et al.* 2006) are major components of NO signalling.

CGMP-DEPENDENT SIGNALING

The evidence that cGMP is an NO signalling intermediate has been obtained in several systems (Neill *et al.* 2003; Delledonne, 2005). Both salt and osmotic stress, two conditions which would both be expected to induce ABA synthesis, induced a rapid increase in the cGMP content of *Arabidopsis* seedlings (Donaldson *et al.* 2004). Using a sensitive radioimmunoassay technique, the cGMP content of pea epidermis and *Arabidopsis* guard cell fragments has been similarly measured as being in the pmol g^{-1} range, and transient increases in cGMP levels following either ABA or SNP treatment have been observed that could be prevented by co-incubation of the treated tissues with PTIO. Further evidence that cGMP can mediate the effects of ABA in stomatal guard cells has resulted from pharmacological work using the cell-permeable cGMP analogue 8-bromo cGMP (8BrcGMP) and inhibitors of NO-sensitive sGC such as 1H-[1, 2, 4] oxadiazolo[4, 3-a]quinoxalin-1-one (ODQ). The ABA- or NO-induced closure of pea (Neill *et al.* 2002) and *Arabidopsis* stomata was inhibited by ODQ. This inhibition could be prevented by co-incubation of the ABA-/NO-stimulated, ODQ-treated guard cells with 8BrcGMP. However, 8BrcGMP alone did not induce stomatal closure. Thus, it would appear that although an elevated level of cGMP is required for effective ABA-induced stomatal closure; additional signalling pathways stimulated by ABA must operate in concert for such an increase to mediate its effects.

In non-plant systems, cADP ribose (cADPR), an agent that mobilizes Ca^{2+} from internal stores, is a downstream messenger of NO. Nicotinamide, a potential inhibitor of cADPR synthesis, blocks ABA- and NO-induced stomatal closure (Neill *et al.* 2002).

Garcia-Mata *et al.* (2003) have also shown that NO-induced intracellular Ca^{2+} release and the regulation of guard cell plasma membrane K^+ and Cl^- channels are mediated by a cGMP- and cADPR-dependent pathway.

cGMP may also signal by binding to and directly activating cyclic nucleotide-gated ion channels (CNGCs) or by similarly activating cGMP-dependent protein kinases. To date, no cGMP-activated plant protein kinases have been identified, and the potential role of CNGCs in guard cell NO signalling awaits clarification.

Though cGMP has been unequivocally identified in various plant tissues (Neill *et al.* 2003) and pharmacological data indicate a role for it during stomatal closure, very little is known of the mechanisms by which cGMP might be turned over in plant cells. The one plant guanylyl cyclase gene (*AtGC1*) cloned from *Arabidopsis* encodes a protein which shows many domain differences from mammalian sGC, lacks a haem-binding motif, and is insensitive to NO (Ludidi and Gehring, 2003). The biological roles of *AtGC1* remain to be described and it may be that plants contain other enzymes capable of cGMP production. Interestingly and indicating that there may well be novel plant guanylyl cyclases awaiting discovery, a recent report has indicated that the *Arabidopsis* brassinosteroid receptor, *AtBRI1*, contains a domain with guanylyl cyclase activity (Kwezi *et al.* 2007). If cGMP is an intracellular plant signal, then mechanisms for its rapid degradation will exist. Plant cells do indeed possess cGMP hydrolysis activity and the *Arabidopsis* genome contains several genes for potential phosphodiesterases including that encoding a putative cGMP phosphodiesterase (Maathuis, 2006).

NO and its related species can oxidize, nitrate, or nitrosylate proteins (Mur *et al.* 2006; Wang *et al.* 2006). Peroxynitrite, formed by the reaction of NO with superoxide, can oxidize proteins on cysteine, methionine, or tryptophan residues or nitrate tyrosine residues to form nitrotyrosine. These post-translational modifications may well turn out to have roles in intracellular signalling and the subsequent physiological effects. For example, recent work indicates a role for protein tyrosine nitration in plant defence responses (Saito *et al.* 2006).

S-nitrosylation is the reversible covalent attachment of NO to the thiol group of cysteine residues forming an *S*-nitrosothiol (SNO) and may well be an ancient highly conserved cell signalling mechanism (Wang *et al.* 2006). Nitrosylation can occur either directly through the interaction of NO and other NO-related species with the cysteine group or indirectly by trans-nitrosylation where the NO is derived from *S*-nitrosglutathione (GSNO) or other *S*-nitrosylated proteins. Some recent studies indicate that this redox-based mechanism plays a pivotal role in plant biology and will, therefore, also be important with regard to NO signalling in guard cells.

S-nitrosylated proteins can be detected using the 'biotin switch assay' which is based on specifically biotin labelling any *S*-nitrosylated cysteines within proteins. This assay is useful for demonstrating the existence of potentially *S*-nitrosylated proteins but, like many assays, may be prone to identifying false positives. However, *S*-nitrosylation can be demonstrated unequivocally using mass spectrometry (Lindermayr *et al.* 2006). The biotin switch assay has been used to show the presence of *S*-nitrosylated proteins in plants and to demonstrate that specific proteins can be *S*-nitrosylated (Lindermayr *et al.* 2005, 2006; Wang *et al.* 2006; Belenghi *et al.* 2007). A large number of potentially *S*-nitrosylated proteins have been identified which include stress-related proteins, redox-related proteins, signalling proteins, cytoskeletal proteins, and proteins involved in photosynthesis and metabolism (Lindemayr *et al.* 2005). Moreover, conserved protein *S*-nitrosylation and GSNO-binding motifs are present in plant proteins (Wang *et al.* 2006), and the effects of *S*-nitrosylation on protein activity and plant physiology are now being addressed (Lindermayr *et al.* 2006; Belenghi *et al.* 2007). It would seem likely that all cells contain nitrosylated proteins and that the spectrum and levels of these, the 'nitrosylome', will alter during NO accumulation in, for example, guard cells after their challenge with ABA. In fact some evidence that this is the case already exists, and data published by Sokolovski and Blatt (2004) suggest that NO regulates outward rectifying K^+ channels in *V. faba* guard cells by *S*-nitrosylation.

DOWNSTREAM SIGNALLING

Alterations in the calcium concentrations of specific cellular compartments and reversible protein phosphorylation are two processes central to cell signalling. Thus, it is not surprising that NO is likely to affect these processes in guard and other cells. ABA and NO activate mitogen-activated protein kinases (MAPKs) (Desikan *et al.* 2004*a*; Pagnussat *et al.* 2004; Zhang *et al.* 2007), and MAPK activation has been suggested as a convergence point for guard cell H_2O_2 and NO signalling during ABA-induced stomatal closure (Desikan *et al.* 2004*a*). NO also activates other protein kinases such as the osmotic stress-activated kinase, NtOSAK, in tobacco (Lamotte *et al.* 2006). NO affects calcium mobilization (Lamotte *et al.* 2006), and Sokolovski *et al.* (2005) demonstrated a requirement for both protein phosphorylation and altered calcium flux in order for normal NO responses to occur in guard cells. Further evidence that NO signalling requires protein dephosphorylation results from the observation that stomata of the *Arabidopsis* ABA-insensitive phosphatase mutant, *abi1*, are insensitive to NO (Desikan *et al.* 2002).

NO AND H_2O_2

It is commonly observed that NO and ROS such as superoxide and H_2O_2 are generated in response to similar stimuli and with similar kinetics. NO and ROS can interact in various ways. For example, they can interact chemically as in the formation of compounds such as peroxynitrite. NO may also affect the activities of enzymes that alter ROS levels, and *vice versa*. Thus, they could both impact either negatively or positively on the same or related signalling pathways and thereby lead to additive and possibly synergistic responses. Stomatal closure in response to ABA is one such example where this occurs. ABA induces the production of H_2O_2. In *Arabidopsis* guard cells this is likely to arise from superoxide generated by the isoforms of NAD(P)H oxidase encoded by the genes *AtrbohD* and *AtrbohF* (Kwak *et al.* 2003; Desikan *et al.* 2004*b*; She *et al.* 2004). Consequently, the simultaneous presence of both superoxide and NO in the same location may lead to the generation of peroxynitrite anions (Neill *et al.* 2003). Peroxynitrite may well have unique signalling properties, but as yet there is no evidence that it affects guard cell movements, and the co-incubation of ABA-treated leaves with a peroxynitrite scavenger does not inhibit the induction of stomatal closure (Bright, 2006).

Lum *et al.* (2002) observed that exogenous H_2O_2 induced NO generation in the guard and other cells of *Phaseolus aureus* leaves. In a chemiluminescence-based assay, the H_2O_2 induced a substantial increase in an apparent NOS-like activity. This increase was reduced by using a potential NOS inhibitor. Interestingly, the H_2O_2-induced NO generation was inhibited by the calcium channel blocker verapamil. Thus, it is possible that Ca^{2+} ions may mediate this effect of H_2O_2. She *et al.* (2004) and He *et al.* (2005) also reported that H_2O_2 induced NO synthesis in *V. faba* guard cells and that again this accumulation could be reduced by a NOS inhibitor. Moreover, they found that both NO generation and stomatal closure stimulated by darkness or UV-B were dependent on guard cell synthesis of H_2O_2. Similarly, in *Arabidopsis*, H_2O_2 induces guard cells to synthesize NO which is required if the induction of stomatal closure is to follow (Bright *et al.* 2006). ABA-induced NO synthesis is dependent on prior H_2O_2 generation in both *V. faba* (Dong *et al.* 2005) and *Arabidopsis* (Bright *et al.* 2006). Removal of the H_2O_2 with antioxidants or inhibition of its synthesis by inhibiting NAD(P)H oxidase activity prevents NO generation and stomatal closure. Similarly, removal of the NO using PTIO prevents either H_2O_2- or ABA-induced stomatal closure. Guard cells of the *AtrbohD/F* double mutant (Kwak *et al.* 2003) also fail to make NO in response to ABA, as do those of the NR-deficient *nia1nia2* mutant. However, H_2O_2 stimulation of guard cell NO accumulation in the *Atnos1* (*Atnoa1*) mutant was as in the wild type (Bright *et al.* 2006). Thus, the requirement of AtNOS1 for NO synthesis must be epistatically upstream of H_2O_2 and its signalling effects. Regardless, the current evidence would suggest that H_2O_2 induces guard cell NO synthesis through the activity of NR. More detailed analysis has demonstrated that only the NIA1 isoform is required, as H_2O_2-stimulated NO generation is as wild type in the *nia2-5* mutant (Bright, 2006). Critically, it has been suggested that the presence of oxidants could increase the sensitivity of the DAF-2DA used in these studies. Thus, the implication is that increased concentrations of H_2O_2 might result in an increase in the fluorescence of this dye without any actual increase in the level of NO (Planchet and Kaiser, 2006*a*, *b*). While this is difficult to discount completely *in vivo*, it would appear that H_2O_2 does not increase the sensitivity of DAF-2DA *in vitro* (Bright *et al.* 2006). Furthermore, DAF-2DA fluorescence does not increase in the guard cells of the *nia1nia2* mutant following ABA or H_2O_2 challenge. The fact that this mutant generates increased H_2O_2 levels in guard cells in response to ABA strongly suggests that H_2O_2 does indeed induce NO production and that it this that is measured by the increased DAF-2DA fluorescence.

It seems that the ABA–H_2O_2–NO cascade is not restricted to guard cells. Zhang *et al.* (2007) have now demonstrated that it also operates in the mesophyll cells of *Zea mays* leaves and that the effects of ABA and H_2O_2 on the induction of antioxidant defences depended on NO generation and could be reduced using cPTIO or L-NAME.

While it is well established that H_2O_2 induces NO synthesis and accumulation, there is also a suggestion that NO may modulate H_2O_2 levels. Both She *et al.* (2004) and He *et al.* (2005) have reported that this was the case in *V. faba* guard cells. However, Dong *et al.* (2005), also working with *V. faba* guard cells, did not observe this phenomenon. NO does not appear to induce H_2O_2 generation in *Arabidopsis* guard cells (Bright *et al.* 2006) or in maize mesophyll cells (Zhang *et al.* 2007). Thus, there is argument as to whether or not this occurs, but feedback mechanisms and autocatalysis are part and parcel of signalling pathways, and it may be that such discrepancies simply reflect differences in the physiological states of the tissues examined.

STOMATAL CLOSURE DURING WATER STRESS

Although Magalhaes *et al.* (2000) reported decreased NO emission following wilting, other studies has shown that water stress causes enhanced NO synthesis and accumulation in guard cells (Leshem and Haramaty, 1996; Garcia-Mata and Lamattina, 2002; Gould *et al.* 2003). Recently such findings have been confirmed by Zhang *et al.* (2007) who showed that polyethylene glycol-induced water stress stimulated NO production in guard cells as a result of increased ABA synthesis. However, the effects of dehydration and turgor loss on the level and emission of NO in and from plants, respectively, certainly requires clarification. ABA synthesis is undoubtedly induced by turgor loss. Thus, the question arises as to whether or not NO plays a direct role in mediating drought-induced stomatal closure. Recent work indicates that NO is not required for rapid water stress-induced stomatal closure. Rapid dehydration of detached *Arabidopsis* leaves caused equally rapid stomatal closure. This was not inhibited by pre-incubation of the leaves in PTIO and occurred in both the *nia1* and *nia1nia2* mutants in the same way as in leaves from wild-type plants. The stomata of the *abi1* mutant do not close at all during this treatment, indicating that the stomatal closure response is active and ABA dependent. Thus, in contrast to the requirement for NO synthesis during ABA-induced stomatal closure in well-hydrated leaves, stomatal closure during rapid wilting may result from ABA signalling that bypasses the need for NO synthesis and perception. It may be different during stomatal closure induced in intact plants by a more gradual reduction in water potential, but it is worth noting that neither the *AtNOS1* nor the *nia1* mutants are wilty. Possibly the effects of NO differ depending on its concentration and the physiological state of the guard cell. For example, Sakihama *et al.* (2003) reported that NO donors actually caused stomatal opening when used at high concentrations. Similar results have been obtained in that, unlike low (0.1–50 µM) concentrations of SNP, high (0.5–1 mM) concentrations failed to cause stomatal closure (Bright, 2006). Another interesting occurrence that has yet to be explained concerns the interaction of ABA, ethylene, and NO signalling in guard cells. ABA, ethylene, and NO all induce stomatal closure when applied alone, but stomatal closure is reduced when either ABA or NO is combined with ethylene (Desikan *et al.* 2006), a situation likely to arise during the wilting of plants. Whichever, NO does appear to have ameliorative benefits in warding off the effects of drought. Garcia-Mata and Lamattina (2001) reported that prolonged SNP treatment provided plants with such beneficial long-term effects. Wheat plants watered with SNP and then exposed to water stress had higher relative water contents than those not receiving the SNP pre-treatment. Presumably the long-term exposure to NO in some way enables plants to retain more water during subsequent dry periods.

STOMATAL MOVEMENTS DURING LIGHT/DARK TRANSITIONS

The inhibition of stomatal closure by cPTIO when bean leaves are transferred into the dark coupled to increased fluorescence of dark-incubated guard cells stained with DAF-2DA indicates that NO is involved in light/dark-regulated stomatal movements (She *et al.* 2004). As also reported for pea leaves (Desikan *et al.* 2004*b*) and indicating the involvement of H_2O_2, this dark-induced stomatal closure could also be inhibited by the addition of ascorbate or exogenous catalase. Adding weight to the concept that NO is involved in dark-induced stomatal closure, stomata of the *Arabidopsis nia1* mutant fail to close after transfer from the light to the dark, and their guard cells show markedly reduced DAF-2DA fluorescence compared with those of the wild type which close after such treatment. Re-opening of *Arabidopsis* stomata on transfer from the dark to the light can be prevented by the addition of ABA or the NO donor SNP (Bright, 2006). The addition of the NO scavenger PTIO effectively prevents this ABA inhibition of re-opening. It is generally considered that the dark-induced cessation of photosynthesis allows the pH gradient between the chloroplast and cytosol to dissipate such that the cellular distribution of ABA changes, with more accumulating in the cytosol. At this point the redistributed ABA may induce stomatal closure directly in the guard cells or indirectly after its movement from the mesophyll to epidermal cells. These data suggest that stomatal closure on transfer from the light to dark involves a linear signalling process from H_2O_2 and NO and that their re-opening, either on transfer to the light or on return to conditions of hydration, may also be partly dependent on the metabolic removal of H_2O_2 and NO. The degree to which stomata close in the dark depends on numerous factors including the stomatal density and leaf age, the position of the plant in its circadian rhythm, the light intensity and quality of the preceding photoperiod, the ambient CO_2 concentration, the relative humidity and soil water availability, and whether the plant operates a C3, C4, or crassulacean acid metabolism (Caird *et al.* 2007). Thus, the involvement of NO with respect to these variables remains to be fully determined.

STOMATAL RESPONSES TO PHYTOPATHOGENIC BACTERIA

Gaining entry into subepidermal tissues is one of the first steps in the bacterial infection of plants, and it is often assumed that openings such as stomata serve as passive ports of entry for such pathogens. However, active stomatal closure has recently been described as part of the innate immunity of plants to bacteria (Melotto *et al.* 2006). For example, in what has been termed a pathogen-associated molecular pattern (PAMP)-induced basal defence mechanism, *Arabidopsis* guard cells are able to perceive *Pseudomonas syringae* cell surface molecules such as lipopolysaccharide (LPS) and flagellin which trigger NO synthesis and result in stomatal closure. Inhibitors of NO synthesis prevent this stomatal closure. In contrast to those of the wild type, stomata of the ABA-insensitive stomata mutant, *ost1-2*, or the ABA-deficient mutant, *aba3-1*, failed to close in response to LPS and flagellin. *Escherichia coli* also induced stomatal closure that was prevented by inhibitors of NO synthesis. Thus, it would appear that the ABA and/or NO signalling mechanism are required for stomatal closure during bacterial challenge. Interestingly, virulent pathogenic bacteria have a mechanism to overcome this barrier and cause stomata to re-open. Using mutants of *P. syringae* which were unable to elicit stomatal re-opening, Melotto *et al.* (2006) identified the phytotoxin coronatine (COR) as the bacterial virulence effector responsible for overcoming this barrier to infection. COR effectively prevented ABA-induced stomatal closure in wild-type *Arabidopsis*, but not in the COR-insensitive mutant, *coi1-20*. COR did not prevent ABA-induced NO synthesis in wild-type plants. Thus, it is likely that COR acts downstream of ABA and NO to prevent stomatal closure (Melotto *et al.* 2006).

NO AS AN ANTIOXIDANT AND PROTECTANT

Exogenous NO, applied *via* the roots in the form of SNP, has been shown to provide some protection against the oxidative damage associated with water stress (Garcia-Mata and Lamattina, 2001). Several other studies have also shown that exogenous NO ameliorates the oxidative stress induced by a range of abiotic conditions including those resulting from the presence of heavy metal ions, salinity, high temperatures, the presence of H_2O_2, dehydration, UV irradiation, and the presence of paraquat (Cheng *et al.* 2002; Hung *et al.* 2002; Uchida *et al.* 2002; Kopyra and Gwóźdź, 2003; Zhang *et al.* 2003, 2006; Hsu and Kao, 2004; Hung and Kao, 2005; Laspina *et al.* 2005; Shi *et al.* 2005; Wang and Yang, 2005; Yu *et al.* 2005; Tian *et al.* 2007). Some of the antioxidant effects of NO may be due to its direct interaction with ROS such as superoxide to form peroxynitrite that might then be scavenged by other cellular processes. In other cases NO may enhance the antioxidant capacity of cells by increasing the activities of antioxidant enzymes such as superoxide dismutase, which converts superoxide to H_2O_2, and catalase and ascorbate peroxidase, which both remove H_2O_2 (Cheng *et al.* 2002; Hung *et al.* 2002; Uchida *et al.* 2002; Kopyra and Gwó d , 2003; Zhang *et al.* 2007).

Whether or not endogenous NO has an antioxidant function is debatable. Guo and Crawford (2005) detected high levels of oxidized proteins and lipids in the *AtNOS1* mutant. *AtNOS1* has enhanced sensitivity to the oxidative stress induced by either methyl viologen or salinity that can be ameliorated by exogenous NO (M Zhao *et al.* 2007; MG Zhao *et al.* 2007). Moreover, in calli of stress-tolerant and stress-sensitive reed ecotypes, heat and salt stress induce endogenous NO generation in the tolerant, but not in the sensitive ecotype. NO enhanced stress tolerance in both ecotypes, but more so in the sensitive ecotype, and depletion of endogenous NO by cPTIO reduced survival in the tolerant ecotype. The enhanced antioxidant enzyme activity and stress tolerance associated with the ability of the tolerant ecotype to make NO was abolished by this cPTIO treatment (Zhao *et al.* 2004; Song *et al.* 2006). Similarly, in *Populus* callus and *Stylosanthes guianensis*, stress tolerance has been associated with endogenous NO synthesis (Zhou *et al.* 2005; F Zhang *et al.* 2007). Given that various stresses induce ABA and, as a consequence NO, and given that NO appears to enhance antioxidant enzyme activity, one might expect a signalling link between ABA, NO, and oxidative stress tolerance. In *S. guianensis*, ABA enhances the activity of antioxidant enzymes, and this is dependent on NO production (Zhou *et al.* 2005). Recent work by Zhang *et al.* (2007) has clearly demonstrated this link. Their data show that the enhanced MAPK activity and antioxidant gene expression and enzyme activity induced in maize leaves by ABA and H_2O_2 are in fact dependent on endogenous NO generation. Thus, the data indicate a key 'ABA–H_2O_2–NO–MAPK–antioxidant survival cycle' and suggest that during water stress ABA has several ameliorative functions that involve NO as a key signalling intermediate and which include the rapid induction of stomatal closure to reduce transpirational water loss and the activation of antioxidant defences to combat oxidative damage .

NO AND CA^{2+}: THE TWO MESSENGERS

Ca^{2+} is a ubiquitous second messenger. The fluctuations in the intracellular levels control an incredible number of processes. The specificity of Ca^{2+} signaling might be related to its extreme compartmentalization and its complex temporal and spatial distribution. The molecular systems for producing signaling events, namely Ca^{2+} channels and transporters, responds to an array of physiological stimuli including extracellular or intracellular messengers, membrane depolarization or hyperpolarization, and mechanical or thermal stimuli. Work on various animal cell types indicate that most of the channels and transporters involved in the control of Ca^{2+} homeostasis seem to be modulated by NO. NO acts either directly on channels and transporters, or indirectly *via.* interconnected secondary signaling cascades.

The list of transporters and Ca^{2+} channels directly regulated by NO by the means of S-nitrosylation on oxidation comprises Ca^{2+} -ATPase, store operated Ca^{2+} chaneels, voltage operated P/Q type and L type Ca^{2+} channels, CNGC and ryanodine receptors (RyRs; Stamler *et al.* 2001). RyRs are large, homotetrameric high conductance Ca^{2+} channels mediating the release of Ca^{2+} from the endoplasmic reticulum (Eu *et al.* 1999). RyRs are the targets of NO because they contain large number of sulfhydryls. For instance the skeletal muscle and cardiac RYRs (RyR1 and RyR2, respectively) contain 100 and 89 cysteines per 560-kDa subunit. Modification of RyR activity by NO might influence key physiological functions such as vasomotor control, muscular contraction, and insulin release. At molecular level, NO interacts with RyRs in a complex way since both activation and inhibition have been reported (Hart and Dulhunty, 2000). At physiological pO_2, RyR1 is activated by S-nitrosylation of a single cysteine residue (Cysteine 3635) located with in CaM binding domain. It is assumed that by S-nitrosylation cysteine 3635 reverses the inhibitory effect of CaM on RyR1 (Sun *et al.* 2001).

The indirect modulation of Ca^{2+} channels by NO is mainly mediated by cGMP. cGMP can mediate ion channel activity by two different mechanisms (1) direct action through its binding to cyclic-nucleotide binding sites of CNGCs, (2) indirect action mediated by PKGs (Ahern *et al.* 2002; Clementi, 1998). PKGs are widespread proteins that controls ion cahannels function, since virtually every type of ion channel contains PKG phosphorylation sites. Accordingly PKG mediated phosphorylation of Ca^{2+} channels, including the inositol 1, 4, 5-triphosphate receptor (IP_3R) and voltage dependent Ca^{2+} channels, has been proposed to account for the stimulatory and inhibitory effects of NO on Ca^{2+} mobilization (Clementi 1998). There is also increasing evidences that PKGs activate ADP-ribosyl cyclase, the enzyme catalyzing the cyclization of NAD^+ to cyclic ADP-ribose (c-ADPR). cADPR is a nucleotide showing Ca^{2+} mobilizing in a wide variety of cells. RyRs appears to be cADPR intracellular target; although it is not clear whether cADPR acts as a direct triggering molecule for the RYR mediated Ca^{2+} release or as a modulating factor. Activation of the cGMP/PKGs/cADPR cascade is a key pathway used by NO to activate RyRs in Sea urchin eggs and in various neuronal cell types.

The efficacy of NO in increasing intracellular Ca^{2+} concentration might have deep repercussions on cell metabolism and adaptation. It has been shown that the combined action of NO and Ca^{2+} lead to the activation of protein kinase-A dependent activation of CREB transcription factor capable of reprogramming the pattern of gene expression in neurons. NO and Ca^{2+} have to act in a very narrow time lapse for the activation to occur. Another important example of the cross talk between NO and Ca^{2+} is provided by Patel *et al.* 1999, who reported that increase in the frequency of Ca^{2+} oscillations in the rat liver cells, induced by the endothelial derived NO, stimulate metabolic activity in hepatocytes together with an increased blood flow resulting from the vasodilating effects of NO. The modulation of the Ca^{2+} signal by the NO is therefore critical for cellular and organ function.

PROTEIN KINASES AS TARGETS OF NO ACTION

During the last few years, an increasing number of protein kinases have been shown to be mediated by NO. NO activate all three mitogen activated protein kinase (MAPK) cascades, namely the stress activated protein kinase/c-Jun N-terminal kinase (SAPK/JNK) cascade, the stress activated p38 activated MAP kinase cascade and the ERK/MAPK cascade. The activation of ERK/MAPK module by NO has important implications. For instance, in stress conditions, this process leads to the inhibition of the proapoptotic protein Bad and to the expression of pro-apoptotic proteins (such as Bcl-2) and enzymes involved in the antioxidant defence. In neuronal cells, No induces the expression of immediate early genes through the ERK/MAPK pathway and plays a role in the synaptic plasticity. NO activates the MAPK cascade *via* the upstream S-nitrosylation of low molecular proteins including $p21^{ras}$, Rac1 and Cdc42. It has also been reported that NO can directly suppress JNK activation by S-nitrosylation, supporting pro-survival mechanisms. Similarly, NO inhibits the autokinase activity of JAK2 and JAK3, through oxidation of crucial dithiols to disulfide. Because JAKs transmit signals from almost all types of cytokine receptors, such inhibition may explain how macrophage produced NO affects the ability of certain lymphoid cells to proliferate in response to cytokine. Finally protein phosphates also represent targets for NO (Stamler *et al.* 2001). By direct interaction with phosphates, either activation or inhibition might occur, which will counteract or prolong activation of protein kinase activity. Taken together, these findings establish a role for protein kinase signaling pathways in the cellular response of NO.

NO AND PROGRAMMED CELL DEATH (PCD)

The most recent major advances in the study of programmed cell death (PCD) in plants include the observation that peptide inhibitors of caspases inhibit the hypersensitive response and NO has been shown to be required for the induction of disease related PCD. Mutant analysis has led to the cloning of the first genes involved in PCD related disease resistance, *LSD1* and *MLO*. Intrinsically programmed cell death is, by now, an accepted requisite for life in multicellular eukaryotes. Extra and intracellular inputs dictate whether cells find themselves in the right place, at the right time in development, and surrounded by the right neighbors (Ashkenazi and Dixit, 1998). If reinforcing microenvironmental signals are perceived, the cell lives and continues about its business. But when spatial and temporal signals clash, and the 'I'm OK, you're OK' monitoring system is perturbed, cells follow an intrinsic pre-set default pathway leading to their own death (Raff, 1992). Cells which beat the system, through mutations disarming the default death pathway(s), can become malignant. Conversely, it is well established that some cells are destined to die as part of normal developmental processes, or can be triggered to die quickly as part of a response to infection. This broad outline was established in animal systems. Recent data suggests that several fundamental aspects of this model can also operate in plants. Possible mechanistic overlap and molecular parallels between cell death control in plants and animals are under investigation. Definition of a 'program' is one which is intrinsically encoded by the plant cell. Hence, mutants who interrupt or induce cell death might define steps in pathways which either respond to, or regulate, cellular homeostasis. This definition, at first glance, could preclude cell death initiated by pathogen-produced toxins. But if that toxin has a specific cellular target, and that target performs a cellular task which is monitored for fidelity, then lack of fidelity in the process could trigger an intrinsic death program (Dangl *et al.* 1996; Greenberg, 1997; Pennell and Lamb, 1997; Wang *et al.* 1996). Programmed cell death (PCD) occurs at many points during plant development, but the most progress to date has been made using the hypersensitive cell death response (HR) to pathogen infection as a model. Whether the HR, per se, stops pathogen growth or whether it is a consequence of the mechanism(s) which does is not clear. It is very often correlated with disease resistance. As well, common disease symptoms — the outcome of a successful infection — also very often include the death of host cells. It is important to consider that the deaths engendered in these two outcomes could be mechanistically different; these differences might reflect the fact that in the HR the plant is controlling its own cellular destiny, whereas during the onset of disease symptoms it is the pathogen which may use interdiction of a normal cellular signaling pathway to further its goal of growth and propagation.

Many, but not all, programmed cell deaths in animals lead to a stereotyped cellular dismantling known as apoptosis. Apoptosis is the outcome of the integration of a wide range of signals, all of which feed into a conserved cysteine proteinase (caspase) cascade (Thornberry and Lazebnik 1998). This cascade is under constitutive negative regulation. If that regulation is removed, the default pathway is activation of caspases and orderly cellular dismemberment. Apoptosis, contrary to vernacular usage, is not synonymous with PCD, as there are other programs which can operate to initiate cell suicide in animal cells (*e.g.* [Schwartz *et al.* 1993]). Some facets of apoptosis are shared between kingdoms. Broadly, signals are transduced, ion fluxes occur, and specific proteinases are activated during plant cell death. Nuclear condensation and DNA cleavage first into large fragments, and subsequently into fragmented 'ladders', is observed in some, but not all cases of pathogen-induced PCD. The lack of apoptotic bodies in plants is not surprising, if their function is to facilitate phagocytosis of their contents by neighboring cells. There is no way for typical apoptotic bodies to pass through the cell wall. A recent study points out yet another difference between plant and animal versions of PCD processes. Tobacco cells pulsed with chemical inducers of PCD has a window within which they can reverse the program, even after chromatin condensation has begun (O'Brien *et al.* 1998). This is in contrast to the animal model where once initiated, there is no turning back from PCD. Proteins in the BCL2-CED9 family can either prevent or trigger PCD in animals (Adams and Cory 1998). No genes encoding BCL2 family members been unambiguously defined in public plant databases. Immunolocalization studies using antibodies to a human BCL2 family member, however, detected a BCL2-like epitope in leaves and meristematic tissues of tobacco and maize. BCL2 family members are mostly localized on the mitochondria in animals (Green and Reed 1998,). As might be expected, the BCL2-like plant protein localizes to mitochondria, nucleus and chloroplasts (Dion *et al.* 1997). An additional tool to uncover genes operative in the *BCL2* control pathway, perhaps including plant genes, was recently described [Xu Q and Reed JC, 1998)—yeast hybrid analysis was used to screen a human library for an inhibitor of BAX (a pro-apoptotic BCL2 family member), and homologues of this inhibitor were found in *Arabidopsis*, mouse, and *C. elegans*, implying possible mechanistic conservation. Overexpression of the animal anti-apoptotic BCL-XL, however, in transgenic tobacco did not inhibit HR (Mittler *at al.*, 1996). Animal viruses often encode proteins which inhibit caspase activation. When overexpressed in tomato and tobacco, the p35 protein from Baculovirus seems to inhibit both pathogen growth and disease symptoms caused by necrotrophic pathogens. Interestingly, growth of obligate biotrophs was not slowed although the eventual disease symptoms these pathogens produce were diminished. This result strongly suggests that induction of conserved mechanisms for apoptosis are used by virulent, necrotrophic pathogens to kill cells upon which they can then feed. Another preliminary study concludes that TMV induced HR was inhibited by p35 overexpression in tobacco and that mutation in p35 which abolish caspase inhibition also abolish this phenotype. A recent demonstration that plants induce a caspase-related proteinase activity preceding HR suggests that, among the many cysteine proteases in plants, at least one has the substrate requirements of an animal caspase (del Pozo and Lam E, 1998). In animals, the CED-4/Apaf1 proteins activate caspases and these are tethered by BCL-2 proteins until required (Zou *et al.*

1997). Enticing homologies between these proteins and a domain in plant disease resistance proteins has been noted (van der Biezen and Jones, 1998), further suggesting structural similarities between plant and animal cell death control.

SIGNAL TRAFFIC TO CONTROL HR CELL DEATH

The signal cascade leading to HR is triggered through recognition of a pathogen avirulence gene product by the appropriate disease resistance (*R*) gene (Hammond-Kosack and Jones, 1996), or by an elicitor of plant defense responses recognized by a specific receptor (*e.g.* [Nürnberger, *et al.* 1994]). Recognition of either type of signal initiates the overall resistance response, in association with an influx of Ca^{2+} ions from the extracellular space, anion fluxes leading to alkalinization of the extracellular milieu, an oxidative burst producing reactive oxygen intermediates (ROIs), defense gene activation, development of local and systemic disease resistance (Hammond-Kosack and Jones, 1996; Yang *et al.* 1997; Scheel, 1998). Measurements of increase in cytosolic calcium support the concept that it plays a very early role in this transduction chain (Xu and Heath, 1998). The overall order of these events seems to reflect the activation of multiple pathways, as inhibition of Ca^{2+} flux or oxidative burst prevents cell death but not defense gene activation (Jabs, *et al.* 1997). An exciting recent development is the identification of a MAP kinase which mediates a variety of input signals, is salicylic acid (SA) inducible, and leads to defense gene activation (Ligternik, *et al.* 1997; Zhang and Klessig, 1997; Zhang *et al.* 1998,). Interestingly, this kinase acts either upstream to or independently of the oxidative burst. This, or another, kinase cascade may end in the phosphorylation of a putative transcription factor for defense response genes (Dröge-Laser, *et al.* 1997). ROI are ordinarily generated during both metabolic and photoactivated processes, and damage the cell *via* uncontrolled oxidation of cellular components. Different ROI are produced within the cell, and, in combination with different sites of production, could drive cell death in a variety of cellular contexts. The oxidative burst initiated by pathogen attack results in superoxide synthesis, which can occur within regions of the cell wall adjacent to the pathogen (Bestwick, *et al.* 1997). Superoxide can spontaneously dismutate to the more stable product hydrogen peroxide (H_2O_2) or it can act as a localized signal molecule. H_2O_2 can serve multiple roles in plant disease resistance: as a direct antimicrobial activity; a component of structural defense through oxidative cross-linking of the cell wall; and as both an intra- and inter-cellular signaling molecule (Lamb and Dixon, 1997). Oxidative burst, in isolation, does not lead to cell death, but high levels of hydrogen peroxide (much greater than those produced during a typical oxidative burst) can kill cultured soybean cells (Levine *et al.* 1994). In addition, mutant bacteria which cannot trigger HR still generate a normal oxidative burst (Glazener, *et al.* 1996) and the oxidative burst is apparently not a sufficient signal for HR in the cowpea–cowpea rust fungus pathosystem (Heath, 1998). Nitric oxide (NO) can act to potentiate PCD in animals (Hausladen and Stamler, 1998) and important recent evidence supports a similar role for NO in plants (Delledone, *et al.* 1998; Durner, *et al.* 1998). If NO production is inhibited in cultured soybean cells or tobacco leaves, the HR is blocked and resistance to avirulent bacteria is moderately attenuated.

Furthermore, exogenous generation of NO and ROI synergistically promote cell death and induce gene expression of both PR1 (SA-dependent) and PAL (SA-independent) defense genes. How do these local events relate to the onset of systemic signals? Recent evidence suggests that ROI may also be rapidly induced at distal uninfected leaves (Alvarez, *et al.* 1998) following a local oxidative burst in infected leaves. The notion here is that rapid systemic signaling can be used as an early warning system which primes the SA-dependent systemic responses. The cell has numerous compounds and enzymes which serve to scavenge ROI before untoward damage can occur. These antioxidants are typically upregulated in times of oxidative stress (*e.g.* high light, ozone exposure). Ascorbate peroxidase (APX) uses ascorbate to detoxify hydrogen peroxide. APX is found throughout the cell, and is believed to be one of the main scavengers of peroxides. Like other antioxidative enzymes, the levels of APX declines in cells undergoing HR (Mittler, *et al.* 1998; Dorey, *et al.* 1997; Fodor, *et al.* 1997). Overproduction of APX in the chloroplast, however, does not protect tobacco against ozone induced damage (Torsethaugen, *et al.* 1997). Are there relationships between the set of signal events leading to HR and any case of developmentally controlled cell death? One highly developed system for analyses of PCD during development is tracheary element (TE) formation in trans differentiating *zinnia* mesophyll cells (Fukuda, 1996). Cell death during TE formation is not apoptotic and does not generate an observable oxidative burst. A calcium influx and perception of extracellular signals, mediated by a serine protease, however, are required for TE cell death (Groover, *et al.* 1997; Groover and Jones, 1998). Thus, most features of this developmental cell death differ from signaling in HR as outlined above.

SALICYLIC ACID (SA) AND THE POTENTIATION OF CELL DEATH

In tobacco and *Arabidopsis*, endogenous SA levels rise following pathogen attack and correlates with expression of pathogenesis-related (*PR*) genes as well as the onset of SAR. Exogenous addition of SA induces *PR* gene expression as well as heightening disease resistance (Ryals, *et al.* 1996). Yet, SA acts downstream of the oxidative burst and SA addition without pathogen triggers neither substantial increase in ROI nor cell death. A variety of experiments strongly support a model whereby SA and ROI (and probably NO) potentiate the overall HR and defense response (Fauth, *et al.* 1996; Kauss and Jeblick, 1996; Kauss and Jeblick, 1995; Shirasu, *et al.* 1997). Transgenic expression of salicylate hydroxylase (*nahG*) under the control of temporally different promoters demonstrates that SA accumulates during pre-necrotic phases of TMV infection in tobacco and that this accumulation is required to curtail viral spread (Mur, *et al.* 1997). SA pretreatment of parsley cell suspensions also potentiates subsequent induction of various defense genes by both elicitor-dependent and -independent modes. SA potentiated, pathogen-induced cell death is also supported by observations of reduced lag time to cell death from eight to four hours (Thulke and Conrath, 1998). The most likely scenario is that ROI production triggers both NO and SA synthesis. Superoxide and NO can combine to form the very dangerous peroxynitrite radical, and high levels of NOS activity could lead to more superoxide production (Dangl, 1998), resulting in an amplification which produces more SA and NO. It is critical to note that amplification mechanisms containing an extracellular component (ROI production or signals emanating from dying cells), must be negatively controlled by desensitization once a sufficient response has been reached. This desensitization could respond to gradients of ROI or SA around HR sites. For example, recent work suggests that although the levels of ROI and SA in live cells around HR sites are sufficient to induce defense gene transcription, they are not enough to trigger HR (Dorey, *et al.* 1998).

MISREGULATION OF HR CELL DEATH IN MUTANTS

A number of mutants misregulate cell death, suggesting that the wild-type function of the genes they define may be in PCD control. They are collectively termed 'lesion mimics' because their phenotypes are reminiscent of either HR or disease symptoms (Buckner, *et al.* 1998). Some lesion mimics map to disease resistance genes, suggesting that misregulation of R function can lead to inappropriate cell death. For example, the *Rp1 R*-gene locus in maize is prone to unequal crossing over that generates new fungal resistance specificities. Derivative alleles exist that cause lesioning in the absence of pathogen (Hu, *et al.* 1996). Some genes whose mutant phenotypes are increased disease resistance and propensity for cell death have been cloned. The *Arabidopsis LSD1* gene is a negative regulator of cell death which responds to a superoxide dependent signal. It encodes a zinc finger protein which may function as a transcriptional regulator (Dietrich *et al.* 1994; Dietrich *et al.* 1997). In barley, the *MLO* gene encodes a novel transmembrane protein (Büschges *et al.* 1997). In *mlo* mutants, defense pathways are primed; this leads to both a low level lesion phenotype and resistance against downy mildew. In contrast, *lsd1* mutants are generally resistant to virulent bacterial and oomycete pathogens. Thus, *MLO* negatively regulates subset defense response components active against one species of pathogen, while *LSD1* negatively regulates a potentially broader set. These two mutants are instructive because they suggest that defense systems are in part pre-existing and active in the absence of negative control. Mutations in negative regulators which interpret activation signal thresholds can lead to default cell death. In addition to negative regulators controlling the extent of HR, there are a number of lesion mimic mutants, mostly dominant, that induce the SA-dependent disease resistance pathway in *Arabidopsis*. These mutations can be sometimes genetically upstream of requirements for SA accumulation in defense gene induction and disease resistance (Hunt *et al.* 1997). Alternatively, they can function in the amplification loop of SA responses such that SA is both required for lesion formation and able to potentiate lesion formation (Weyman *et al.* 1995). There are also a number of constitutive disease resistant mutants which do not make lesions. In particular the *dnd* mutation is resistant to virulent pathogens in the absence of HR and makes a smaller and less obvious HR when challenged with avirulent pathogen (Yu *et al.* 1998). This phenotype is in keeping with classic examples of systemic tobacco mosaic virus resistance in tobacco and suggests that the potential normal requirement for HR in either local or systemic resistance can be bypassed. Do all mutants that trigger PCD define genes whose wildtype function is either in pathogen recognition or response pathways? *Les22* in maize was recently cloned and found to encode an enzyme in the porphyrin pathway (Hu *et al.* 1998). Mutation in this enzyme causes a build up of a photoactivatable intermediate, which generates ROI in response to high light. Additionally, this pathway seems to impinge on heme formation, which is required for catalase and APX function, resulting in an inability to detoxify the ROI produced. If leaves are protected from high light until they are fully developed, thereby lowering flux through this pathway, lesioning disappears. Thus, ROI resulting from metabolic perturbation of pathways can trigger PCD and yet not be involved in defense. The maize *LLS1* gene is a negative regulator of cell death, and encodes an enzyme believed normally to degrade a phenolic signal. The absence of this enzyme leads to cell death (Gray *et al.* 1997). Neither *Les22* nor *lls1* mutants induce disease resistance pathways. Mechanistically, this class of mutants suggests that cells sense metabolic perturbations and can follow a default pathway, normally under negative control, to cell death. Alternatively, these mutants result in accumulation of toxic compounds which then kill the cell. Lesion mimics can also result from transgene overexpression. Several recent papers illustrate how these may trigger the disease resistance pathway much as the lesion mimics upstream of SA accumulation described above. Transgenic tobacco plants with reduced levels of catalase form lesions and express PR genes in response to high light, and this death increases protection against TMV and Pseudomonas (Chamnongpol *et al.* 1996; Chamnongpol *et al.* 1998; Takahashi *et al.* 1997) in an SA-dependent manner (Du and Klessig 1997). Expression of the light driven proton pump bacterio-opsin in tobacco results in increased systemic SA levels, PR gene expression, and resistance to several pathogens (Mittler, *et al.* 1995). Expression of the same gene in potato imparted resistance to certain fungal isolates, but unexpectedly increased susceptibility to Potato Virus X (Abad *et al.* 1997). Hence, inducible artificial lesion mimics resulting from transgene expression may not be a panacea for disease resistance in agriculture, but may prove useful in certain cases.

Measurement of NO

Measurement of NO in biological specimens is difficult both because of the small amounts present (usually less than nanomolar) and the lability of NO in the presence of O_2. This instability is partially ameliorated by adding superoxide dismutase to the medium in which NO is to be measured (Archer SL and Cowan NJ, 1991). Superoxide dismutase (2000-4000 units) enhances the detection of NO from isolated pulmonary artery rings, presumably by decreasing the amount of superoxide anion present and thereby preventing the destruction of NO. NO is commonly measured by the chemiluminescence reaction with acid reflux (Palmer RMJ *et al.* (1987); Rees DD *et al.* (1989); Myers P *et al.* (1990); Amezcua JL *et al.* (1989); Chen WZ *et al.*(1989)) or without acidification (Menon N *et al.* 1989), by the diazotization of sulfaniic acid assay (Ignarro LJ *et al.* (1987); Tracey WR *et al.* (1990); Martin W *et al.* 1988; Schmidt HHHW *et al.* 1988), by electron paramagnetic resonance (EPR) with nitroso or hemoglobin traps (Arroyo C and Kohno M 1991; Greenberg, SS *et al.* 1990), and by spectrophotometric measurement of the conversion of reduced to met-hemoglobin (Kelm M *et al.* 1988; Kelm M and Schrader J 1988). These assays use one of three detection strategies: they chemically trap NO and measure either adduct (nitrosyl hemoglobin, NO-Hb), the conversion of reduced to methemoglobin, or the light emission (chemiluminescence) that occurs when ozone interacts with NO.

CHEMILUMINESCENCE ASSAY

This assay was originally developed to measure NO as an atmospheric pollutant (Zafiriou OC and McFarland M 1980), but when performed with acid and K! it is also an assay for nitrite (Aoki T 1990). Detection of NO is based on the observation that ozone interacts with NO to generate light. Luminescence, measured by a sensitive photomultiplier tube (PMT), is directly proportional to NO levels. The assay takes advantage of the low solubility of NO in aqueous solutions by measuring NO in the gas phase, whether the original sample was a gas or a liquid. If the specimen is liquid, the NO is driven into the gas phase by bubbling the solution with an inert gas under vacuum conditions, a process called "stripping?'. The gas above the liquid is called the "head space?'. Chung and Fung (1990) documented the predisposition of NO to exist as a gas. They found that when they injected NO into a physiological buffer, 83% immediately appeared in the head space. The specimen (gas or liquid) is injected into a glass purge chamber previously evacuated to 3 mm Hg to remove O_2 and stabilize NO. If the specimen is gaseous (Brien J *et al.* 1991) one can simply enter the sample into the reaction chamber without the need for stripping. Several components are necessary to measure NO using chemiluminescence. In addition to a PMT and ozone generator, one must have a sample chamber in which NO can be held, and if necessary stripped into the gas phase. A glass purge chamber (Sievers Research, Inc., Boulder, and Cob.) that has a fit in its floor to permit stripping of liquid specimens is useful for this purpose. This purge chamber has an inflow line for delivery of an inert gas to the fit and an outflow line to carry the gas in the head space to the reaction chamber. NO is "stripped" from the liquid to the gas phase by bubbling the specimen with an inert gas such as helium (12 ml/mm x 60 s).

Aoki T (1990) showed that the rate of stripping the liquid samples is an important determinant of NO measurement; flow rates of less than 8 mi/mm cause underestimation of NO due to incomplete stripping. If one is dealing with a gaseous specimen the stripping phase is omitted, greatly simplifying the assay. Once stripping is complete, the gaseous specimen is drawn through a "needle valve" into a reaction chamber. The opening size in the needle valve is adjusted to make the NO signal appear as a brief spike and avoid entrainment of foam from the sample. In the reaction chamber, ozone (generated internally by electrical discharge) is mixed with NO in front of a cooled, red-sensitive PMT. It is critical to avoid aspiration of foam or bubbles into the reaction chamber as this can coat the PMT and diminish its sensitivity. This precaution is especially important in dealing with biological specimens because proteinaceous material foams dramatically during stripping, limiting the assay's usefulness in specimens containing albumin or blood.

Assay calibration

Gaseous NO or aqueous solutions containing various amounts of NO (20-1000 pmol) are used to calibrate the photomultiplier tube. A calibration curve is produced by plotting the signal from the PMT (mV) vs. the known NO dose. This calibration curve is used to estimate the amount of NO present in biological specimens. The integration time, the period of time during which photons are counted before a value is presented and counting resumes, can be varied to optimize the assay for weak vs. strong NO signals. The optimal integration time should be short enough so the counts do not saturate the tube but long enough to detect small signals. Whereas an integration time of 0.06 is adequate for most experiments, use of an integration time of 0.25 s results in a more linear NO/chemiluminescence curve and greater sensitivity, albeit at the price of increased noise. A NO dose/chemiluminescence curve must be constructed for each integration time.

Sensitivity/detection threshold

Zafiriou and McFarland Aoki, T. (1990) reported that the chemiluminescence assay is sensitive to 10- M NO. However, most investigators find the sensitivity threshold for detection of NO is 20-50 pmol. Several authors have noted that the chemiluminescence/NO curve are linear between NO doses of 300 and 3000 pmol. The sensitivity of the assay for authentic NO is not dramatically affected by the use of reducing agents or acids if done properly, for unlike measurements in biological models, O_2 can be excluded and there should be little signal amplification due to recruitment of nitrite or other oxidation products.

Specificity

The specificity of the chemiluminescence assay depends on whether the sample is acidified or refluxed under strong reducing conditions (*e.g.*, in the presence of KI or Nal) before to measurement. If the assay is performed without acidification or Nal, the signals achieved in biological models are much smaller (usually by 1 log order) than those produced in the presence of acid and Nal (Menon N *et al.* 1991). In the absence of Nal and acid, chemiluminescence is almost exclusively due to NO. The few other substances that could elicit chemiluminescence (*e.g.*, H2S) are either nonvolatile or do not occur in biologic models. To be detected in this assay a molecule must first enter the head space, a fact that provides a good deal of specificity. NO has a high partition coefficient and tends to enter the head space whereas nitrite and nitroso thiols will remain in solution. Once in the gas phase, a substance must be capable of interacting with ozone to produce light. This requires that the substance have unsaturated or strongly polar groups (*e.g.*, sulfides, amines), characteristics that are rare in small molecules. Thus NO's unique chemical properties greatly enhance the specificity of the assay. Most authors report that the assay is entirely specific; however, it has been found that DMSO at high doses can cause a chemiluminescent signal. This is a potential problem because DMSO is used as a solvent for many drugs, including the endothelial-dependent vasodilator A23187. This problem can be resolved by interposing a red filter, which absorbs light at the wavelengths emitted by the interaction of ozone and sulfoxides, in front of the PMT and limiting the dose of DMSO used to less than 0.1 ml/I0 ml sample.

An alternative means of proving that a chemiluminescence signal is due to NO is to scrub the head space by passing the gas through a solution of 1 M Fe^{2+} before it reaches the reaction chamber. Fe^{2+} interacts with NO and reduces the signal strength several hundred fold if the signal was due to NO. A number of features of the chemiuminescence assay merit discussion.

DIAZOTIZATION ASSAY

This assay is the standard technique for measuring inorganic nitrite. It is discussed as a NO assay only because it has been used as such in the literature. This two-step assay is based on the observation that the adduct of nitroxides and sulfanilic acid interacts with N-(1-naphthyl) ethylenediamine, generating a product that is readily monitored by spectrophotometry.

Methods

The basic assay is designed to measure single samples, but it can be modified to give "on-line" results (Tracey WR *et al.* 1990). In the basic assay the sample is placed in a vessel containing 4 N HCI. An aliquot is withdrawn and incubated for 10 mm with a mixture of 2 N HC1 and sulfanilic acid (2 mg/ml). N- (1-naphthyl) ethylenediamine (1 mg/ml) is added, and after 30 mm absorbance at 548 nm is measured. The NO level can be calculated by comparing the observed change in absorbance to a NO/absorption calibration curve.

Utilization

Tracey *et al.* (1990) evaluated the ability of an automated version of the diazotization assay to detect authentic NO and attempted to correlate measured NO with EDRF activity in a parallel bioassay. In these studies the test fluid (either authentic NO or EDRF produced by stimulating cultured endothelial cells with bradykinin BK, adenosine triphosphate ATP, or A23187) was split, so that part was delivered to a bioassay (a denuded aortic ring) and the remainder to a spectrophotometer. The diazotization assay was performed en route to the spectrophotometer. Perfusate destined for the spectrophotometer (1.2 ml/min) was mixed first with 2 M HC1 and then sulfaniic acid (0.7 mg/ml) was added. This mixture was passed through a 3.5 mm delay coil and then N-(l-naphthyl) ethylenediamine (0.7 mg/ml) was added. All reagents were infused at 0.3 ml/mm. After passing through a 6 mm delay coil, the mixture entered the spectrophotometer and the absorbances at 548 nm were compared to baseline (the absorption measured with all reagents present, but before adding EDRF or NO). The detection threshold of the diazotization assay is roughly 10 to 106 M NO. In this assay, absorption at 548 nm should increase solely as a function of NO dose. In fact, when using this assay on authentic NO solutions, Tracy *et al.* (1990) found the magnitude of signal varied depended on the duration of NO bubbling used to prepare the authentic NO solution. The longer the solutions were bubbled with NO, the greater the signal noted by the diazotization assay. In contrast, the duration of bubbling had no effect on the vasodilator potency of the NO solution in bioassay. They then removed the NO from the test solutions by bubbling it with an inert gas and made the elegant observation that bioassay activity was, as expected, severely diminished; surprisingly, NO levels estimated from the diazotization assay were unaltered. They correctly interpreted these findings as evidence that a nonvolatile, hemodynamically inactive substance, probably NO, was accounting for much of the NO signal in the diazotization assay. It appears that despite the use of nominally O_2-free solutions, the small contaminating amounts of O_2 caused conversion of NO to NO, which was biologically inactive but was converted to NO during the diazotization assay by the acid pH.

Of even more concern, the administration of ATP, A23187, and BK decreased tension in aortic rings from 40 to 50% (as did authentic NO), but the EDRF failed to produce any signal in the diazotization assay. This indicates that either the diazotization assay is insufficiently sensitive to detect the amount of NO produced under these conditions or the endothelial relaxing factor in these experiments was not NO. It appears that the diazotization assay is not adequately sensitive to measure NO at submicromolar concentrations and is best considered a measure of nitrite and mixed nitroxides.

ELECTRON PARAMAGNETIC RESONANCE (EPR) ASSAY

NO is a gaseous, paramagnetic molecule with the unpaired electron in the _ orbital. The basic principle of detecting radicals by EPR spectroscopy is that at a discrete amount of energy (microwave frequency) and magnetic field strength, unpaired electrons are promoted to higher energy levels; relaxation from this state produces a characteristic spectrum. NO and NO_2 cannot be studied by simple EPR as the relaxation time of the stimulated electron to ground state is too rapid to be detected (Maples K *et al.* 1991). Spin traps are compounds that interact with the unstable radicals, producing a more stable adduct that can be detected by EPR. The spin traps that can be used to detect NO include the nitroxide compounds and hemoglobin. Nitroxide traps are relatively stabilized by the resonance of the unpaired electron. There are two types of nitroxide spin traps, the nitrones and the nitroso compounds. The nitroso compounds usually provide the best information about the identity of a radical, as the radical is added directly to the nitroso nitrogen, thereby increasing the information in the hyperfine splitting parameters (Buettner GR 1987). However, oxygen-based adducts of the nitroso compounds are unstable and one must usually rely on the nitrones for such studies. The identity of the radical, defined by EPR, is characterized by the magnitude and multiplicity of the hyperfine splitting of the spin adduct's spectra. The amount of radical present is proportional to the magnitude of the signal.

Methods for EPR using nitroxide traps

A commercially available spectrometer is used. The magnetic field is usually set at or near 330 mT, with a microwave power of 8-10 mW, frequency of 9 GHz, and modulation amplitude of 0.1 mT. Experiments need to be performed in darkness to prevent photolysis of the spin traps. When NO interacts with the nitroso compounds (*e.g.*, 2-methyl-2-nitrosopropane (MNP) and 3, 5, -dibromo-4-nitrosobenzene, DBNBS), a triplet signal is formed that can be characterized by its line width and hyperfine coupling constant.

Utilization

Although in principle this technique is simple, the attempt to detect NO using traps which themselves have - NO moieties is problematic. Arroyo and Kohno (1991) compared the specificity of multiple spin traps for NO and concluded that the nitroso spin traps (MNP, DBNBS) are better suited for the identification of NO-rebated signals than the nitrones (*e.g.*, 5, 5 - dimethyb-pyrrobine-N-oxide (DMPO). The spectra generated by the nitrones in the presence of NO appear to result from hydrolysis of the probes and provide an inaccurate estimate of NO levels. In fact, nitrones are unstable at acid pH and tended to yield NO-type spectra even in the absence of NO.

Hemoglobin is also useful as a spin trap because nitrosyl hemoglobin (NO-Hb) is readily measured by EPR. NO interacts with hemoglobin, producing both NO-Hb and methemoglobin (the NO-Hb interacts with oxygen, producing further methemoglobin) (Henry Y *et al.* 1991). The 77 K spectra of NOHb has a characteristic three-line, hyperfine signal that is not seen when hemoglobin is exposed to other nitro compounds (*e.g.*, NO, azide). A super-hyperfine structure (triplet of a triplet) is also detectable. The g values for NO-Hb are 2.060, 2.010, and 2.005, with values of 26.8 and 6.4 G. An advantage of using hemoglobin as a spin trap technique is the commercial availability of hemoglobin bound to agarose (Sigma, St. Louis, Mo.). This gel can be applied to columns suitable for EPR spectrometry, as described by Wennmalm *et al.* (1990). Quartz columns (length 15 cm, ID 3.8 mm) are lined with a hemoglobmn-agarose gel (created by dissolving 400 l of Hb-agarose in 1 ml of H_2O) and then excess material is washed free with 10 ml of H_2O. The hemoglobin is reduced (manifest as a change in color from brown to purple) by the addition of dithionite (10 mg/I ml H_2O). The column can then be exposed to the experimental specimen and analyzed by EPR at 77 K. Like the chemiluminescence assay, the EPR NO-Hb assay can measure NO itself, or by adding dithionite or acid to the specimen, can measure combined NO and nitrite. Greenberg *et al.* (1990) studied EDRF activity and NO production in femoral arterial rings using the N0-Hb EPR technique. The threshold for detection of authentic NO was I nM (spectra obtained at 77 K using a spectrometer operating at 9.42 GHz with 20 mW power and 100 kHz modulation frequency). They attempted to detect NO production in response to acetylcholine by using this technique, but found no signal in the effluent from acetylcholine-treated femoral arteries even though ACH and the authentic NO had produced similar relaxation in the bioassay. They concluded that ACH-induced vasodilatation was not mediated by NO in the canine femoral artery. The conclusion that ACH-induced vasodilatation does not (always) occur through production of NO is supported by studies of bovine and rat pulmonary arteries using the chemiluminescence assay. Wang *et al.* (1991) noted increased levels of NO-Hb in rats with endotoxic shock. They enhanced EPR-detection of NOHb by pooling the blood from several animals to increase the signal intensity. The signal-to-noise ratio was improved by subtracting the EPR spectra from the blood of control animals. They reported that levels of NO-Hb did not correlate with vascular tone, an observation made previously in studies that failed to find a correlation between dilator potency of the nitro vasodilator drugs and the blood levels of NO-Hb. The authors discuss the hypothesis that only that fraction of NO that reaches the lumen is available for interaction with reduced hemoglobin. The assay may therefore not have access to the pool of NO that determines vasodilator efficacy, namely, that which reaches the smooth muscle. However, this theoretical criticism applies potentially to all assays. One advantage of the N0-Hb assay over the chemibuminescence assay is the ability of the former to measure NO in blood, and unlike the diazotization assay, acidification is unnecessary, although acidification of the specimen will convert the NO-Hb EPR assay to a nitrite assay. Wennmalm *et al.* (1990) used the N0-Hb complex, as detected by EPR, to measure nitrite in humans after forearm ischemia. They added dithionite to specimens, there by converting nitrite to NO, and then measured EPR spectra using a spectrometer (microwave frequency 9.17 GHz, power 8 mW). Spectra were scanned from 3000 to 3500 gauss with a scan rate of 250 G/min. They established a "normal" venous nitrite level in humans of 0-0.6 mM and noted that ischemia greatly increased this level.

Westenberger *et al.* (1990) detected the NO adduct of Hb in the venous blood of rats with endotoxic shock using EPR (microwave power 10 mW at 9.20 GHz, modulation amplitude 8 G). In sepsis, NO-Hb accounted for as much as 0.8% of all hemoglobin. They found the typical three-line hyperfine splitting of 1.75 mT at g' 2.012 and 1.6 mT at g' 2.609 "characteristic of an unpaired electron coupled to nitrogen in a ferrous heme-NO complex". The authors identified a separate EPR signal at g' 6 that corresponded to the ferric heme in methemoglobin. Maples *et al.* (1991) found a linear relationship between the amount of NO inhaled and the intensity of the NO-Hb EPR signal in broncho-alveolar lavage fluid from human volunteers who intermittently inhaled low doses of NO (1.5-4 ppm, 20 mm every other day x 6). Unfortunately the equipment required to perform EPR is expensive and considerable expertise is necessary. In contrast, the chemiluminescence assay is easier to learn and much less costly. Furthermore, the interaction between NO and hemoglobin is pH dependent. Liu *et al.* (1987) showed that at alkaline pH there was significant impairment of EPR detection of NO-Hb adduct due to inhibition of the interaction between NO and Hb (hyperfine coupling was unchanged). Hemoglobin forms a fairly stable adduct with NO that is readily detected by EPR (Doyle MP and Hoekstra JW 1981). For those with the necessary equipment and expertise, EPR, particularly using hemoglobin as a trap, is a useful means of assaying NO.

METHEMOGLOBIN SPECTROPHOTOMETRY ASSAY

Unlike the EPR assay, this technique is based on the rapid oxidation of reduced hemoglobin (Fe^{2+}) to methemoglobin (Fe^{3+}) by NO rather than on the detection of NO-Hb. NO's oxidation of reduced hemoglobin occurs in less than 100 ms, suggesting the assay is sufficiently dynamic to study the time course of NO synthesis. The threshold for detection of NO using this assay is I nM.

Methods

Commercially available hemoglobin is 50-95% methemoglobin (Matsunaga K and Furchgott RF 1988) and must be reduced with a molar excess of dithionite. Once reduced, the hemoglobin gradually reoxidizes to methemoglobin. Although some authors have then purified the mixture by gel chromatography using a Sephadex column, this is not essential. NO is detected by observing the characteristic shift in the Soret absorbance peak of hemoglobin from 433 nm to 406 nm. Accumulation of oxyhemoglobin is not a problem in this assay due to the much higher affinity of hemoglobin for NO than O_2. Furthermore, oxyhemoglobin can be detected by its Soret peak (416 nm). Conversion of reduced to methemoglobin can be measured as the extinction difference at 401 and 411 nm (the isobestic point) (Schror, K *et al.* 1991).

Utilization

Ignarro *et al.* (1987) detected the production of NO or a related nitrosyl product from A23187-stimulated aortic endothelial cells by using this assay. They found that estimates of NO synthesis provided by the diazotization and methemoglobin assays were comparable. Kelm *et al.* (1988) measured the production of EDRF/NO by porcine thoracic aortic endothelial cells using a "difference spectrophotometric" variant of this assay. They used 2-4 ml of endothelial cells grown on beads as a source of EDRF and perfused this column of beads with medium containing 4 iM HbO_2. Methemoglobin production was measured as the extinction difference between the absorption maximum and isobestic point of reduced vs. met-hemoglobin. These authors measured NO in real time using a dual wavelength (401 and 411 nm), double-beam spectrophotometer equipped with a flow-through cell. They were able to document the production of NO beginning 15 s after stimulation of endothelial cells with bradykinin (peak production at I min). Kostic and Schrader (Kostic M and Schrader J 1992) used this assay to indicate that post ischemic myocardial hyperemia is due at least in part to increased production of NO. Schror *et al.* (1991) demonstrated increased NO-Hb in response to nitro vasodilators in the Langendorff rabbit heart. The advantages of the measurement of NO by monitoring methemoglobin production include the wide availability of spectrophotometers, avoidance of sample acidification, and the relative stability of the methemoglobin. Furthermore, as the reduced hemoglobin is present from the beginning of an experiment, NO synthesis can be measured continuously, removing the uncertainty as to when to sample for NO. However, unlike the non acidified chemiluminescence assay, which gains specificity from the volatility and water insolubility of NO, this assay might detect other nitrosyl groups in the sample.

GAS CHROMATOGRAPHY

Although NO can be detected by standard gas chromatography techniques, this assay is much less sensitive than chemiluminescence or EPR assays. Pai *et al.* (1987) used a Varian gas chromatograph with a Porapak Q column and electron capture detector to measure enzymatical by generated NO. They found that gas chromatography was much less sensitive than chemiluminescence assay, a difference that they attributed part to the sample processing methodology required for gas chromatography.

ALTERNATE NITRIC OXIDE ASSAYS

Mass spectroscopy

Gustafsson *et al.* (1991) recently used mass spectroscopy to measure NO in the breath of guinea pigs. They trapped NO as nitroso thio proline by bubbling the expirate through a degassed aqueous solution of thioproline (15 nM). The nitroso thioproline was extracted with ethyl acetate, and dried and derivatized by N-methyl-N-(tertbutyldimethylsilyl) trifluroacetamide in acetonitnile. Samples were placed on a CP-Sil-5-CB column interfaced with a mass spectrometer operating in the positive chemical ionization mode with isobutane. The derivatizedform of nitrosothioproline had a retention time of 4.76 mm. It is not clear whether this assay is more sensitive than measuring breath chemiuminescence. Indeed, in the same experiment breath NO was exposed to Fe (II) sulfate and the resulting nitrite was measured by chemiluminescence. The principle of trapping NO in degassed water for subsequent analysis, regardless of the technique, appears to a reasonable strategy for measuring exhaled NO. In addition, these experiments noted that in the rabbit, inhalation of physiologic levels of hypoxia (6% O_2) acutely decreased the amount of expired NO from normoxic levels of approximately 24 to 20 ppb.

MICROELECTRODE NO ASSAYS

There are currently two amperometric methods available for measuring NO from intact tissues and single cells. Shibuki (1990) recently reported the measurement of NO release in the cerebellum using a micro electrode. He modified a miniature O_2 electrode by sealing its fire-polished 150-250 tm tip with a thin chloroprene rubber seal so that only low molecular weight gases could enter. By introducing a platinum cathode within the pipette and holding it at positive voltage, it was possible to detect NO, a readily oxidized gas, by its oxidation at the electrode's surface. Pipettes were fibbed with 30 mM NaC1 and 0.3 mM HC1 (pH 3.5), and a Teflon-coated platinum wire was placed as close as possible to the membrane. A 200 im silver wire was the anode. The cathode was kept 0.9 V from the anode and the anode was grounded. The principle of the assay is that if the anode is grounded there should be no current. There was a linear relationship between current and NO concentration *in vitro* over a small range (1-3 tM NO). Using this electrode, Shibuki was able to measure NO release in rat cerebellar tissue elicited by electrical stimulation of the white matter. The author claims a nanomolar NO sensitivity for the technique with a rapid response time (seconds). If others duplicate this technique it may permit measurement of local production of NO in tissues.

Luis A.J. Mur *et al.* 2005 demonstrate the use of laser photoacoustic detection (LPAD) as a highly sensitive method to detect in planta nitric oxide (NO) production from tobacco (*Nicotiana tabacum*). The specificity of the photoacoustic signal for NO when adding gas or the NO donor, sodium nitroprusside, on injection into plant leaves, was demonstrated by its abolition with O_3 ($NO + O_3 \rightarrow NO_2 + O_2$). The utility of the LPAD method was shown by examination of a nonhost hypersensitive response and a disease induced by *Pseudomonas syringae* (*P. s.*) pv *phaseolicola* and *P. s.* pv *tabaci* in tobacco.

NO in plants

NO involved in various aspects of plant growth and development (Beligni and Lamattina 2000, Gouvea *et al.* 1997, Leshem *et al.* 1998) action of NO in plants has been mainly accomplished by exogenous application of NO donors in a wide range of plant systems. The concentration of NO release and the reducing power (Ramamurthi and Lewis1997). Various role of nitric oxide in growth and developmental processes, such as :

SEED GERMINATION

Exogenous application of NO and related nitrogen oxides have been reported as stimulators of seed germination and break dark imposed dormancy (Keenley and Fotheringham 1997, Giba, 1998). Seeds of *Paulownia tormentosa* need long periods of light to germinate. However, if seeds are supplied with 10 mM KNO_3, a single red light exposure was sufficient to induce germination (Grubisic and Konjevic, 1990). SNP and some other organic and inorganic nitrates also stimulated germination with a similar efficacy (Grubusic *et al.* 1992). The dark imposed dormancy of lettuce seeds could also be broken by either 100μM SNP or 100μM SNAP to the same extent as both GA_3 (Beligni and Lamattina, 2000). Thus, stimulatory effects of nitrate compounds in plants could be due to their reduction to NO and that phytochrome regulated dormancy breakage in photoblastic seeds was closely connected with redox changes (Grubusic *et al.* 1992). These effects of nitrite on seed germination suggest that soil nitrite is one of the factors determining seed germination (Neill *et al.* 2003).

CELL GROWTH AND DIFFERENTIATION

In the life cycle of a plant, light is known to inhibit the elongation growth of hypocotyls and internodes. NO donors are reported to significantly reduce hypocotyls growth and internode elongation in three different plants species (*Arabidopsis thaliana, Lactuca sativa and Solanum tuberosum)* grown in dark or under conditions of low light intensity (Beligni and Lamattina, 2000). Low micromolar concentrations of NO increased the rate of leaf expansion in pea and elongation of maize root segments (Leshem and Haramaty1996; Gouvea *et al.* 1997). NO also exerts effects on cellular differentiation, especially xylogenesis, and is currently regarded as a signal molecule in plant cell differentiation. Programmed cell death (PCD) is one of the most distinctive characterstics exhibited by the differentiating xylem in plants and is coordinated with the processes of secondary wall formation and lignification (Groover and Jones, 1999). NO has been shown to affect plant peroxidases involved in cell wall lignification in the xylem of *Zinnia* seedlings and *in vitro Zinnia* cell culture system. NO production remains characteristic of vascular tissue (predermined to die). The scavenging of NO by 2-phenyl-4, 4, 5, 5- tetramethyl imdazoline-1-oxyl-3-oxide (PTIO) inhibits tracheary element differentiation and increase cell viability (Gabaldon *et al.* 2005). In the second system, Gabaldon *et al.* (2005) used mesophyll cells of *Zinnia* leaves for studying transdifferentiation of cells into tracheary elements.

Cell respiration: High concentration of NO affects mitochondrial functionality in plant cells by reducing cell respiration due to strong inhibition of the cytochrome pathway (Zottini *et al.* 2002). They observed that treatment of suspension cultures of carrot cells with SNP (1 mM) decreased cell viability and cPTIO prevented the killing effect of NO, thus demonstrating that NO is responsible for cell lethality. In plants, in addition to cytochrome oxidase, a second terminal oxidase, termed the alternative oxidase (AOX) is also present. Unlike cytochrome pathway, which is coupled to oxidative phosphorylation, electron transport is shunted into the alternative pathway resulting in a nearly normal rate of respiratory O_2 uptake (Miller and Day, 1996). According to Millar and Day (1996), AOX pathway may play a role in NO tolerance in higher plants by lowering respiration dependent production of ROS under conditions where NO is overproduced (nitrosative stress).

PHOTOSYNTHESIS

Air pollution studies in plants have shown that NO inhibits photosynthetic CO_2 assimilation (Wellburn, 1990). Since NO is degraded to nitrite (or nitrate) in an aqueous solution under ambient air, But studies of Takahashi and Yamasaki (2002) suggest that nitrite is involved in the mechanism of NO dependent inhibition of photophosphorylation, and confirm that NO molecule itself has a suppressive effect on photophosphorylation by inhibiting electron transport activity in photosystem II. According to these authors, inhibitory effect of NO is due to the NO induced dissociation of bicarbonate from the specific sites of thylakoids. The sensitivity of photosynthesis against NO depends on the local concentration of bicarbonate within thylakoid membranes. NO also inhibited induced pH formation across the thylakoid membrane. This NO-induced inhibition was found to be temporary as the activity was completely recovered by removing NO.

Emission of NO from plants has been linked with the accumulation of nitrite in many cases (Mallic *et al.* 2000 a,b; Klepper, 1990, 1991).Nonetheless, nitrite is accumulated in the cells when photosynthetic electron transport (Shingles *et al.* 1996) or nitrogen metabolism (Mallic *et al.* 2000a,b) is not operational. The activities of nitrogen metabolism and carbon assimilation are interdependent (Huppe and Turpin, 1994). From these studies, it appears that NO acts a transmembrane messenger to alarm dysfunction nitrogen assimilation so that photosynthetic activity is down regulated.

CELL METABOLISM

NO also affect the functioning of cellular organelles like chloroplasts (Takahashi and Yamasaki, 2002), mitochondria (Zottini *et al.* 2002) and subcellular organelles like peroxisomes (del Rio, 2002). Chloroplast and mitochondria are important cell organelles for energy transduction in the processes of photosynthesis and respiration, respectively NO exerts its effect at various levels.

ROOT ORGANOGENESIS

Exogenously applied NO donors, such as sodium nitrite (SN), sodium nitroprusside (SNP), nitrosoglutathione (NGLU) and nitrocystein (NCYS) are reported to induce elongation of maize root tip segments in a dose- dependent manner (Gouvea *et al.* 1997). These workers suggested that NO is a putative elongation inducer and that IAA and NO donors conceivably share common steps in the signal transduction pathway, since both elicited the same plant response. Pagnussat *et al.* 2002 also demonstrated dependence of auxin- induced adventitious root formation in cucumber explants on NO. IAA treatment in cucumber explants induced a transisents increase in the level of endogenous NO in the basal region of the hypocotyls where new root meristems develop. Moreover, NO has been found effective in inducing adventitious root formation in woody species (Lamattina *et al.* 2003). Stohr (2001) reported involvement of plasma membrane- bound enzyme the nitrite: NO reductase in the formation of NO in plant roots. The presence of NO is also reported in root nodules (Mathieu *et al.* 1998). EPR analysis of soybean plants grown in the absence of nitrate showed the presence of leghemoglobin-NO (Lb-NO) complexes in root nodules. The concentration of Lb-NO complexes in root nodules. The concentration of Lb-NO complexes varied with age, being higher in younger nodules and almost absent in senescent ones (Mathieu *et al.* 1998) they suggested that NO binding to heme group of hemoglobin is essential to have a role in nitrogen fixation . Possibly, Lb-NO could react with O^{2-} and / or H_2O_2 to avoid ROS damage to the nodules and controlled levels of Lb-NO complex in active nodules could result in an equilibrium between oxy-Lb mediated oxygen transport and ROS formation. But according to Nathan (1992), NO exerts deletrious effect on the rate of nitrogen fixation. However, significance of the effect of NO during symbiosis is still far from completely understood.

FLOWERING

He *et al.* (2004) provided pharmacological, physiological, molecular and genetic data to demonstrate that NO represses the photoperiod and autonomous floral pathways in *Arabidopsis thaliana*. Treatment of *A. thaliana* plants with NO donor (SNP) enhanced the vegetative growth and significantly delayed flowering. Studies revealed that NO affects the genes that control both the autonomous and environmentally sensitive pathways that control flowering. NO suppressed CONSTANS and GIGANTEA gene expression and enhanced FLOWERING LOCUS C expression indicating thereby that NO regulates the photoperiod and autonomous pathways. Several genes affected by NO (CAB, CO and GI) are also regulated by light, the primary effect of NO may be on light signaling, and the effect on flowering may be secondary.

PIGMENT SYNTHESIS

NO donors are reported to stimulate de-etiolation and increase chlorophyll content in potato, lettuce, *Arabidopsis* and dark brown wheat seedlings (Beligni and Lamattina, 2000). Short term pulse of white light or red light cause 2-3 fold increase in the chlorophyll content of NO treated seedlings over control ones suggesting thereby a putative role for NO in de-etiolation (Beligni and Lamattina 2000). The magnitude of increase in chlorophyll was further elevated when seedlings were subjected to abiotic and biotic stresses, suggesting thereby a possible connection between NO, light and stress signaling pathways in plants. NO also increased chlorophyll content in leaves of pea, particularly in guard cells (Leshem *et al.* 1996) and retarded chlorophyll loss in potato leaves infected with *phytophthora infestans*. NO donors are reported to significantly increase chlorophyll content and chloroplast membrane density in maize plants growing under conditions of very low iron concentration (Granziano *et al.* 2002). It is well known fact that iron deficiency impairs chlorophyll biosynthesis and chloroplast development (Thoiron *et al.* 1997). It seems quite likely that enhanced chlorophyll content and vigor of N- fertilized crops is the result of increased availability of NO. It also enhances betacynanin synthesis in *Amaranthus caudatus*. The synthesis of anthocyanin and betacynanin is promoted by cytokinins and red light (Bamberger and Mayer 1990, Piatelli *et al.* 1969). Use of NOS inhibitors and NO scavengers blocked the Cytokinin induced pigment synthesis suggesting that NO is required to accomplish cytokinin function.

STOMATAL CLOSURE

ABA is an endogenous antitranspirants that induces stomatal closure, thereby leading to water conservation. Recently, NO has been identified as a mediator of guard cells ABA signaling (Neill *et al.* 2002a). Scavenging of NO or inhibition of NO synthesis reduces ABA induced stomatal closure. Garcia-Mata and Lamattina (2001) showed that exogenous application of NO as SNP reduced transpiration and induced stomatal closure in several species, such as *Vicia faba*, *Salpichera* and *Tradescantia*. NO- induced stomatal closure is likely to be a common phenomenon. The ABA induced stomatal closure was completely checked when epidermal peelings of pea leaves were treated with PTIO, a scavenger of NO (Garcia-Mata and 2002). They suggested involvement of NO as an endogenous signaling intermediate in ABA induced stomatal closure.

PEROXISOMAL FUNCTIONALITY

Peroxisom are subcellular organelles with oxidative type of metabolism, biochemical, immunological and immunocytochemical data provide evidence that plant peroxisomes contain NOS activity (per NOS) which is calcium-dependent, constitutively expressed, and immunorelated with the mammalian NOS (Barroso *et al.* 1999). The presence of NOS in peroxisomes, suggests that these organelles are a cellular source of NO. Inside peroxisomes, NO could react with oxygen radicals generated in peroximal matrix by xanthin oxidase to form powerful oxidants peroxynitrite which could regulate conversion of xanthin dehyrogenase into superoxide- generating xanthin oxidase (del

Rio *et al.* 1992). The level of measured NO was linearly dependent on the amount of peroximal protein (Corpus *et al.* 2004). NO can also react with reduced GSH to form S-nitrosoglutathione (GSNO), a reactive nitrogen oxide species (Wink *et al.* 1995). GSNO has been reported as a powerful inducer of defense genes (Durner *et al.* 1998).Presence of NOS in plant peroxisomes implies that these organelles could also have a function as a source of signal molecules like NO, superoxide radicals, hydrogen peroxide and GSNO.

SENESCENCE AND PROGRAMMED CELL DEATH

Cell death is an essential process during the plant's life cycle. Two modes of cell death are described in plants: Programmed cell death (PCD) and Necrosis. Two examples of PCD in plants, senescence and cell death associated with the hypersensitive response (HR); demonstrate range of forms that PCD can take place in plants. Senescence is a highly regulated process during which new metabolic pathways are activated and others are turned off (Pedroso *et al.* 2000).

PCD is genetically controlled and has characteristic feature of the apoptotic cell death in animals. These features include cell shrinkage, chromatin condensation and DNA fragmentation.

Necrosis, also a type of cell death results from severe and persistent trauma and is considered not to be genetically orchestrated (Loak, 2001).

Delledone *et al.* (2001) have shown that the interaction between NO and ROS can determine whether or not PCD is determined by NO: superoxide ratio. If level of superoxide is greater than NO, then NO reacts with superoxide to form peroxynitrite which does not results in PCD. On the other hand, Leshem *et al.* (1998) observed lower NO concentration in mature fruits than in green fruits, and senescing flowers compared with fresh flowers. To confirm this, the authors applied exogenous NO to flowers, fruits and vegetables and observed that NO delayed senescence and extended the post harvest life in all cases. If level of NO is greater than superoxide, then NO reacts with H_2O_2 to induce cell death. Treatment of tobacco cells with both H_2O_2 and NO together induced substantial increase in cell death with PCD characteristics (Neill *et al.* 2002 b).

TARGETING OF POLLEN TUBES TO OVULES

Prado *et al.* (2008) have investigated the role of NO in the process of pollen–pistil communication, using *Arabidopsis in-vivo* and lily *semi-vivo* assays. Cross-pollination between wild-type and AtNOS1 plants shows that the mutation affects the pistil tissues in a way that is compatible with abnormal pollen tube guidance. The presence of NO in an asymmetric restricted area around the micropyle is demonstrated. The pollen–pistil interaction transcriptome indicated a time-course-specific modulation of transcripts of AtNOS1 and two Nitrate reductases (nr1 and nr2), which collectively triggered a putative NO signaling pathway. It was also inferred that NO was necessary for micropyle targeting to occur. Activation of calcium influx in pollen tubes partially rescued normal pollen tube morphology, suggesting that this pathway is also dependent on Ca^{2+} signaling. A role of NO in modulating Ca^{2+} signaling was further substantiated by direct imaging the cytosolic free Ca^{2+} concentration during NO-induced reorientation, where two peaks of Ca^{2+} occurred. Taken together, these results provided evidence for the participation of NO signaling events during pollen–pistil interaction. Of special relevance, NO seems to directly affect the targeting of pollen tubes to the ovule's micropyle by modulating the action of its diffusible factors.

NITRIC OXIDE IN PLANTS AND CROSS TALK WITH VARIOUS PLANT HORMONES

Growth and development of plants is under the control of natural mobile biomolecules called plant hormones, which are transported from cell to cell and organ to organ to signal and coordinate various processes. Apparently, NO has emerged as an important bioactive molecule involved in regulation of a wide range of processes. Nitric oxide (NO) research is a new exciting field in plant biology. Yaacov Leshem, the pioneer in the study of nitric oxide in plants, has proposed that this bioactive molecule should be classified as a phytohormone which may function as a gaseous endogenous plant growth regulator (Leshem, 2001) have postulated NO to be a non-traditional regulator of plant growth.

Several studies have reported that NO can regulate processes related to plant growth and development, some of which are summarized in Table 1.

Exogenous application of different plant hormones has been reported to trigger endogenous production of NO (Lamattina *et al.* 2003), which could alleviate many stress symptoms in plants by contributing to cellular homeostasis. The small size and high diffusion rate of NO through biological membranes mean that NO fits the premises that hormones are easily transported.

NO AND INDOLE ACETIC ACID (IAA)

NO synthesis has also been found to be induced by auxin in cucumber roots (Pagnussat *et al.* 2002). The dependence of NO in auxin induced adventitious root development reflects the possibility that IAA and NO might share common steps in signal transduction pathway. A plasma membrane bound enzyme, nitrite: NO reductase

TABLE 1: Biological activities of nitric oxide in plants, Beligni and Lamattina, 2000.

Tissue or organ	*Physiological action*	*Species*	*Optimum concentrations*
Seeds	Induction of germination	*Paulownia tormentosa,* Lettuce,California chaparral(*Emmenanthe penduliflora)*	10^{-6}
	Inhibition of respiration after imbibitions	Soybean	10^{-6}
	Inhibition of aleuronic cell death	Barley	10^{-6}

Roots	Elongation	Maize	10^{-10}
	Induction of adventitious and lateral root formation	Cucumber, lavender (*Lavendula spp.*)	10^{-9} to 10^{-5}, 10^{-6}
Tuber	Tuberization	Potato	10^{-6}
Hypocotyls	Inhibition of growth in low light conditions	Lettuce, *Arabidopis thaliana*	10^{-6}
Stems	Inhibition of internodal growth in low light conditions	Potato	10^{-6}
Leaves	Induction of de-etiolation	Wheat, barley	10^{-6}
	Delay of senescence	Pea	$5\text{-}10^{-6}$
	Stoma closure	Wheat, *Vicia faba*	10^{-6}
	Leaf expansion	Peas	$5\text{-}10^{-6}$
	Induction of defensive responses	*Arabidopsis*, Tobacco	$2\text{-}10^{-6}$
	Inhibition of cell death	Potatoes	10^{-6}

(Ni-NOR) was reported in the formation of NO in plant roots (Stoher *et al.* 2001). But it remains to be tested whether NO is also involved in other IAA-mediated plant physiological responses.

NO AND GIBBERELEIC ACID (GA)

NO donors also reported to promote germination of photoblastic seeds in the dark to the same extent as both GA treatment or 5 min pulse of while light. But these seeds also germinate in light in presence of NO scavenger, suggesting that their mode of action to stimulate germination is different (Beligni & Lamattina, 2000).

NO AND CYTOKININS

Cytokinins have been shown to induce NO synthesis in tobacco parsley and *Arabidopsis* cell cultures (Tun *et al.* 2001,Chory *et al.* 1994). Some effects of cytokinins can be mimicked by NO. The synthesis betacyanins in *Amaranthus* seedlings is promoted by cytokinins and red light (Bamberger & Mayer, 1990). NO mimicked cytokinins action on pigment synthesis, suggesting that it act downstream of cytokinins to promote the response, or that NO is necessary to accomplish its function (Lamattina *et al.* 2003). NOS inhibitors and NO-scavenger block the effect of this hormone on betacyanin synthesis (Scherer & Holk, 2000). A novel role of cytokinins in the induction of programmed cell death has been recently proposed. It is possible that this cytokinins induced response is mediated by NO. It also triggers several responses in which hormone and phytochrome interact or act separately to give same response. Exogenous application of cytokinins inhibits hypocotyls elongation in manner similar to light treatment (Chory *et al.* 1994).

NO AND ABA

The recent discovery of NO involvement in ABA-induced stomatal closure represents a significant development in NO research. Studies of Garcia-Mata & Lamattina, (2002) in *Pisum sativum* and *Vicia faba* demonstrated at ABA induced NO production was essential for stomatal closure as removal of NO with scavenger PTIO substantially inhibited ABA-induced closure. However,. ABA did not induce NO synthesis in *Arabidopsis* suspension cultures, indicating tissue specificity (Tun, *et al.*).

NO AND ETHYLENE

Ethylene, a gaseous hormone, plays an active role in plant responses from germination, cell expansion to stress responses and fruit ripening. The effect of NO and ethylene during maturation and senescence is antagonistic and an inverse relationship between endogenous content of these two gaseous molecules has been observed during fruit ripening (Leshem & Pinchasov, 2002). NO application extends post harvest life of horticulture produce by inhibiting ethylene production (Leshem *et al.* 1998). The administration of sildenafil, an inhibitor of cGMP-degrading phosphodiestrases, results greater inhibition of ethylene production and indicates that c GMP could be involved in the NO-induced inhibition of ethylene metabolism.

NO AND SALICYLIC ACID

NO action in plants is closely linked to salicyclic acid (SA). SA functions at multiple nodes in the defense signaling network to potentiate a variety of plant defense responses including oxidative burst. NO burst has been reported in tobacco and soybean plant tissues treated with a virulent bacterial pathogens or elicitors (Clarke *et al.* 2000 Pedroso *et al.* 2000). NO acts synergistically with SA as well as ROS to potentiate defense and according to song and Goodman, (2001) NO are at least partially required for function of SA.

FUNCTIONS OF NITRIC OXIDE IN BIOTIC STRESS

Indications that NO is involved in signalling defence responses during plant-pathogen interactions have been well documented in many experiments. The challenge of a pathogen very often leads to the induction of HR. There is some evidence that NO plays a key signalling role during HR, next to the accumulation of ROS and SA. Hypersensitive cell death *via* NO is a typical example of programmed cell death (PCD). It was shown that NO-donor treatment of plant tissue initiates chromatin condensation and DNA fragmentation. Moreover, NO-provoked cell death may be inhibited by animal caspase-1 inhibitor. Although studies confirm caspase activity in plants, and transgenic plants with an over expression of a caspase inhibitor-protein p35 and Op-IAP show the inhibition of HR, A plant cellular responses to infection are often correlated with rapid, localized cell death termed as HR at the site of infection. This is also said as (PCD) in the sense that it can be mimicked by plant mutant in the absence of pathogen. This rapid responses is followed by the acquisition of resistance to a range of pathogen at sides away from the original infection termed as systemic acquired resistance (SAR) (Alvarz *et al.* 1998; Hammond-Kosack & Jones, 1996). The responses to plant attack by the pathogens involves the activation of a wide variety of defenses in order to prevent pathogen attack which include the increased expression of various defense genes, such as that for phenyl-alanine ammonia lyase (PAL). PAL is the first enzyme in phenyl propanoid pathway involved in synthesis of phytoalexins with antimicrobial activities. They First induction of the involvement of NO in disease resistance came from the work on potato tuber (Noretake *et al.* 1996). They observed that treatment of the potato tuber tissue with NOC-18 (NO-donor) provoked an accumulation of phytoalexin riehitin, an endogenous antibiotic compound. The effect NOC-18 was counteracted by PTIO and free radial scavenger Tiron (Noritake *et al.* 1996). NO also protected chlorophyll levels in potato leaves infected with pathogen *Phytophthora infestans* (Beligni *et al.* 1997). Durner *and* Delledonne *et al.* (1998) demonstrated a regulatory role for NO as a plant defense signal against pathogen infection. They demonstrated that NO collaborates with ROS to trigger transcriptional activation of plant defense genes and HR. Injection of the animal NO synthase, its substrate and cofactor to tobacco leaves, a common marker for defense-gene transcription called PR-1 was activated through the accumulation of SA, an intermediate required for many disease resistance responses. SA is a crucial signal during plant-pathogen interactions and NO acts synergistically with SA as well as with ROS to potentiate defense response (Song and Goodman 2001). In *Arabidopsis* suspension cells, exogenous NO induced cell death at concentrations similar to those generated by cells challenged by avirulent bacteria (Clarke *et al.* 2000). In soybean and tobacco cell suspensions a simultaneous increase in NO and H_2O_2 activated cell death, whereas, an independent increase of only one of the above-mentioned factors induced cell death only slightly (Tuteja *et al.* 2004; Pinto *et al.* 2002) Moreover, cytological observations showed that either administration of NO-donors or a change in H_2O_2 level has no effect on elicitation of HR in infected oat cells, although both molecules were required for onset of death in neighbouring cells (Tada *et al.* 2004). The mechanism by which NO and H_2O_2 interact in killing is still largely unknown. However, chemical reaction between NO and H_2O_2 produce either singlet oxygen or hydroxyl radicals (Noronha *et al.* 1993), which can cause cell death. Reaction of NO with O_2 produces a highly toxic molecule for animal cells, peroxynitrite, mediating apoptosis. For plants ONOO is relatively non-toxic (Dellodone *et al.* 2001). However, it was found that in tobacco leaves ONOO induces PR-1 accumulation (Durner *et al.* 1999) and protein nitration modulating cell redox state. Early NO production, known as NO burst seems to be closely dependent on the genetic makeup of the plant (R genes) and of the pathogen (avr) (Bennett *et al.* 2005). As a confirmation of this fact, prompt nitric oxide production (30-45 min after inoculation) was recorded in non-compatible systems of P. s. pv. *Phaseolicola*-tobacco and P. s. pv. Tomato-*Arabidopsis.* This early "NO burst" directly preceded H_2O_2 generation and occurred approximately 6 h before the appearance of visible HR-type cell death symptoms. As a confirmation lack of NO emission was observed in plants inoculated with a mutant of the avirulent bacterium (hrp), incapable of supplying the avr protein to the plant (Mur *et al.* 2005). Phytoalexin accumulation is another event connected with plant resistance, in which NO seems to be engaged (Able *et al.* 2003). Treatment of potato tubers with exogenous NO stimulated the accumulation of rishitin. Additionally, the effect of inhibited synthesis of this compound was observed after the application of an NO-scavenger (Noritake *et al.* 1996). Independently, biosynthesis of specific phytoalexins was observed after NO treatment of soybean cotyledons (Modolo *et al.* 2002). NO may also participate in the onset of systemic acquired resistance (SAR), in tobacco exogenous NO induces the accumulation of salicylic acid-playing a fundamental role in SAR (Durner *et al.* 1998). Activation of PR-1 protein, obtained *via* NO, occurs with the participation of SA, since as it was shown in transgenic plants unable to accumulate SA (NahG), a similar effect was not observed. Moreover, disease spots caused by TMV on leaves pretreated with NO were considerably more reduced in comparison to those on transgenic ones. The application of inhibitors specific for animal NOS or NO scavengers reduced SAR (Song *et al.* 2001) Thus; these results suggest an important role of NO in the induction of a distal signalling network leading to enhance SAR in tobacco. Studies of effects of NO inhibitors in genetically well defined host pathogen interaction in response to *Pseudomonas syringe* showed promotion of disease infestation and bacterial growth. This clearly demonstrates that NO plays a key role in disease resistance in plants. Likewise, O-infiltration of NOS inhibitors L-NNA (N^W–nitro-L-arginine) or PBITU (5, 5'-1, 3-phenyl-bis(1, 2-ethanediyl)-bis- isothiourea) promoted the spreading of chlorosis accompanied by enhanced bacterial growth in leaves of *Arabidopsis* ecotype Col-O (Delledonne *et al.* 1998). A pathogen inducible NOS (iNOS) has been recently identified in tobacco and *Arabidopsis* (Chandok *et al.* 2003). They observed that plant iNOS has similar kinetic properties to its animal counterpart and also requires tetrahydrobiopterin, FAD, NADPH, Ca^{2+}, and calmodulin for its activity *in vitro*. Most of the available on NO detection during plant-pathogen interaction come from studies of infections by biotrophic pathogens. Rapid accumulation of NO has been observed in soybean and *Arabidopsis* suspension cultured cells in response to avirulent bacteria (Delledonne *et al.* 1998, Clarke *et al.* 2000); in tobacco leaves treated with fungal elicitor cryptogein (Foissner *et al.* 2000) and in potato tubers treated with an elicitor from *Phytophthora infestans* (Yamamoto *et al.* 2003). A direct contact of avirulent crown rust fungus with oat plants also induces production of NO at an early stage in the defense response (Tada *et al.* 2004). Further experiments are needed to examine the course of metabolism of NOS inhibitors or NO scavengers in plants to further assess the ability of NO to induce the SAR response.

FUNCTIONS OF NITRIC OXIDE IN ABIOTIC STRESS

The most important factors limiting the crop productivity are environmental stresses. At whole plant level, the effect of any stress is usually perceived as a decrease in photosynthesis and growth which is associated with alterations in carbon and nitrogen metabolism (Loggini *et al.* 1999). At molecular level, negative effect of stress is associated with generation of reactive oxygen species (ROS) which initiate several oxidative destructive processes and also triggers various signaling pathways (Neill *et al.* 2002 b, Varnova *et al.* 2002). Thus maintenance of appropriate ROS levels might represent survival response. Recently, NO has been found to be involved in diverse biological pathways where it has been proved to be protective against damages provoked by oxidative stress (Beligni and Lamattina, 1999). Favourable balance between ROS/NO is important for healthy plant growth (Bolwell 1999). Evidences showing that plant response to such stressor as drought, high or low temperature, salinity, heavy metals and oxidative stress, is regulated by NO.

DROUGHT STRESS

The major environmental constraints that limit crop productivity and performance is drought stress. It is highly desired to understand the underlying cellular processes that ameliorate the injurious effects of drought stress and conserve water in crop plants. Water deficit is associated with the accumulation of the plant hormone ABA which is found to induce NO synthesis in guard cells. Application of NO donors has been shown to ameliorate injurious effects of drought stress in several plant species (Garcia-Mata and Lamattina, 2001). NO donor reduced transpiration rates and stomatal aperture by 35, 30, and 65 % respectively, in *Tradescantia, Salpickroa organifolia* and *Vicia faba* (Garcia- Mata and Lamattina, 2001). This NO induced stomatal closure was correlated with increase in water content as a result of reduction in transpiration rate, stomatal aperture and ion leakage to the extent found in untreated leaves. ABA synthesis in wheat roots in response to water deficit was much higher in the presence of NO donors and ROS, which suggests synergistic action of ROS and NO (Zhao *et al.* 2001) as a confirmation of this fact, ABA accumulation under stress was blocked after the administration of ROS scavengers and a NOS inhibitor. The accumulation of NO also proved to be necessary in stomata of *Vicia faba* during ABA induced closure of stomata (Garcia-Mata *et al.* 2002) Close functional cooperation between signal molecules of ABA, H_2O_2 and NO was also confirmed by the recent reports. In the cited study it was shown that stress-induced ABA accumulation along with NO and the accompanying reaction of stomatal closure in *Arabidopsis* is additionally dependent on H_2O_2 synthesis (Desikan *et al.* 2006).

TEMPERATURE STRESS

NO participates also in plant response to high and low temperature stress. For example, high temperature treatment of lucerne cells resulted in an increase of NO synthesis, whereas, the application of exogenous NO increased cold tolerance in tomato, wheat and maize (Neill *et al.* 2003) Most probably the observed effects were related with the antioxidative action of NO, which elevates negative effects caused by the intensification of peroxidative metabolism in thermal stress (Neill *et al.* 2002). Application of NO donors mediated chilling resistance in tomato, wheat and corn (Lamattina *et al.* 2001). In *Brassica* species, SNP treatments ameliorated injurious effects of chilling stress imposed on the leaf disc under *in vitro* conditions by modulating the regulation of generation of peroxide radicals and activities of anti oxidants enzymes (Setia *et al.* 2005). Stress tolerance induced by NO reflects its antioxidants properties *via* suppression on high levels of ROS that accumulate following exposure to heat or chilling stress (Neill *et al.* 2002 b).

Very recently Malik *et al.* (2007) demonstrated that cool temperature ($32^0C/12^0C$) delayed initiation of fibers but addition of NO caused a shift in redoxbalance from a reduced state to an oxidative state in the initiating fibers of cotton ovules. Thus, IAA was sprayed to cause initiation of fibers at low temperature. NO is also shown to regulate the auxin degrading system during various phases of fiber growth. The role of NO in early initiation of fibers and their prompt elongation is inferred. NO also affected the activity of acid invertase, preventing sucrose accumulation during the elongation phase. Total sugars decreased with low temperature but NO restored their content. NO also promoted the activity of α-galactosidase indicating restoration of fiberwall synthesis during the elongation.

HEAVY METALS STRESS

Pretreatment of lupin (*Lupinus lutens*) seedlings for 24 h with 10 µM SNP resulted in efficient reduction of detrimental effect of heavy metal like cadmium and lead (Kopyra and Gwozdz, 2003). The protective effect of NO may be at least partly due to stimulation of SOD activity and / or direct scavenging of the superoxide anion.

SALINITY STRESS

NO-donor treatment of rice seedlings caused the effect of loss minimization (Uhida *et al.* 2002). In this case after the application of NO a stronger growth of plants was observed, along with the maintenance of appropriate PS II activity, an increase of antioxidative enzymatic activity and the expression of specific salinity stress resistance genes. On the other hand, prolonged stress conditions may result in an over production of NO and NO-derived products, leading to a range of specific responses, known as nitrosative stress. Recently, (Valderrama *et al.* 2007) indicated that in olive leaves salinity stress induced the production of RNS, *i.e.* NO, GSNO and RSNO, and in consequence, a rise in tyrosine-nitrated proteins, which are good markers of nitrosative stress.

UV-B RADIATION

Mackerness *et al.* 2001 showed also the participation of NO in plant response to UV-B radiation, demonstrating post-stress induction of CHS expression, an increase in NOS-type enzymatic activity and an elevation of NO level. Results of a study by Shi *et al.* 2005 suggest that NO may effectively protect plants against UV-B radiation, most probably through increased activity of antioxidative enzymes. It needs to be mentioned that NO-donor treatment of potato tubers prior to UV-B irradiation resulted in the development of almost 50% more healthy leaves in comparison to plants not subjected to NO.

OZONE STRESS

NO may also interact with atmospheric pollutant ozone. Treatment of *Arabidopsis* plants with ozone induced NOS activity that preceded accumulation of SA and cell death (Rao and Davis, 2001).

WOUNDING

According to Orozco-Cardenas and Ryan, 2002 demonstrated that injury itself does not induce NO synthesis; however, the application of exogenous NO inhibits the process of NO generation and the expression of wounding inducible genes during pathogenesis.

PHOTO-OXIDATIVE STRESS

Oxidative stress within chloroplast originates due to light dependent O_2 reduction. Artifically ROS can be generated with bipyridinum herbicides like diquat and paraquat (Dodge, 1994). These herbicides act at photosystem I as artificial electron acceptor (Beligni and Lamattiana, 2002). Their oxidation produces ROS which when produced in high amounts provokes severe damage in almost every cellular component. In natural conditions, superoxide is formed when electron transport within photosynthetic chain occurs in excessive amounts related to $NADP^+$/NADPH ratio. During a photo-oxidant condition, ROS concentration in the cell can rise to such toxic levels to cause severe damage. Recently, NO has been reported as an important ROS scavenger, able to reduce the extent of chlorophyll loss, ion leakage, necrosis and defoliation caused by herbicides (Beligni and Lamattina, 1999; Dodge, 1994). Beligni and Lamattina (2002) observed marked reduction in diquat- triggered lipid peroxidation, ribulose 1, 5 bis- phosphate carboxylase/oxygenase (rubisco) and DI protein by two nitric oxide donors, SNP (100μM) and SNAP (200μM) in potato leaf pieces and isolated chloroplast . NO donors also protected the RNA against oxidative damage. These effects may, at least in part, be explained by its ability to directly scavenge ROS.

CONCLUSION

Clearly much remains to be discovered about NO action in plants. NO is undoubtedly made by plant cells and has a range of biological activities. Therefore, it would seem likely that the processes by which NO is made and removed are subject to regulation. Even though the details remain to be resolved, it is clear that various stimuli can increase the rate of NO production and that altering NO turnover in cells, either by modulating its production or removal, does have biological effects. NO is evolved from plants and the rate of evolution can be dramatically increased in response to various stimuli, but the physiological significance of such evolution is not clear. Increased NO evolution probably reflects increased cellular NO generation, but whether NO derived from different intracellular sources is evolved at different rates or from different cells is not known. There is clearly much to explore about NO.

Overall, much more work needs to be done before assigning the term plant hormone to NO. The endogenous variation of NO during the plant life cycle need to be monitored whether NO acts directly or indirectly through changes in the balance of plant hormones also needs to be determined. A mutational analysis of NO- regulated growth in *Arabidopsis* and the gene cloning of mutant phenotypes should be powerful tools to dissect the cross talk between NO and other chemical messenger. In the mean time, NO will surely go on surprising us with novel functions.

REFERENCES

Able AJ 2003 Role of reactive oxygen species in the response of barley to necrotrophic pathogens *Protoplasma* **221** 137–143

Adams JM Cory S 1998 The Bcl-2 protein family: arbiters of cell death *Science* **281** 1322-1326

Ahern GP Klyachko VA Jackson MB 2002 cGMP and S-nitrosylation: two routs for modulation of neuronal excitability by NO *Trends Neuro Sci* **25** 510-517

Alvarez ME Pennell RI Meijer PJ Ishikawa A Dixon RA and Lamb C 1998 Reactive oxygen intermediates mediate a systemic signal network in the establishment of plant immunity *Cell* **92** 773-784

Amezcua JL Palmer RMJ de Souza BM and Moncada S 1989 Nitric oxide synthesized from L-arginine regulates vascular tone in the coronary circulation of the rabbit *Br J Pharmacol* **97** 1019-1024

Anderson L Mansfield TA 1979 The effects of nitric oxide pollution on the growth of tomato *Environmental Pollution* **20** 113-121

Aoki T 1990 Continuous flow determination of nitrite with membrane separation/ chemiluminescence detection *Biomed Chromatogr* **4** 128-130

Archer SL and Cowan NJ 1991 Acetylcholine causes endothelium dependent vasodilatation but does not stimulate nitric oxide production by rat pulmonary arteries or elevate endothelial cytosol calcium concentrations *Circ Ret* **68** 1569-1581

Arnaud N Murgia I Boucherez J Briat J-F Cellier F Gaymard F 2006 An iron-induced nitric oxide burst precedes ubiquitin-dependent protein degradation for *Arabidopsis* AtFer1 ferritin gene expression *Journal of Biological Chemistry* **281** 23579–23588

Arroyo C and Kohno M 1991 Difficulties encountered in the detection of nitric oxide (NO) by spin trapping techniques A cautionary note *Free Radic Ret Commun* **14** 145-155

Ashkenazi A Dixit V 1998 Death receptors: signaling and modulation *Science* **281** 1305-1308

Ball L Accotto GP Bechtold U et al 2004 Evidence for a direct link between glutathione biosynthesis and stress defence gene expression in *Arabidopsis The Plant Cell* **16** 2448–2462

Bamberger E and Mayer AM 1990 Effect of Kinetin on formation of red pigment in seedlings of *Amaranthus retroflexus Science* **131** 1094-1095

Barroso J Corpas FJ Carreras A *et al* 2006 Localization of S-nitrosoglutathione and expression of S-nitrosoglutathione reductase in pea plants under cadmium stress *Journal of Experimental Botany* **57** 1785–1793

Barroso JB Corpas FJ Carrweas A Sandalio LM Valderrama R Palma JM Lupianez JA and del Rio LA 1999 Localization of nitric-oxide synthase in plant peroxisomes *J Biol Chem* **274** 36729-36733

Belenghi B Romero-Puertas MC Vercammen D Brackenier A Inze D Delledonne M Van Breusegem F 2007 Metacaspase activity of *Arabidopsis thaliana* is regulated by S-nitrosylation of critical cystein residue *J Biol Chem* **282** 1352-1358

Beligni MV and Lamattina L 1999 Is nitric oxide toxic or protective? *Trends Plant Science* **4** 299-

Beligni MV and Lamattina L 2000 Nitric oxide stimulates seed germination and de-etiolation and inhibits hypocotyls elongation three light-inducible responses in plants *Planta* **210** 215-221

Beligni MV and Lamattina L 2002 Nitric oxide interferes with plant photo-oxidative stress by detoxifying reactive oxygen species *Plant Cell Environment* **25** 737-748

Beligni MV Lamattina L 2001a Nitric oxide in plants: the history is just beginning *Plant Cell and Environment* **24** 267-278

Beligni MV Lamattina L 2001b Nitric oxide: a non-traditional regulator of plant growth *Trends Plant Sci* **6** 11 508-509

Bestwick CS Brown IR Bennett MHR Mansfield JW 1997 Localization of hydrogen peroxide accumulation during the hypersensitive reaction of lettuce cells to *Pseudomonas syringae pv phaseolicola Plant Cell* **9** 209-221

Bethke PC Badger MR Jones RL 2004 Apoplastic synthesis of nitric oxide by plant tissues *The Plant Cell* **16** 332–341

Bethke PC Gubler F Jacobsen JV Jones RL 2004 Dormancy of Arabidopsis seeds and barley grains can be broken by nitric oxide *Planta* **219** 847–855

Bethke PC Libourei IG Jones RL 2006 Nitric Oxide reduces seed dormancy in *Arabidopsis Journal of Experimental Botony* **57** 17-526

Boccara M Mills CE Zeier J Anzi C Lamb C Poole RK Delledonne M 2005 Flavohaemoglobin HmpX from Erwinia chrysanthemi confers nitrosative stress tolerance and affects the plant hypersensitive reaction by intercepting nitric oxide produced by the host *The Plant Journal* **43** 26–237

Bolwell GP 1999 Role of reactive oxygen species and NO in plant defense responses *Curr Opin Plant Biol* **2** 287-294

Boon EM Davis JH Tran R Karow DS Huang SH Pan D Miazgowicz MM Mathies RA Marletta MA 2006 Nitric oxide binding to prokaryotic homologus ofthe soluble guanylate cyclase 1 H-NOX domain *Journal of Biological Chemistry* **281** 1892–21902

Bourgis F Roje S Nuccio ML *et al* 1999 S-methylmethionine plays a major role in phloem sulfur transport and is synthesized by a novel type of methyltransferase *The Plant Cell* **111** 485–1498

Brien J McLaughlin B Nakatsu K and Marks G 1991 Quantitation of nitric oxide formation from nitro vasodilator drugs by chemiluminescence analysis of headspace gas *j PharmacoL Methods* **25** 19-27

Bright J 2006 Nitric oxide signalling in *Arabidopsis thaliana* guard cells Bristol UK: University of the West of England PhD thesis

Bright J Desikan R Hancock JT Weir IS Neill SJ 2006 ABA-induced NO generation and stomatal closure in *Arabidopsis* are dependent on H_2O_2 synthesis *The Plant Journal* **45** 113–122

Buckner B Janick-Buckner D Gray J Johal GS 1998 Cell-death mechanisms in maize *Trends Plant Sci* **3** 218-223

Buettner GR 1987 Spin Trapping: ESR parameters of spin adducts *Free Radic Biol & Med* **3** 259-303

Büschges R Hollricher K Panstruga R Simons G Wolter M Frijters A van Daelen R van der Lee T Diergaarde P Groenendijk J *et al* 1997 The barley Mlo gene: a novel control element of plant pathogen resistance *Cell* **88** 695-706

Caird MA Richards JH Donovan LA 2007 Night-time stomatal conductance and transpiration in C3 and C4 plants *Plant Physiology* **143** 4–10

Chamnongpol S Willekens H Langebartels C Van Montagu M Inzé D Van Camp W 1996 Transgenic tobacco with a reduced catalase activity develops necrotic lesions and induces pathogenesisrelated expression under high light *Plant J* **10** 491-504

Chamnongpol S Willekens H Moeder W Langebartels C Sanderman HJ Van Montagu M Inze D Van Camp W 1998 Defense activation and enhanced pathogen tolerance induced by H_2O_2 in transgenic tobacco *Proc Natl Acad Sci* USA **95** 5818-5823

Chandok MR Ytterberg AJ Van Wijk KJ and Klessing DF 2003 The pathogen-inducible nitric oxide synthase (iNOS) in plants is a variant pf the P protein of the glycine decarboxylase complex *Cell* **113** 469-482

Chen WZ Palmer RMJ and Moncada S 1989 Release of nitric oxide from the rabbit aorta] *Vasc Med BioL* **1** 1-6

Cheng F Hsu S Kao CH 2002 Nitric oxide counteracts the senescence of detached rice leaves induced by dehydration and polyethylene glycol but not by sorbitol *Plant Growth Regulation* **38** 265–272

Chory J Reinecke D Sim S Washbun T and Brenner M 1994 A role for cytokinins in de-etiolation of *Arabidopsis* det mutants have an altered response to cytokinins *Plant Physiol* **104** 339-347

Chung S and Fung H 1990 Identification of the subcellular site for nitroglycerin metabolism to nitric oxide in bovine coronary smooth muscle cells *Pharmacol Exp The'* **253** 614-619

Clarke A Desikan R Hurst RD Hancock JT Neill S J 2000 NO way back: nitric oxide and programmed cell death in *Arabidopsis thaliana* suspension cultures *Plant J* **24** 667-677

Clementi E 1998 Role of Nitric Oxide and its intracellular signaling pathways in the control of calcium homeostasis *Biochem Pharmacol* **55** 713-718

Conrath U Amoroso G Köhle H Sültemeyer DF 2004 Non-invasive online detection of nitric oxide from plants and some other organisms by mass spectrometry *The Plant Journal* **38** 1015–1022

Corpas FJ Barroso JB Carreras A *et al* 2004a Cellular and subcellular localization of endogenous nitric oxide in young and senescent pea plants *Plant Physiology* **136** 2722–2733

Corpas FJ Barroso JB Carreras A Valderrama R Palma JM Leon AM Sandalio LM del Rio LA 2006 Constitutive arginine-dependent nitric oxide synthase activity in different organs of pea seedlings during plant development *Planta* **224** 246–254

Corpas FJ Barroso JB del Rio LA 2004b Enzymatic sources of nitric oxide in plant cells: beyond one protein-one function *New Phytologist* **162** 246–248

Crawford NM 2006 Mechanisms for nitric oxide synthesis in plants *Journal of Experimental Botany* **57** 471–478

Crawford NM Galli M Tischner R Heimer YM Okamoto M Mack A 2006 Response to Zemojtel *et al*: plant nitric oxide synthase: back to square one *Trends in Plant Science* **11** 526–527

Dangl JL 1998 Plants just say NO to pathogens *Nature* **394** 525- 527

Dangl JL Dietrich RA Richberg MH 1996 Death don't have no mercy: cell death programs in plant–microbe interactions *Plant Cell* **8** 1793-1807

Davidson EA Kingerlee W 1997 A global inventory of nitric oxide emissions from soils *Nutrient Cycling in Agroecosystems* **48** 37–50

De Pinto MC Tommasi F and de Gara L 2002 Changes in the antioxidant systems as part of the signaling pathway responsible for the programmed cell death activated by nitric oxide and reactive oxygen species in tobacco Bright-yellow 2 cells *Plant Physiol* **130** 698-708

Dean JV Harper JE 1986 Nitric oxide and nitrous oxide production by soybean and winged bean during the *in vivo* nitrate reductase assay *Plant Physiology* **82** 718–723

del Pozo O Lam E 1998 Caspases and programmed cell death in the hypersensitive response *Curr Biol* **8** in press

Del Rio Corpas FJ Sandalio LM Palma MG and barroso JB 2002 Reactive oxygen species antioxidants systems and nitric oxide peroxisomes *J Exp Bot* **53** 1255-1272

Del Rio LA Corpas FJ Sandalio LM Palma JM Gomez M Barroso JB 2002 Reactive oxygen species antioxidant systems and nitric oxide in peroxisomes *Journal of Experimental Botany* **53** 1255–1272

Delledone M Xia Y Dixon RA Lamb CJ 1998 Nitric oxide functions as a signal in plant disease resistance *Nature* **394** 585-588

Delledonne M Xia Y Dixon RA Lamb C 1998 Nitric oxide functions as a signal in plant disease resistance *Nature* **394** 585-588

Delledonne M Zeier J Marocco A Lamb C 2001 Signal interactions between nitric oxide and reactive oxygen intermediates in the plant hypersensitive disease resistance response *Proceedings of the National Academy of Science* USA **98** 13454–13459

Delledonne M 2005 NO news is good news for plants *Current Opinion in Plant Biology* **8** 390–396

Desikan R Cheung MK Bright J Henson D Hancock JT Neill SJ 2004a ABA hydrogen peroxide and nitric oxide signalling in stomatal guard cells *Journal of Experimental Botany* **55** 205–212

Desikan R Cheung M-K Clarke A Golding S Sagi M Fluhr R Rock C Hancock J Neill S 2004b Hydrogen peroxide is a common signal for darkness and ABA-induced stomatal closure in *Pisum sativum* L *Functional Plant Biology* **31** 913–920

Desikan R Griffiths RHancock J Neill SJ 2002 A new role for an old enzyme: nitrate reductase-mediated nitric oxide generation is required for abscisic acid-induced stomatal closure in *Arabidopsis thaliana Proceedings of the National Academy of Sciences* USA **99** 16314–16318

Desikan R Last K Harrett-Williams R Tagliavia C Harter K Hooley R Hancock JT Neill SJ 2006 Ethylene-induced stomatal closure in *Arabidopsis* occurs via AtrbohF-mediated hydrogen peroxide synthesis *The Plant Journal* **47** 907–916

Diaz M Achkor H Titarenko E Carmen Martinez M 2003 The gene encoding glutathione-dependent formaldehyde dehydrogenase/GSNO reductase is responsive to wounding jasmonic acid and salicylic acid *FEBS Letters* **543** 136–139

Dietrich RA Delaney TP Uknes SJ Ward EJ Ryals JA Dangl JL 1994 *Arabidopsis* mutants simulating disease resistance response *Cell* **77** 565-578

Dietrich RA Richberg MH Schmidt R Dean C Dangl JL 1997 A novel zinc-finger protein is encoded by the *Arabidopsis* lsd1 gene and functions as a negative regulator of plant cell death *Cell* **88** 685-694

Dion M Chamberland H St-Michel C Plante M Darveau A Lafontaine JG Brisson LF 1997 Detection of a homologue of bcl-2 in plant cells *Biochem Cell Biol* **75** 457-461

Dodge A 1994 Herbicide action and effects on detoxification processes in Causes of Photoactive Stress and Amelioration of Defense Sytems in Plants pp 219-236 eds CH Foyer and PM Mullineaux (CRC Boca Raton FL : USA)

Donaldson L Ludidi N Knight MR Gehring C Denby K 2004 Salt and osmotic stress cause rapid increases in *Arabidopsis thaliana* cGMP levels *FEBS Letters* **569** 317–20

Dong Lu Zhang X Jiang J An GY Zhang LR Song CP 2005 NO may function in the downstream of H_2O_2 in ABA-induced stomatal closure in *Vicia faba* L *Journal of Plant Physiology and Molecular Biology* **31** 62–70

Dordas C Hasinoff BB Igamberdiev AU Manac'h N Rivoal J Hill RD 2003 Expression of a stress-induced hemoglobin affects NO levels produced by alfalfa root cultures under hypoxic stress *Plant J* **35** 763–770

Dordas C Rivoal J Hill RD 2003 Plant haemoglobins nitric oxide and hypoxic stress *Annals of Botany* **91** 173–178

Dorey S Baillieul F Pierrel M-A Saindrenan P Fritig B Kauffmann S 1997 Spatial and temporal induction of cell death defense genes and accumulation of salicylic acid in tobacco leaves reacting hypersensitively to a fungal glycoprotein *Mol Plant–Microbe Interact* **10** 646-655

Dorey S Baillieul F Saindrenan P Fritig B Kauffmann S 1998 Tobacco class I and II catalases are differentially expressed during elicitorinduced hypersensitive cell death and localized acquired resistance *Mol Plant–Microbe Interact* **11** 321-334

Doyle MP and Hoekstra J W 1981 Oxidation of nitrogen oxides by bound dioxygen in hemoproteins *j Inorg Biochem* **14** 351-358

Dröge-Laser W Kaiser A Lindsay WP Halkier BA Loake GJ Doerner P Dixon RA Lamb CJ 1997 Rapid stimulation of a soybean protein serine-threonine kinase which phosphorylates a novel bZIP DNAbinding protein G/HBF-1 during the induction of early transcription-dependent defenses *EMBO J* **16** 726-738

Du H Klessig DF 1997 Role for salicylic acid in the activation of defense responses in catalase-deficient tobacco *Mol Plant–Microbe Interact* **7** 922-925

Durner J and Wendehenne D and Klessing DF 1998 Defense gene induction in tobacco by nitric oxide cyclic GMP and cyclic ADP- ribose *Proc Natl Acad Sci* USA **95** 10328-10333

Durner J Klessig DF 1999 Nitric oxide as a signal in plants *Current Opinion in Plant Biology* **2** 369-374

Eu JP Xu L Stamler JS Meissner G 1999 Regulation of ryanodine receptors by reactive nitrogen species *Biochem Pharmacol* **57** 1079-1084

Fauth M Merten A Hahn MG Jeblick W Kauss H 1996 Competence for elicitation of H_2O_2 in hypocotyls of cucumber is induced by breaching the cuticle and is enhanced by salicylic acid *Plant Physiol* **110** 347-354

Feechan A Kwon E Yun B Wang Y Palls JA Loake G 2005 A central role for S-nitrosothiols in plant disease resistance *Proceedings of the National Academy of Sciences* USA **102** 8054–8059

Floryszak-Wieczorek J Milczarek G Arasimowicz M Ciszewski A 2006 Do nitric oxide donors mimic endogenous NO-related response in plants? *Planta* **224** 1363–1372

Fodor J Gullner G Adam AL Barna B Kömives T Király Z 1997 Local and systemic responses of antioxidants to Tobacco Mosaic Virus infection and to salicylic acid in tobacco: role in systemic acquired resistance *Plant Physiol* **114** 1443-1451

Foissner ID Wendehenne D Langebartels C Dürner J 2000 *In vivo* imaging of an elicitor-induced nitric oxide burst in tobacco *Plant J* **23** 817–824

Fukuda H 1996 Xylogenesis: initiation progression and cell death *Annu Rev Physiol Plant Mol Biol* **47** 299-325

Furchgott R 1988 In: Vasodilation: Vascular Smooth Muscle Peptides Autonomic nerves and Endothelium (PVanhoutte Ed) Raven Press NY 401–414

Furchgott RF and Zawadzki J V 1980 The obligatory role of the endothelium in the relaxation of arterial smooth muscle by acetylcholine *Nature* **288** 373-376

Gabaldon C Gomez RL Pedreno MA and Barcelo AR 2005 Nitric oxide production by the differentiating xylem of *Zinnia elegans* *New Phytol* **165** 121-130

Garcia-Mata C and Lamattina L 2001 Nitric oxide induces stomatal closure and enhances the adaptive plant responses against drought stress *Plant Physiol* **126** 1196-1204

Garcia-Mata C and Lamattina L 2002 Nitric oxide and abscisic acid cross talk in guard cells *Plant Physiol* **128** 790–792

Garcia-Mata C Gay R Sokolovski S Hills A Lamattina L Blatt MR 2003 Nitric oxide regulates K^+ and Cl^- channels in guard cells through a subset of abscisic acid-evoked signaling pathways *Proceedings of the National Academy of Sciences* USA **100** 11116–11121

Gladwin MT Schechter AN Kim-Shapiro DB *et al* 2005 The emerging biology of the nitrite anion *Nature Chemical Biology* **1** 308–314

Glazener JA Orlandi EW Baker CJ 1996 The active oxygen response of cell suspensions to incompatible bacteria is not sufficient to cause hypersensitive cell death *Plant Physiol* **110** 759-763

Gould KS Lamotte O Klinguer A Pugin A Wendehenne D 2003 Nitric oxide production in tobacco leaf cells: a generalized stress response? *Plant Cell and Environment* **26** 1851–1862

Govea CM CP Souza JF Magalhes AC and Martins IS 1997 NO- releasing substances that induce growth elongation in maize root segments *Plant Growth Regul* **21** 183- 187

Gray J Close PS Briggs SP Johal GS 1997 A novel suppressor of cell death in plants encoded by the Lls1 gene of maize *Cell* **89** 25-32

Graziano M Beligni MV and Lamattina L 2002 Nitric oxide improves internal iron avilalability in plants *Plant Physiol* **130** 1852-1859

Graziano M Lamattina L 2005 Nitric oxide and iron in plants: an emerging and converging story *Trends in Plant Science* **10** 4-8

Green DR Reed JC 1998 Mitochondria and apoptosis *Science* **281** 1309-1312

Greenberg JT 1997 Programmed cell death in plant–pathogen interactions *Annu Rev Plant Physiol Plant Mol Biol* **48** 525- 545

Greenberg SS Wilcox DE and Rubanyi GM 1990 Endothelium-derived relaxing factor released from canine femoral artery by acetylcholine cannot be identified as free nitric oxide by electron paramagnetic resonance spectroscopy *Circ Rex* **67** 1446-1452

Groover A DeWit N Heidel A Jones A 1997 Programmed cell death of plant tracheary elements differentiating *in vitro Protoplasma* **196** 197-211

Groover A and Jones AM 1999 Tracheary element differentiation uses a novel mechanism coordinating programmed cell death and secondary cell wall synthesis *Plant Physiol* **119** 375-384

Groover A Jones AM 1998 Tracheary element programmed cell death *Plant Physiol* **112** 1765-1782

Grubisic D and Konjevic R 1990 Light and nitrate interaction in phytochrome-controlled germination of *Paulownia tormentosa Planta* **181** 239-243

Grubisic D and Konjevic R 1992 The effect of organic nitrates in phytochrome controlled germination of *Paulownia tormentosa seeds Photochem Photobiol* **56** 629-632

Guo F Okamoto M Crawford NM 2003 Identification of a plant nitric oxide synthase gene involved in hormonal signaling *Science* **302** 100–103

Guo FQ Crawford NM 2005 Arabidopsis nitric oxide synthase1 is targeted to mitochondria and protects against oxidative damage and dark-induced senescence *The Plant Cell* **17** 3436–3450

Guo FQ 2006 Response to Zemojtel *et al* plant nitric oxide synthase: At NOS1 is just the beginning *Trends in Plant Science* **11** 527–528

Gupta KJ Stoimenova M Kaiser WM 2005 In higher plants only root mitochondria but not leaf mitochondria reduce nitrite to NO *in vitro* and *in situ Journal of Experimental Botany* **56** 2601–2609

Gustafsson L Leone A Persson M Wikiund N and Moncada S 1991 Endogenous nitric oxide is present in the exhaled air of rabbits guinea pigs and humans *Biochem Biophys Rex Commun* **181** 852-857

Hammond- Kosack KF and Jones RDG 1996 Resistance gene dependent plant defence responses *Plant Cell* **8** 1773-1791

Hammond-Kosack KE Jones JDG 1996 Inducible plant defense mechanisms and resistance gene function *Plant Cell* **8** 1773-1791

Hari P Raivonen M Vesala T Munger JW Pilegaard K Kulmala M 2003 Ultraviolet light and leaf emission of NO_X *Nature* **422** 134

Harper JE 1981 Evolution of nitrogen oxide(s) during *in vivo* nitrate reductase assay of soybean leaves *Plant Physiology* **68** 1488–1493

Hart JDE and Dulhunty AF 2000 Nitric Oxide activates or inhibits skeletal muscle ryanodine receptors binding on its concentration membrane potential and ligand binding *J Memb Biol* **173** 227-236

Hausladen A Stamler JS 1998 Nitric oxide in plant immunity *Proc Natl Acad Sci* USA **95** 10345-10347

He JM Xu H She XP Song XG Zhao WM 2005 The role and the interrelationship of hydrogen peroxide and nitric oxide in the UV-B-induced stomatal closure in broad bean *Functional Plant Biology* **32** 237–247

He Y Tang RH Hao Y Stevens RD Cook CW Ahn SM Jing L Yang Z Chen L Guo F Fiorani F Jackson RB Crawford NM and Pei ZM 2004 Nitric oxide represses the *Arabidopsis* floral transition *Science* **305** 1968-1971

Heath MC 1998 Involvement of reactive oxygen species in the response of resistant (hypersensitive) or susceptible cowpeas to the cowpea rust fungus *New Phytol* **138** in press

Henry Y Ducrocq C Drapier J Servent D Pellat C and Guissani A 1991 Nitric oxide a biological effector electron paramagnetic resonance detection of nitrosyl-iron-protein complexes in whole cells *Eur Biophys j* **20** 1-15

Hsu YT Kao CH 2004 Cadmium toxicity is reduced by nitric oxide in rice leaves *Plant Growth Regulation* **42** 227–238

Hu G Richter TE Hulbert SH Pryor T 1996 Disease lesion mimicry caused by mutations in the rust resistance gene rp1 *Plant Cell* **8** 1367-1376

Hu G Yalpani N Briggs SP Johal GS 1998 A porphyrin pathway impairment is responsible for the phenotype of a dominant lesion mimic mutant of maize *Plant Cell* **10** 1095-1105

Hu X Neill SJ Tang Z Cai W 2005 Nitric oxide mediates gravitropic bending in soybean roots *Plant Physiology* **137** 663–670

Hung KT Chang CJ Kao CH 2002 Paraquat toxicity is reduced by nitric oxide in rice leaves *Journal of Plant Physiology* **159** 159–166

Hung KT Kao CH 2005 Nitric oxide counteracts the senescence of rice leaves induced by hydrogen peroxide *Botanical Bulletin of Academia Sinica* **46** 21–28

Hunt MD Delaney TP Dietrich RA Weyman KB Dangl JL Ryals JA 1997 Salicylate-independent lesion formation in *Arabidopsis* lsd mutants *Mol Plant–Microbe Interact* **10** 531-536

Huppe HC and Turpin DH 1994 Integration of carbon and nitrogen metabolism in plant and algal cells *Annu Rev Plant Physiol MolBiol* **45** 577-607

Ignarro L Buga G Wood K Byrns R Chadhuri G 1987 Endothelium derived relaxing fadtor produced and released from artery and vein is nitric oxide *Proceedings of the National Academy of Science* USA **84** 9265–9269

Jabs T Colling C Tschöpe M Hahlbrock K Scheel D 1997 Elicitor- stimulated ion fluxes and reactive oxygen species from the oxidative burst signal defense gene activation and phytoalexin synthesis in parsley *Proc Natl Acad Sci* USA **94** 4800-4805

Jasid S Simontacchi M Bartoli CG Puntarulo S 2006 Chloroplasts as a nitric oxide cellular source Effect of reactive nitrogen species on chloroplastic lipids and proteins *Plant Physiology* **142** 1246–1255

Karow DS Pan D Tran R Pellicena P Presley A Mathies RA Marietta MA 2004 Spectroscopic characterization of the soluble guanylate cyclase-like heme domains from *Vibrio cholerae* and *Thermoanaerobacter fengcongensis Biochemistry* **43** 10203–10211

Kauss H Jeblick W 1996 Influence of salicylic acid on the induction of competence for H_2O_2 elicitation Comparison of ergosterol with other elicitors *Plant Physiol* **111** 755-763

Kauss H Jeblick W 1995 Pretreatment of parsley suspension cultures with salicylic acid enhances spontaneous and elicited production of H_2O_2 *Plant Physiol* **108** 1171-1178

Keeley JE and Fortheringham CJ 1997 Trace gas emission and smoke- induced seed germination *Science* **276** 1248- 1250

Kelm M and Schrader J 1988 Nitric oxide release from the isolated guinea pig heart *EurJ Pharmacol* **155** 317-321

Kelm M Feelisch M Spahr R Piper HM Noack E and Schrader J 1988 Quantitative and kinetic characterization of nitric oxide and EDRF release from cultured endothelial cells *Biochem Biophys Res Commun* **154** 237-244

Klepper L 1990 NO_x evolution by soybean leaves treated with salicylic acid and selected derivatives Pesticide *Biochemistry and Physiology* **39** 43–48

Klepper LA 1979 Nitric oxide (NO) and nitrogen dioxide emissions from herbicide-treated soybean plants *Atmospheric Environment* **13** 537–542

Klessig DF Durner J Noad R *et al* 2000 Nitric oxide and salicylic acid signaling in plant defense *Proceedings of the National Academy of Sciences* USA **97** 8849–8855

Kopyra M and Gwozdz EA 2003 Nitric oxide stimulates seed germination and counteracts the inhibitory effect of heavy metals and salinity on root growth of *Lupinus luteus Plant Physiol Biochem* **41** 1011-1017

Kostic M and Schrader J 1992 Role of nitric oxide in reactive hyperemia of the guinea pig heart *Circ Rex* **70** 208-212

Kwak JM Mori IC Pei ZM Leonhardt N Torres MA Dangl JL Bloom RE Bodde S Jones JD Schroeder JI 2003 NADPH oxidase AtrbohD and AtrbohF genes function in ROS-dependent ABA signalling in *Arabidopsis EMBO Journal* **22** 2623–2633

Kwezi L Meier S Mungur L Ruzvidzo O Irving H Gehring C 2007 The *Arabidopsis thaliana* Brassinosteroid Receptor (AtBRI1) contains a domain that functions as a guanylyl cyclase *in vitro* In: Public Library of Science ONE 2:e449 oi:101371/journalpone0000449

Lamattina L Garcia-Matta C M Graziano M Pagnussat G 2003 Nitric Oxide: The versatility of an extensive signal molecule *Ann Rev Plant Biol* **54** 109-136

Lamb C Dixon RA 1997 The oxidative burst in plant disease resistance *Annu Rev Physiol Plant Mol Biol* **48** 251-275

Lamotte L Courtois C Barnavon L Pugin A Wendehenne D 2005 Nitric oxide in plants: the biosynthesis and cell signalling properties of a fascinating molecule *Planta* **221** 1–4

Lamotte O Courtois C Dobrowolska G Besson A Pugin A Wendehenne D 2006 Mechanisms of nitric oxide induced increase of free cytosolic Ca^{2+} concentration in Nicotiana plumbaginifolia cells *Free Radical Biology and Medicine* **40** 1369–1376

Laspina NV Groppa MD Tomaro ML Benavides MP 2005 Nitric oxide protects sunflower leaves against Cd-induced oxidative stress *Plant Science* **169** 323–330

Laxalt AM Beligni M V Lamattina L 1997 Nitric oxide preserves the level of chlorophyll in potato leaves infected by *Phy- tophthora infestans European Journal of Plant Pathology* **103 643-651**

Lesham Y Haramaty E 1996 The characterization and contrasting effects of the nitric oxide free radical in vegetative stress and senescence of *Pisum sativum Linn foliage Journal of Plant Physiology* **148** 258–263

Leshem Y Y 1996 Nitric oxide in biological systems *Plant Growth Regul* **18** 155-159

Leshem Y 2001 Nitric Oxide in Plants (London UK: Kluwer Academic Publishers)

Leshem YY and Pincasov Y 2002 Noninvasive photoacustic spectroscopic determination of relative endogenous nitric oxide and ethylene content stoichiometry during the ripening of strawberries *Fragaria anannasa* (Duch) and avocados *Persea americana* (Mill) *J Exp Bot* **51** 1471-1473

Leshem YY Willis RBH and Ku VVV 1998 Evidence for the function of the free radical gas nitric oxide (NO) as an endogenous regulating factor in higher plants *Plant Physiol Biochem* **36** 825-833

Levine A Tenhaken R Dixon R Lamb CJ 1994 H_2O_2 from the oxidative burst orchestrates the plant hypersensitive disease resistance response *Cell* **79** 583-593

Ligternik W Kroj T zur Nieden U Hirt H Scheel D 1997 Receptor- mediated activation of a MAP kinase in pathogen defense of plants *Science* **276** 2054-2057

Lindermayr C Saalbach G Bahnweg G Durner J 2006 Differential inhibition of *Arabidopsis* methionine adenosyltransferases by protein S-nitrosylation *Journal of Biological Chemistry* 2811–7

Lindermayr C Saalbach G Durner J 2005 Proteomic identification of S-nitrosylated proteins in *arabidopsis Plant Physiology* **137** 921–930

Liu M Huynh B Payne W Peck HJ Dervartanian D and Legall J 1987 Optical EPR and M{246} ssbauer spectroscopic studies on the NO derivatives of cytochrome cdl from *Thiobacilluc deni{128}rficanx Eur j Biochem* **169** 253-258

Liu X Miller MJS Joshi MS Thomas DD Lancaster JR Jr 1998 Accelerated reaction of nitric oxide with O_2 within the hydrophobic interior of biological membranes *Proceedings of the National Academy of Sciences* USA **95** 2175–2179

Loake G 2001 Plant cell death: unmasking the gatekeepers *Curr Biol* **11** R1028- R1031

Ludidi N Gehring C 2003 Identification of a novel protein with guanylyl cyclase activity in *Arabidopsis thaliana Journal of Biological Chemistry* **278** 6490–6494

Luis AJ Mur I Edi Santosa Lucas JJ Laarhoven Nicholas J Holton Frans JM Harren and Aileen R Smith 2005 Laser Photoacoustic Detection Allows in Planta Detection of Nitric Oxide in

Tobacco following Challenge with Avirulent and Virulent *Pseudomonas syringae* Pathovars *Plant Physiology* **138** 1247-1258

Lum HK Butt YKC Lo SCL 2002 Hydrogen peroxide induces a rapid production of nitric oxide in mung bean (*Phaseolus aureus*) Nitric Oxide: *Biology and Chemistry* **6** 205–213

Maathuis JM 2006 cGMP modulates gene transcription and cation transport in *Arabidopsis* roots *The Plant Journal* **45** 700–711

Magalhaes JR Monte DC Durzan D 2000 Nitric oxide and ethylene emission in *Arabidopsis* thaliana *Physiology and Molecular Biology of Plants* **2** 117–127

Mallick N Mohna FH Rai LC and Soeder CJ 2000a Evidence for the non involvement of nitric oxide synthase in nitric oxide production by the green algae *Scendesmus Obliquus J Plant Physiol* **156** 423-426

Mallick N Mohna FH Rai LC and Soeder CJ 2000b impact of physiological stresses on nitric oxide formation by green alga *Scendesmus Obliquus J Microbial Biotechnol* **10** 300-306

Mansfield TA 2002 Nitrogen oxides: old problems and new challenges In: Air pollution and plant life—Bell JNB Treshow M eds Chichester UK: John Wiley and Sons Ltd 119–133

Maples K Sandstrom T Su Y and Henderson R 1991 The nitric oxide/heme protein complex as a biologic marker of exposure to nitrogen dioxide in humans rats and in vitro models *Am j llespir Cell Mol Biol* **4** 538-543

Martin W Smith J Lewis MJ and Henderson AH 1988 Evidence the inhibitory factor extracted from bovine retract or penis is nitrite whose activated derivative is stabilized nitric oxide *Br j Pharmacol* **93** 579-585

Mathieu C Moreau S Frendo P Puppo A and Davis MJ 1998 Direct detection of radicals in intact soybean nodules: Presence of Nitric oxide- Leghemoglobin complexes Free Radical *Biol Med* **24** 1242-149

Matsunaga K and Furchgott RF 1988 Interactions of light and sodium nitrite in producing relaxation of rabbit aorta *J PharmacoL Exp Ther* **248** 687-695

Melotto M Underwood W Koczan J Nomura K He SY 2006 Plant stomata function in innate immunity against bacterial invasion *Cell* **126** 969–980

Menon N Patricza J Binder T and Bing R 1991 Reduction of biological effluents in purge and trap micro reaction vessels and detection of endothelium-derived nitric oxide (edno) by chemiluminescence *j MoL Cell CardioL* **23** 389-393

Menon N Wolf A Zehetgruber M and Bing R 1989 An improved chemiluminescence assay suggests non nitric oxide mediated action of lysophosphatidylcholine and acetylcholine *Proc Soc Exp Biol Med* **191** 316-319

Millar AH and Day DA 1996 Nitric oxide inhibits the cytochrome oxidase but not the alternative oxidase of plant mitochondria *FEBS Lett* **398** 155-158

Millar TM Stevens CR Benjamin N Eisenthal R Harrison R Blake DR 1998 Xanthine oxidoreductase catalyses the reduction of nitrates and nitrite to nitric oxide under hypoxic conditions *FEBS Letters* **427** 225–228

Mitchell DE Madore MA1992 Patterns of assimilate production and translocation in muskmelon (*Cucumis melo L) Plant Physiology* **99** 966–971

Mittler R Feng X Cohen M 1998 Post-translational suppression of cytosolic ascorbate peroxidase expression during pathogeninduced programmed cell death in tobacco *Plant Cell* **10** 461-473

Mittler R Shulaev V Lam E 1995 Coordinated activation of programmed cell death and defense mechanisms in transgenic tobacco plants expressing a bacterial proton pump *Plant Cell* **7** 29-42

Mittler R Shulaev V Seskar M Lam E 1996 Inhibition of programmed cell death in tobacco plants during a pathogen-induced hypersensitive response at low oxygen pressure *Plant Cell* **8** 1991-2001

Modolo LV Augusto O Almeida IMG Magalhaes JR Salgado I 2005 Nitrite as the major source of nitric oxide production by *Arabidopsis thaliana* in response to *Pseudomonas syringae FEBS* Letters **579** 3814–3820

Modolo LV Augusto O Almeida IMG Pinto-Maglio CAF Oliveira HC Seligman K Salgado I 2006 Decreased arginine and nitrite levels in nitrate reductase-deficient *Arabidopsis thaliana* plants impair nitric oxide synthesis and the hypersensitive response to *Pseudomonas syringae Plant Science* **171** 34–40

Morot-Gaudry-Talarmain Y Rockel P Moureaux T Quilleré I Leydecker MT Kaiser WM Morot-Gaudry JF 2002 Nitrite accumulation and nitric oxide emission in relation to cellular signalling in nitrite reductase antisense tobacco *Planta* **215** 708–715

Munemasa S Oda K Watanabe-Sugimoto M Nakamura Y Shimoishi Y Murata Y 2007 The coronatine-insensitive 1 mutation reveals the hormone signalling interaction between abscisic acid and methyl jasmonate in *Arabidopsis* guard cells Specific impairment of ion channel activation and second messenger production *Plant Physiology* **143** 1398–1407

Mur LAJ Bi Y-M Darby RM Firek S Draper J 1997 Compromising early salicylic acid accumulation delays the hypersensitive response and increases viral dispersal during lesion establishment in TMVinfected tobacco *Plant J* **12** 1113-1126

Mur LAJ Craver TLW Prats E 2006 NO way to live; the various roles of nitric oxide in plant–pathogen interactions *Journal of Experimental Botany* **57** 489–505

Mur LAJ Santosa IE Laarhoven LJJ Holton NJ Harren FJM Smith AR 2005 Laser photoacoustic detection allows in planta detection of nitric oxide in tobacco following challenge with avirulent and virulent *Pseudomonas syringae* pathovars *Plant Physiology* **138** 1247–1258

Murad F Arnold WP Mittal CK Braughler JM 1979 Properties and regulation of guanylate cyclase and some proposed functions for cyclic GMP Adv Cyclic Nucleotide Res **11** 175–204

Myers P Minor RJ Guerra RJ Bates J and Harrison D 1990 Vasorelaxant properties of the endothelium-derived relaxing factor more closely resemble 5-nitroso cysteine than nitric oxide *Nature (London)* **345** 161-163

Nathan C 1992 Nitric oxide as a secretory product of mammalian cells *FASEB J* **6** 3051-3064

Neill SJ Desikan R Clarke A Hurst RD and Hancock JT 2002b Hydrogen peroxide and nitric oxide as signaling molecules in plants *J Exp Bot* **53** 1237-1242

Neill SJ Desikan R Clarke A and Hancock JT 2002a Nitric oxide is a novel component of abscisic acid signaling in stomatal guard cells *Plant Physiol* **128** 13-16

Neill SJ Desikan R and Hancock JT 2003 Nitric oxide signaling in plants *New Phytol* **159** 11-35

Neill SJ Desikan R Hancock JT 2003 Nitric oxide signalling in plants *New Phytologist* 15911–35

Noronha Dutra AA Epperlein MM Woolf N 1993 Reaction of nitric oxide with hydrogen peroxide to produce potentially cytotoxic singlet oxygen as a model for nitric oxide-mediated killing *FEBS Lett* **321** 1993 59–62

Nürnberger T Nennstiel D Jabs T Sacks WR Hahlbrock K Scheel D 1994 High-affinity binding of a fungal oligopeptide elicitor to parsley plasma membranes triggers multiple defense responses *Cell* **78** 449-460

O'Brien IEW Baguley BCMurray BG Morris BAM Ferguson IB 1998 Early stages of the apoptotic pathway in plant cells are reversible *Plant J* **13** 803-814

Ohwaki YKawagishi-Kobayashi M Wakasa K Fujihara S Yoneyama T 2005 Induction of class-1 non-symbiotic haemoglobin genes by nitrate nitrite and nitric oxide in cultured rice cells *Plant Cell Physiology* **46** 324–331

Orozco-Cardenas ML and Ryan CA 2002 Nitric oxide negatively modulates wound signaling in tomato plants *Plant Physiol* **130** 487–493

Pagnussat G Simontachi M Puntarulo S and Lamattina L 2002 Nitric oxide is required for root organogenesis *Plant Physiol* **129** 954-956

Pagnussat GC Lanteri ML Lombardo MC Lamattina L 2004 Nitric oxide mediates the indole-acetic acid activation of a mitogen-activated protein kinase cascade involved in adventitious root formation *Plant Physiology* **135** 279–286

Pai T Payne W and LeGall J 1987 Use of a chemiluminescence detector for quantitation of nitric oxide produced in assays of denitrifying enzymes *Analyt Biochem* **166** 150-157

Palmer RMJ Ferridge AG and Moncada S 1987 Nitric oxide release accounts for the biological activity of endothelium-derived relaxing factor *Nature (London)* **327** 524-526

Parinov S Sevugan M Ye S Yang W-C Kumaran M Sundaresan V 1999 Analysis of flanking sequences from Dissociation insertion lines: a database for reverse genetics in *Arabidopsis Plant Cell* **11** 2263–2270

Patel S Robb-Gaspers LD Stellato KA Shon M Thomas AP 1999 Coordination of calcium signaling by endothelial derived nitric oxide in the intact liver *Nature Cell Biol* **1** 467-471

Pedroso MC Magalhaes JR and Durzan D 2000 A Nitric oxide burst precedes apoptosis in angiosperm and gymnosperm callus cells and foliar tissues *J Exp Bot* **51** 1027-1037

Pennell RI Lamb CJ 1997 Programmed cell death in plants *Plant Cell* **9** 1157-1168

Perazzolli M Dominici P Romero-Puertas MC Zago E Zeier J Sonoda M Lamb C Delledonne M 2004 *Arabidopsis* non-symbiotic hemoglobin AHb1 modulates nitric oxide bioactivity *The Plant Cell* **16** 2785–2794

Perazzolli M Romero-Puertas MC Delledonne M 2006 Modulation of nitric oxide bioactivity by plant haemoglobins *Journal of Experimental Botany* **57** 479–488

Piatelli M Giudici de Nicola M and Castrogiovanni V 1969 Photocontrol of amaranthin synthesis in *Amaranthis tricolor Phytochemistry* **8** 731-736

Planchet E Gupta KJ Sonoda M Kaiser WM 2005 Nitric oxide emission from tobacco leaves and cell suspensions: rate-limiting factors and evidence for the involvement of mitochondrial electron transport *The Plant Journal* **41** 732–743

Planchet E Kaiser WM 2006a Nitric oxide (NO) detection by DAF fluorescence and chemiluminescence: a comparison using abiotic and biotic NO sources *Journal of Experimental Botany* **57** 3043–3055

Planchet E Kaiser WM 2006b Nitric oxide production in plants *Plant Signaling and Behaviour* **1** 46–51

Raff M 1992 Social controls on cell survival and cell death *Nature* **356** 397-400

Ramamurthi A and Lewis RS 1997 Measurment and modeling of nitric oxide release rates for nitric oxide donors *Chem Res Toxicol* **10** 408-413

Rao MV and Davis KR 1999 Ozone- induced cell death occur via distinct mechanisms in *Arabidopsis:* The role of salicyclic acid *Plant J* **17** 603-609

Rees DD Palmer RMJ and Moncada S 1989 Role of endothelium-derived nitric oxide in the regulation of blood pressure *Proc Natl Acad Sci USA* **86** 3375-3378

Rockel P Strube F Rockel A Wildt J Kaiser WM 2002 Regulation of nitric oxide (NO) production by plant nitrate reductase *in vivo* and *in vitro Journal of Experimental Botany* **53** 103–110

Ryals JL Neuenschwander UH Willits MC Molina A Steiner H-YHunt MD 1996 Systemic acquired resistance *Plant Cell* **8** 1809-1819

Saito S Yamamoto-Katou A Yoshioka H Doke N Kawakita K 2006 Peroxynitrite generation and tyrosine nitration in defense responses in tobacco BY-2 cells *Plant and Cell Physiology* **47** 689–697

Sakamoto A Sakurao SH Fukunaga K Matsubara T Ueda-Hashimoto M Tsukamoto S Takahashi M Morikawa H 2004 Three distinct *Arabidopsis* hemoglobins exhibit peroxidase-like activity and differentially mediate nitrite-dependent protein nitration *FEBS Letters* **572** 27–32

Sakihama Y Murakamai S Yamasaki H Involvement of nitric oxide in the mechanism for stomatal opening in *Vicia faba* leaves *Biologia Plantarum* 2003 **45** 117–119

Sasakura F Uchiumi T Shimoda Y Suzuki A Takenouchi K Higashi S Abe M 2006 A class 1 hemoglobin gene from Alnus firma functions in symbiotic and non-symbiotic tissues to detoxify nitric oxide *Molecular Plant–Microbe Interactions* **19** 441–450

Scheel D 1998 Resistance response physiology and signal transduction *Curr Opin Plant Biol* **1** 305-310

Scherer GFE and Holk A 2000 NO donors mimic and NO inhibitors inhibit cytokinin action in betalain accumulation in *Amaranthus Caudatus Plant Growth Regul* **32** 345- 350

Schmidt H H H W Nau H Wittfoht W Gerlach J PrescherK EKleinM M Niroomand Fand Bohme E 1988 Arginine is a physiologic precursor of endotheliumderived nitric oxide *Eur J PharmacoL* **154** 212-216

Schror K Forster S and Woditsch I 1991 On-line measurement of nitric oxide release from organic nitrates in the intact coronary circulation *Naunyn-Schmiedebergs Arch Pharmacol* **344** 240-246

Schwartz LM Smith SW Jones MEE Osbourne BA 1993 Do all programmed cell deaths occur via apoptosis? *Proc Natl Acad Sci* USA **90** 980-984

Setia N Setia RC and Aggarwal M 2005 Amelioration of temperature effects by NO in leaves of oil seed *Brassica Species J PlSci Res* **21** 191-194

She XP Song XG He JM 2004 Role and relationship of nitric oxide and hydrogen peroxide in light/dark-regulated stomatal movement in *Vicia faba Acta Botanica Sinica* **46** 1292–1300

Shi S Wang G Wang Y Zhang L Zhang L 2005 Protective effect of nitric oxide against oxidative stress under ultraviolet-B radiation *Nitric Oxide* **13** 1–9

Shibuki K 1990 An electrochemical microprobe for detecting nitric oxide release in brain tissue *Neurosci Rex* **9** 69-76

Shimoda Y Nagata M Suzuki A Abe M Sato S Kato T Tabata S Higashi S Uchiumi T 2005 Symbiotic rhizobium and nitric oxide induce gene expression of non-symbiotic haemoglobin in Lotus japonicus *Plant and Cell Physiology* **46** 99–107

Shingles R ROH MH and Mc Carty RE 1996 Nitrite transport in chloroplast inner envelope vesicles I Direct measurement of proton- linked transport *Plant Physiol* **112** 1375-1381

Shirasu K Nakajima H Rajasekhar VK Dixon RA Lamb CJ 1997 Salicylic acid potentiates an agonist-dependent gain control that amplifies pathogen signals in the activation of defense mechanisms *Plant Cellb* 261-270

Sokolovski S Blatt MR 2004 Nitric oxide block of outward-rectifying K^+ channels indicates direct control by protein nitrosylation in guard cells *Plant Physiology* **136** 4275–4284

Sokolovski S Hills A Gay R Garcia-Mata C Lamattina L Blatt MR 2005 Protein phosphorylation is a prerequisite for intracellular Ca^{2+} release and ion channel control by nitric oxide and abscisic acid in guard cells *The Plant Journal* **43** 520–529

Song F Goodman RM 2001 Activity of nitric oxide is dependent on but is partially required for function of salicylic acid in the signaling pathway in tobacco systemic acquired resistance Mol *Plant Microbe Interact* **12** 1458–1462

Song L Ding W Zhao M Sun B Zhang L 2006 Nitric oxide protects against oxidative stress under heat stress in the calluses from two ecotypes of reed *Plant Science* **171** 449–458

Stamler JS Lamas S Fang FC 2001 Nitrosylation: prototypic redox based signaling mechanism *Cell* **106** 675-678

Sto¨hr C Strube F Marx G Ulrich WR Rockel P 2001 A plasma membrane-bound enzyme of tobacco roots catalyses the formation of nitric oxide from nitrite *Planta* **212** 835–841

Stohr C Stremlau S 2006 Formation and possible roles of nitric oxide in plant roots *Journal of Experimental Botany* **57** 463–470

Sun J Xin C Eu JP Stamler JS Meissner G 2001 Cysteine- 3635 is responsible for skeletal muscle ryanodine receptors modulation by NO *Proc Natl Acad Sci* USA **98** 11158-111162

Tada Y Mori T Shinogi T and Yao N 2004 Nitric oxide and reactive oxygen species do not elicit hypersensitive cell death but induce apoptosis in the adjacent cells during the defense response of oat *Mol Plant Microbe Interact* **17** 245–253

Takahashi H Chen Z Du H Liu Y Klessig DF 1997 Development of necrosis and activation of disease resistance in transgenic tobacco plants with severely reduced catalase levels *Plant J* **11** 993-1005

Takahashi S and Yamasaki H 2002 Reversible inhibition of photophosphorylation in chloroplasts by nitric oxide *FEBS Lett* **512** 145-148

The Nobel Prize in Physiology and Medicine1998http://nobelprizeorg/ nobel_prizes/medicine/ laureates/1998/ [Last visited date: 080906]

Thoiron S Pascal N and Briat JF 1997 Impact of iron deficiency and iron re supplly during the early stages of vegetative development in maize (*Zea mays L) Plant Cell Environ* **20** 1051-1060

Thornberry NA Lazebnik Y 1998 Caspases: enemies within *Science* **281** 312-1316

Thulke O Conrath U 1998 Salicylic acid has a dual role in the activation of defence-related genes in parsley *Plant J* **14** 35-42

Tian QY Sun DH Zhao MG Zhang WH 2007 Inhibition of nitric oxide synthase (NOS) underlies aluminium-induced inhibition of root elongation in *Hibiscus moscheutos New Phytologist* **174** 322–331

Torsethaugen G Pitcher LH Zilinskas BA Pell EJ 1997 Overproduction of ascorbate peroxidase in the tobacco chloroplast does not provide protection against ozone *Plant Physiol* **114** 529-537

Tracey WR Linden J Peach MJ and Johns RA 1990 Comparison of spectrophotometric and biological assays for nitric oxide (NO) and endothelium-derived relaxing factor (EDRF): nonspecificity of the diazotization reaction for NO and failure to detect EDRF *J PharmacoL Exp TIler* **252**922-928

Trevaskis B Watts RA Andersson CR Llewellyn DJ Hargrove MS Olson JS Dennis ES Peacock WJ 1997 Two hemoglobin genes in Arabidopsis thaliana: the evolutionary origins of leghemoglobins *Proceedings of the National Academy of Sciences* USA **94** 12230–12234

Tun NN Holk A and Schere FE 2001 Rapid increase of NO release in plant cell cultures induced by cytokinin *FEBS Lett* **509** 174-176

Uchida A Jagendorf AT Hibino T Takabe T Takabe T 2002 Effects of hydrogen peroxide and nitric oxide on both salt and heat stress tolerance in rice *Plant Science* **163** 515–523

Valderrama R Corpas FJ Carreras A Fernandez-Ocana A Chaki M Luque F Gomez-Rodriguez MV Colmenero-Varera P del Rio LA Barroso JB 2007 Nitrosative stress in plants *FEBS Letters* **581** 453–461

Van der Biezen EA Jones JDG 1998 Homologus between plant resistance gene products and regulators of cell death in animals *Curr Biol* **8** R226-R227

Vanin AF Svistunenko DA Mikoyan VD Serezhenkov VA Fryer MJ Baker NR Cooper CE 2004 Endogenous superoxide production and the nitrite/nitrate ratio control the concentration of bioavailable free nitric oxide in leaves *Journal of Biological Chemistry* **279** 24100–24107

Vranova E Inze D and Van Breusegem F 2002 Signal transduction during oxidative stress *J Exp Bot* **53** 1227-1236

Wang H Li J Bostock RM Gilchrist DG 1996 Apoptosis: a functional paradigm for programmed plant cell death induced by a hostselective phytotoxin and invoked during development *Plant Cell* **8** 375-391

Wang Q Jacobs J DeLco J Kruszyna H Kruszyna R Smith R and Wilcox D 1991 Nitric oxide hemoglobin in mice and rats in endotoxic shock *Life Sci* **49** 55-6O

Wang R Guegler K LaBrie ST Crawford NM 2000 Genomic analysis of a nutrient response in Arabidopsis reveals diverse expression patterns and novel metabolic and potential regulatory genes induced by nitrate *The Plant Cell* **12** 1491–14–509

Wang R Tischner R Gutierrez RA Hoffman M Xing X Chen M Coruzzi G Crawford NM 2004 Genomic analysis of the nitrate response using a nitrate reductase-null mutant of *Arabidopsis Plant Physiology* **136** 2512–2522

Wang Y Yun BW Kwon EJ Hong JK Yoon JY Loake GJ 2006 S-nitrosylation: an emerging redox-based post-translational modification in plants *Journal of Experimental Botany* 57 1777–1784

Wang YS Yang ZM 2005 Nitric oxide reduces aluminium toxicity by preventing oxidative stress in the roots of *Cassia tora L Plant and Cell Physiology* **46** 1915–1923

Wellburn AR 1990 why are atmospheric oxides of nitrogen usually phytotoxic and not alternarive fertilizers? *New Phytol* **115** 395-429

Wendehenne D Pugin AKlessig DF Durner J 2001 Nitric oxide: comparative synthesis and signaling in animal and plant cells *Trends in Plant Science* 6177–183

Wennmalm A Lanne B and Petersson A 1990 Detection of endothelial-derived relaxing factor in human plasma in the basal state and following ischemia using electron paramagnetic resonance spectrometry *Analyt Biochem* **4** 128-130

Westenberger U Thanner S Ruf H Gersonde K Sutter G and Trentz O 1990 Formation of free radicals and nitric oxide derivative of hemoglobin in rats during shock syndrome *Free Radic Rex Commun* **11** 167-178

Weyman K Hunt M Uknes S Neuenschwander U Lawton K Steiner H-Y Ryals J 1995 Suppression and restoration of lesion formation in *Arabidopsis* lsd mutants *Plant Cell* **7** 2013-2022

Wildt J Kley D Rockel A Rockel P Segschneider HJ 1997 Emission of NO from several higher plant species *Journal of Geophysical Research* **102** 5919–5927

Wilkinson J Crawford NM 1991 Identification of the *Arabidopsis* CHL3 gene as the nitrate reductase structural gene NIA2 *The Plant Cell* **3** 461–471

Wilkinson JQ Crawford NM 1993 Identification and characterization of a chlorate-resistant mutant of *Arabidopsis thaliana* with mutations in both nitrate reductase structural genes NIA1 and NIA2 *Molecular and General Genetics* **239** 289–297

Wink D A Cook JA Pacelli R Liebmann J Krishne M C and Mitchell J B 1995 Nitric oxide (NO) protects against cellular damage by reactive oxygen species *ToxicolLett* **82\83** 221-226

Xu H Heath M 1998 Role of calcium in signal transduction during the hypersensitive response caused by basidiospore-derived infection of the *Cowpea Rust fungus Plant Cell* **10** 585-597

Xu Q Reed JC 1998 Bax Inhibitor-1 a mammalian apoptosis suppressor identified by functional screening in yeast *Molecular Cell* **1** 337-346

Yamamoto A Katou S Yoshioka H Doke N and Kawakita K 2003 Nitrate reductase a nitric oxide-producing enzyme induction by pathogen signals *J Gen Plant pathol* **69** 218-229

Yang Y Shah J Klessig DF 1997 Signal perception and transduction in plant defense responses Genes Dev **11** 1621-1639

Yu CC Hung KT Kao CH 2005 Nitric oxide reduces Cu toxicity and Cu-induced NH_4^+ accumulation in rice leaves *Journal of Plant Physiology* **162** 1319–1330

Yu I-C Parker J Bent AF 1998 Gene-for-gene disease resistance without the hypersensitive response in *Arabidopsis dnd1 mutant Proc Natl Acad Sci* USA **95** 7819-7824

Yu X Sukumaran S Marton L 1998 Differential expression of the *Arabidopsis* NIA1 and NIA2 genes *Plant Physiology* **116** 1091–1096

Zafiriou OC and McFarland M 1980 Determination of trace levels of nitric oxide in aqueous solution *Analyl Chem* **52** 1662-1667

Zeidler D Zahringer U Gerber I Dubery I Hartung T Bors W Hutzler P Durner J 2004 Innate immunity in *Arabidopsis thaliana:* lipopolysaccharides activate nitric oxide synthase (NOS) and induce defense genes *Proceedings of the National Academy of Sciences* USA **101** 15811–15816

Zemojtel T Fröchlich A Palmieri MC *et al* 2006 Plant nitric oxide synthase: a never-ending story? *Trends in Plant Science* **11** 524–525

Zhang A Jiang M Zhang J Ding H Xu S Hu X Tan M 2007 Nitric oxide induced by hydrogen peroxide mediates abscisic acid-induced activation of mitogen-activated protein kinase cascade involved in antioxidant defense in maize leaves *New Phytologist* **175** 36–50

Zhang F Wang Y Yang Y Wu H Wang Di Liu J 2007 Involvement of hydrogen peroxide and nitric oxide in salt resistance in the calluses from *Populus euphratica Plant Cell* and *Environment* **30** 775–785

Zhang H Shen WB Xu LL2003 Effects of nitric oxide on the germination of wheat seeds and its reactive oxygen species metabolisms under osmotic stress *Acta Botanica Sinica* **45** 901–905

Zhang S Du H Klessig DF 1998 Activation of the tobacco SIP kinase by both a cell wall-derived carbohydrate elicitor and purified proteinaceous elicitins from *Phytophthora spp Plant Cell* **10** 435-449

Zhang S Klessig DK 1997 Salicylic acid activates a 48-kD MAP kinase in tobacco *Plant Cell* **9** 809-824

Zhang Y Wang L Liu Y Zhang Q Wei Q Zhang W 2006 Nitric oxide enhances salt tolerance in maize seedlings through increasing activities of proton-pump and Na^+/H^+ antiport in the tonoplast *Planta* **224** 545–555

Zhao L Zhang F Guo J Yang Y Li B Zhang L 2004 Nitric oxide functions as a signal in salt resistance in the calluses from two ecotypes of reed *Plant Physiology* **134** 849–857

Zhao M Zhao X Wu Y Zhang L 2007 Enhanced sensitivity to oxidative stress in an *Arabidopsis* nitric oxide synthase mutant *Journal of Plant Physiology* **164** 737–745

Zhao MG Tian QY Zhang WH 2007 Nitric oxide synthase-dependent nitric oxide production Is associated with salt tolerance in *Arabidopsis Plant Physiology* **144** 206–217

Zhao Z Chen G and Zhang C 2001 Interaction between reactive oxygen species and nitric oxide in drought- induced abcisic acid synthesis in root tips of wheat seedlings *Aust J Plant physiol* **28** 1050-1061

Zhou B Guo Z Xing J Huang B 2005 Nitric oxide is involved in abscisic acid-induced antioxidant activities in Stylosanthes guianensis *Journal of Experimental Botany* **56** 3223–3228

Zottini M Formentin E Scattolin M Carimi F LoSchiavo F and Terzi M 2002 Nitric oxide affects plant mitochondrial functionality *in vivo FEBS Lett* **515** 75-78

Zou H Henzel W Liu X Lutschg A Wang X 1997 Apaf-1 a human protein homologous to C elegans CED-4 participates in cytochrome c-dependent activation of caspase-3 **Cell 90** 405-414

11

PRIONS: THE LETHAL PROTEINS-REVISITED

C.P.MALIK, CHITRA WADHWANI, VARSHA CHATURVEDI AND YAKSHA SINGH

In a recent review Wadhwani and Malik (2007) discussed various aspects of Prions -the lethal proteins. The present discussion is aimed at looking at the prion diseases from a molecular point of view, taking into consideration the genetic as well as the transmissible and sporadic aspects of the diseases. The present appraisal intends to not only critically analyze the prion particle, ending in identifying the progress that needs to be made in the future, and above all, the major uncertainty of variant CJD.

INTRODUCTION

Prusiner (1981) coined the name Prion agents that cause a novel type of fatal brain diseases. Prions are Proteinaceous infectious agents of about 27 Kilo Daltons which (almost certainly) do not have a nucleic acid genome. It was reported to be present in amyloid fibrils in the brain. They resist inactivation by procedures that modify nucleic acid but respond to agents that disrupt proteins. Prion diseases are often called spongiform encephalopathies because of the post mortem appearance of the brain with large vacuoles in the cerebellum and cortex. Research on prion diseases has recently accelerated because the causative agents are not viruses or bacteria or fungi or parasites. The mounting experimental evidence has generated great interest in what appears to be a totally new kind of mechanism of disease. The main goal of present review is to understand the various aspects of these unique proteins partly. The prions are infectious proteins and on account of lacking a nucleic acid genome appear to be a Protein Life form More recently, Prions have come to the attention of a wide audience because they are the causative agents of neurological disorders such as BSE (Bovine Spongiform Encephalopathy) whose possible transmission to human has caused great concern in U.K and elsewhere. The exact nature of this etiological agent has been and still is a focus of scientific debate.

THE PRION PARTICLE

Perhaps, the best current working definition of a prion is a proteinaceous in acid (Prusiner, 1997). A wealth of data supports the infectious particle that lacks nucleic acid (Prusiner, 1997). A wealth of data supports the contention that scrapie prions are devoid of nucleic acid and seem to be composed exclusively of a modified isoform of PrP, designated PrPSc. The normal cellular PrP, denoted PrPC, is converted into PrPSc through a process whereby a portion of its α-helical and coil structure is refolded into β sheet (Pan *et al.* 1993). This structural transition is accompanied by profound changes in the physicochemical properties of the PrP. While PrPC is soluble in no denaturing detergents, PrPSc is not. PrPC is readily digested by proteases, whereas PrPSc is partially resistant (Oesch *et al.* 1985). Because prions appear to be composed entirely of a protein that adopts an abnormal conformation, it is not unreasonable to think of prions as infectious proteins (Pan *et al.* 1993; Telling *et al.* 1996). But we hasten to add that we still cannot eliminate the possibility of a small ligand bound to PrPSc as an essential component of the infectious prion particle. In a more broad view, prions are elements that impart and propagate variability through multiple conformers of a normal cellular protein. The species of a particular prion is encoded by the sequence of the chromosomal PrP gene of the mammal in which it last replicated. In contrast to pathogens with a nucleic acid genome that encode strain-specific properties in genes, prions seem to encipher these properties in the tertiary structure of PrPSc (Bessen and Marsh, 1994; Telling *et al.* 1996; Prusiner, 1997). The discovery that mutations of the PrP gene caused dominantly inherited prion diseases in humans linked the genetic and infectious forms of prion diseases and presented another hurdle for investigators who contain used to argue that prion diseases are caused by viruses. More than 20 mutations of the PrP gene are now known to cause the inherited human prion diseases, and no significant genetic linkage has been established for five of these mutations (Hsiao *et al.* 1989) (for review, see Prusiner, 1997). The prion concept readily explains how a disease can manifest as a heritable as well as an infectious illness. Moreover, the hallmark common to all of the prion diseases whether sporadic, dominantly inherited, or acquired by infection is that they involve the aberrant metabolism of the prion protein (Prusiner, 1991). Although PrPSc is the only known component of the infectious prion particles, these unique pathogens share several phenotypic traits with other infectious entities such as viruses. Because some features of the diseases caused by prions and viruses are similar, some scientists have difficulty accepting the existence of prions despite a wealth of scientific data supporting this concept (Chesebro and Caughey, 1993; Manuelidis and Fritch, 1996; Lasme´ zas *et al.* 1997; Chesebro, 1998).

PrP gene, PrPc Protein and its Isoform

Prion protein gene has been located on chromosome 20. It bears two exons separated by a single intron. Exon I and II are transcribed and the two RNA's ligated into a single mRNA coding for PrP protein. The latter is a cell surface glycoprotein. Normal Protein is a precursor of infectious several posttranslational events to become Prion Protein (PrP 27-30) which causes scrapie in sheep. In humans PrP is translated as a 253 amino acid protein that is translocated into the Endoplasmic reticulum (ER), where signal sequences at both amino (an ER entry signal) and carboxy (a glycosyl phosphatidylinositol) terminal ends are cleaved to produce mature PrP of 208 amino acids. In the ER, the GPI anchor is attached and two asparagine-linked glycosylation sites are core glycosylated. PrP then passes to the Golgi apparatus for trimming and modification of the sugars prior to transport to the exofacial membrane of the cell. In certain cases, PrP may exist in a transmembrane topology that may be specifically linked to the pathogenesis of the disease. PrP molecules arising from mutated genes probably do not adopt the disease conformation as soon as they are synthesized. Otherwise people become sick in early childhood. The mutation in the PrP gene renders the resulting proteins susceptible to flipping from an α-helical to β-sheet shape. Presumably it takes time until one of the molecules spontaneously flips over and still more time for disease PrP to accumulate and damage the brain enough to cause symptoms. In sucrose gradients, infectivity co migrates with rod shaped particles called amyloid fibrils found by electron microscopy in infected brains. Polyacrylamide gel electrophoresis of the proteins in the infectious fraction from sucrose gradients has revealed a single protein termed the prion protein (PrP) with a relative molecular weight 0f 27-30 kDa. Further analysis has revealed that the scrapie PrP or PrPSc is a glycoprotein. A great advance in the field came with the derivation of a monoclonal antibody that recognizes PrPsc. In addition to the prion associated protein that is readily detected in infected tissue, a cross-reacting normal cellular analogue (PrPc) has also been identified using this antibody. The latter protein has homologues in many animal species including mammalian and most likely avian. Purification and partial sequence analysis of PrPSc allowed the design of synthetic oligonucleotide DNA probes complementary to all possible codons that can encode the PrP amino acid sequence. These probes in turn, have led to the cloning of mRNA's encoding PrP and detection of such mRNA's in both normal and scrapie infected hamster tissues. Because DNA's of both normal and infected cells contain a single copy of a gene for this mRNA, it appears that PrP is a component of normal brain tissue that remains closely associated with scrapie agent infectivity during extensive purification. The amino acid sequence of the normal and scrapie PrP is the same and PrPc expressed in uninfected cells is the precursor for PrPSc. The ability of the PrPSc to form aggregates of fibers in the brain distinguishes it from the normal protein. Moreover, the normal protein is slightly smaller and more easily digested by proteinase K than is PrPSc The conformational differences between the two proteins distinguish them and are determined most likely at the plasma membrane or during transfer to lysosomes. It is the accumulation of PrPSc that causes brain cell damage, but whether PrPSc is the causative agent or a consequence of infection is still not known. It appears unlikely that the mRNA is itself the genome of the prion, since none of this RNA has been detected in spleens of infected mice where high levels of infectivity are found. The length of time for development of disease probably reflects the interval needed for production of this abnormal prion protein shape and recruitment of normal PrP into its unusual form. In experiments involving transgenic mice, a mutation was introduced into the DNA coding for the normal PrPSc so that the defect in this protein found associated with GSS was expressed. The animals developed scrapie-like lesions in the brain. Whether or not an infectious agent is also made is under study. In other experiments, transfer of the gene for the hamster PrPSc to transgenic mice made the animals highly susceptible to the hamster adapted scrapie strain. Disease involved the hamster (and not mouse) PrP. When, however, these mice were inoculated with the mouse scrapie strain, the brain pathology showed the presence of mouse PrP. These observations support the role of the protein in nerve cell pathology but do not resolve if PrP is a component of the causative agent or is required for its replication. Finally, it is interesting to note that, using anti-PrP antisera, researchers have found the PrP antigen in amyloid fibers not only in Scrapie infected brains but also in human brains of CJD victims. However, the role of this protein, if any, in the infectivity or pathogenesis of these diseases has yet to be proved, and morphologically similar structures have been observed in noninfectious neuropathies, including Alzheimer's disease. However, the fibers in the amyloid of Alzheimer's disease differ in amino acid sequence from these PrP.

Sterilization

Infectious particles possessing nucleic acid are dependent upon it to direct their continued replication. Prions however, are infectious by their effect on normal versions of the protein. Therefore, sterilizing prions involves the denaturation of the protein to a state where the molecule is no longer able to induce the abnormal folding of normal proteins. However, prions are generally quite resistant to denaturation by proteases, heat, radiation, and formalin treatments, although their infectivity can be reduced by such treatments. Prions can be denatured by subjecting them to a temperature of 134 degrees Celsius for 18 minutes in a pressurised steam autoclave. Ozone sterilization is currently being studied as a potential method for prion deactivation. Renaturation of a completely denatured prion to infectious status has not yet been achieved, however partially denatured prions can be renatured to an infective status under certain artificial conditions.

PRIONS SHOW THEIR GOOD SIDE

Zamponi and colleagues in a study published in *Journal of Cell Biology* 5 May, 2008, have inferred that normally functioning prions prevent neurons from working themselves to death. Diseases such as mad cow result when the prion protein adopts an abnormal conformation. This infectious form creates a template that induces normal copies of the protein to misfold as well. Scientists have long assumed that prions must also have a beneficial side but have been unable to pinpoint any such favorable traits.

In their work, the authors found that mice lacking the prion protein had overactive brain cells. Their neurons responded longer and more vigorously to electrical or drug-induced stimulation than did neurons that had normal prion protein. This hyperactivity eventually led to the neurons' death. The results might help explain why misfolded prions cause dementia: in the wrong conformation, the prion can no longer protect brain cells from deadly overexcitement.

INFECTIOUS AND NON-INFECTIOUS PRIONS HAVE CLEAR DIFFERENCES IN MOLECULAR STRUCTURES

Prions have been identified as the cause of "mad cow" disease (BSE). The culprits are "incorrectly folded" proteins that can "infect" healthy proteins. The molecular bases for such prion diseases are not yet fully understood. Why are some proteins infectious while others are not?

A team headed by Beat Meier (ETH Zurich, Switzerland) and Raimon Sabaté (University of Bordeaux, France) has examined two different forms of a prion-forming protein domain by means of NMR spectroscopy. In the journal *Angewandte Chemie 2008*, the researchers report that the infectious and non-infectious forms differ markedly in their molecular structure.

These are proteins that can fold in different ways. Pathogenic prions are dangerous because they can convert physiological, non-pathogenic molecules into the diseased form. Often, prions largely consist of β-sheet structures. These are accordion-like folded protein ribbons that can easily aggregate into thread-like structures (amyloid fibrils).

The research team took on the prion-forming domain of the fungal protein HET-s. At a pH value of 7—under physiological conditions—this domain forms infectious fibrils. In acidic solution, at pH 3, it also forms fibrils, but these are not infectious.

By using nuclear magnetic resonance spectroscopy (NMR), the team was able to get a closer look at this protein. NMR allows for the evaluation of interactions of the nuclear spins of specific atomic nuclei with each other and their chemical surroundings, which gives information about the structure and dynamics of molecules and molecular fragments. The spectra of the pH 7 and pH 3 versions of the prion differ significantly. Both are mainly arranged in the rigid β-sheet structure, but up close the structures diverge widely. Particularly striking is the fact that the infectious pH 7 form has highly flexible loops in addition to the rigid domains. These are absent from the non-infectious pH 3 prions.

The lack of infectiousness of the pH 3 fibrils is thus related to the fact that their molecular structure is significantly different from that of the fibrils formed at physiological pH as shown in fig:1

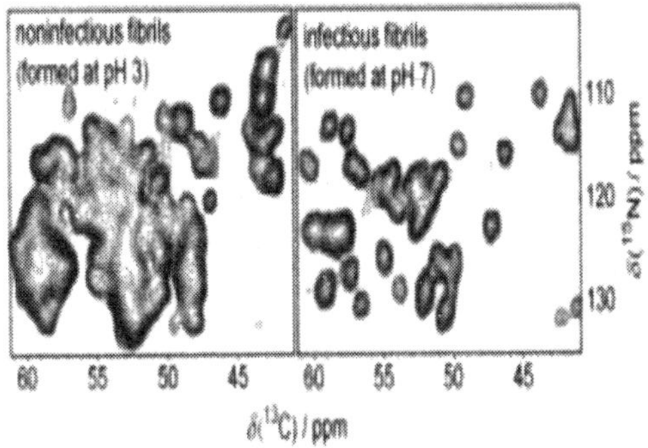

Fig.1. Magnetic resonance studies find clear differences between the structures of infectious and non-infectious prions.

PRION FINGERPRINTS DETECTED WITH GLOWING MOLECULE

Peter Nilsson *et al.* (2007) at Linköping University in Sweden, among other institutions found an effective and sensitive new method for detecting and characterizing prions, the infectious compounds behind diseases like mad cow.

Mad cow disease (BSE), which has caused the death of more than 200,000 cattle and 165 people in the U.K., has now abated. But other prion disorders are on the rise, and there is concern that new strains will infect humans. Prions are not readily transmittable from species to species, but once they have broken through the species barrier they can rapidly adapt and become contagious within the species. Intensive work is now underway to find new, more sensitive methods for detecting these potentially deadly protein structures and distinguish between various strains.

Methods are based on a fluorescent molecule, a so-called conjugated polymer. The research team infected genetically identical laboratory mice with BSE, scrapie (which afflicts sheep), and CWD (chronic wasting disease or "mad elk disease," which is epidemic in the central U.S.) for several generations in a row. Gradually new strains of prions emerge, making the diseases more fatal to the mice. Tissue samples from mice were examined using the fluorescent molecule, which seeks out and binds with prions. This is signaled by a shift in color. By tweaking the molecule, the team has been able to get it to show different colors depending on the structure of the prion each prion strain emits its own optical fingerprint.

This is an important difference compared with other techniques used to find prions, such as antibodies and the well-known stain Congo red.

The technique has also proven to work well on tissue sections from dead animals, such as cows infected with BSE. Now the scientists want to move on and look for alternative sample-taking methods for diagnosing and tracking prion diseases in humans in early stages.

The method would then be useful for screening blood products, since there is a risk that people can be carriers of prions without having any symptoms of disease. In the U.K. it was discovered that 66 people had received blood from blood donors who were infected with the human form of BSE (a variant of Creutzfeldt-Jakob disease, vCJD), and among them, four individuals have been shown to be infected (source: Health Protection Agency, Jan. 2007). We can directly see the structure of the prions and thereby deduce the disease, Nilsson developed the technique has been applying the technology to prion diseases. Prions and other defectively folded proteins give rise to similar disorders, such as Alzheimer's,

ACCELERATED METHOD TO DETERMINE INFECTIOUS PRION STRAINS

According to Lasmézas *et al. PLos ONE* (2009) some prion strains are pathogenic for humans and some are not, it's vital that we know the difference when we find them in the field and when we study them in the laboratory, The new method for distinguishing among various strains combines a transgenic mouse model with a rapid and sensitive cell-based procedure, There are about 20 prion strains known in mouse models. Different varieties of prion strains are identified in mouse models according to incubation time, clinical symptoms, and localization of brain lesions, accelerated incubation. A transgenic mouse line called Tga20 plays an important role, because it succumbs rapidly to prion disease .The prions primarily target the brainstem and the thalamus of this transgenic mouse, explaining why these animals have a shorter incubation time than their normal counterparts.

INFECTIOUS, TEST TUBE-PRODUCED PRIONS CAN JUMP THE 'SPECIES BARRIER'

Castilla *et al* (2009) in, *Cell* have shown that they can create entirely new strains of infectious proteins known as prions in the laboratory by simply mixing infectious prions from one species with the normal prion proteins of another species. Unlike conventional infectious microorganisms, the infectious agent in the case of prion diseases consists exclusively of a misfolded form of the prion protein.

These researchers now find that prion strains produced by combining normal hamster proteins with infectious mouse proteins can infect hamsters and *vice versa.* Although they are both rodents, prions from one of the two species normally don't readily infect the other, a common phenomenon amongst prions known as a species barrier.

The novel prions they produced not only look different, but they also produce symptoms in the animals that differ from any known strain found in nature they report. It is Surprising that the new barriers are crossed and new prions arise. There is the potential for a large variety of new infectious prions—some of which may have dramatic effects.

The infectious agent is a protein with a different shape from the normal protein. Those misfolded and misshapen proteins can spread by causing normal protein to change their shape. Those aberrant forms band together, forming fibrils.

Soto's *et al. in Science* (2005) have reported the generation of infectious prions by amplification of prion misfolding in the test tube. In those experiments, they used a technology called protein misfolding cyclic amplification (PMCA) that mimics some of the fundamental steps involved in the replication of infectious prions in living animals, but at an accelerated rate. The method involves placing small quantities of infectious prions with large quantities of the normal protein from the same species together, allowing the infectious form to imprint on the normal form and thereby replicate itself.

Now, they have demonstrated that the same method can generate new strains when infectious prions from one species are mixed with normal prion proteins from another species. The finding provides conclusive evidence that the imprinting of disease-causing prions on normal forms can overcome species barriers, and doesn't require any other infectious agent.

The new method might provide insight into the risk that other prion diseases could spread from one species to another, For instance, scientists don't know whether chronic wasting disease, a condition now on the rise amongst deer in some parts of the U.S., can be transmitted to humans or not.

Test tube studies like this one might help answer that question, and-- in the case that the deer prions can make the leap—such studies may inform scientists about what those prions might look like,. By studying any new prion strains created in mice with the human prion protein, scientists might also gain insight into the potential symptoms associated with those diseases.

"The data demonstrate that PMCA is a valuable tool for the investigation of the strength of the barrier between diverse species, its molecular determinants, and the expected features of the new infectious material produced," the researchers concluded. "Finally, our findings suggest that the universe of possible prions is not restricted to those currently known but that likely many unique infectious foldings of the prion protein may be produced and that one of the sources for this is cross-species transmission."

PRION DISEASES

A fifty-two year old man stepped into his shower, but could not recognize anything at all. Even his soap was a stranger. One year later, after losing his ability to speak and see he died. Shortly after his death, scientists proclaimed that he had died of variant Creutzfeldt -Jakob disease (Klein *et al.* 2007).

The prion diseases are fatal neurodegenerative disorders of humans and animals that are important because of their impact on public health and because they exemplify a novel mechanism of infectivity and biological information transfer. These diseases are caused by conformational conversion of a normal host glycoprotein (PrP^C) into an infectious isoform (PrP^{Sc}) that is devoid of nucleic acid. These diseases, grouped under the category of transmissible spongiform encephalopathies (TSEs), are the result of the misfolding of the normal host encoded cellular prion protein (PrP^c), to its pathogenic isoform, the scrapie protein (PrP^{sc}) in a post-translational conversion (Prusiner *et al.*1996).The PRNP gene, located on the short arm of human chromosome 20, is responsible for the creation of the normal prion protein (Prusiner *et al.* 1994).

Several polymorphisms occur in the PRNP gene, although the most important one is at codon 129, since it appears to affect the susceptibility to prion diseases (Belay *et al.*1992) Prion diseases have shown to afflict animals as well as humans (Chakra borty *et al.* 2005).

Prion diseases are fatal neurodegenerative disorders of humans and animals that are important because of their impact on public health and because they exemplify a novel mechanism of infectivity and biological information transfer. These diseases are caused by conformational conversion of a normal host glycoprotein (PrP^C) into an infectious isoform (PrP^{Sc}) that is devoid of nucleic acid.

In several cases, little physical data about the infectious agent exist, and the classification is essentially based on similarities in epidemiologic and pathologic findings. Both genetic and sporadic forms of the diseases have been categorized in Table 1.

Table 1: The Prion Diseases, the hosts and mechanism of pathogenesis

Diseases	*Host*	*Mechanism of Pathogenesis*
Kuru	Fore people	Infection through ritualistic cannibalism
iCJD.	Humans	Infection from prion-contaminated HGH, dura mater grafts, etc
vCJD	Humans	Infection from bovine prions?
fCJD	Humans	Germline mutations in PrP gene
GSS	Humans	Germline mutations in PrP gene
FFI C	Humans	Germline mutation in PrP gene (D178N, M129)
sCJD	Humans	Somatic mutation or spontaneous conversion of PrP^{c} into PrP^{Sc}?
FSI	Humans	Somatic mutation or spontaneous conversion of PrP^{c} into PrP^{Sc}?
Scrapie	Sheep	Infection in genetically susceptible sheep
BSE	Cattle	Infection with prion-contaminated MBM
TME	Mink	Infection with prions from sheep or cattle
CWD	Mule deer, elk	Unknown
FSE	Cats	Infection with prion-contaminated beef
Exotic ungulate encephalopathy M	Greater kudu, nyala, oryx	Infection with prion-contaminated MBM

Abbreviations: BSE, bovine spongiform encephalopathy; CJD, Creutzfeldt-Jakob disease; sCJD, sporadic CJD; fCJD, familial CJD; iCJD, iatrogenic CJD; vCJD, (new) variant CJD; CWD, chronic wasting disease; FFI, fatal familial insomnia; FSE, feline spongiform encephalopathy; FSI, fatal sporadic insomnia; GSS, Gerstmann-Straussler-Scheinker disease; HGH, human growth hormone; MBM, meat and bone meal; TME, transmissible mink encephalopathy.

PRION DISEASES IN ANIMALS AND HUMANS

Prion diseases, also called spongiform encephalopathies, are a group of closely related, fatal neurodegenerative disorders that affect mammals, including cows, sheep, and deer, as well as humans. Different strains of the infectious agent, called a prion, cause mad cow disease, chronic wasting disease, and different forms of scrapie and human Creutzfeldt-Jacob disease, Mad cow disease has had devastating consequences for bovine livestock populations, particularly in Europe, and for humans who have consumed contaminated beef products. There have been more than 200 recorded human fatalities worldwide due to mad cow disease. Creutzfeldt-Jacob disease, a low-incidence but always fatal disease, affects humans in all countries. Prions consist mainly or entirely of an abnormal form of a normal cellular protein. They multiply by converting their normal counterparts into a likeness of themselves, which may aggregate to form deposits called amyloid. Accumulation of different kinds of amyloid plays a role in a wide range of neurodegenerative diseases, including Alzheimer's and Parkinson's diseases.

Most mammalian species develop this disease such as the Scrapie in sheep, Transmissible Mink Encephalopathy (TME) in mink, Chronic Wasting Disease (CWD) in mule deer, Bovine Spongiform encephalopathy (BSE) in cows. Humans are also susceptible to several Prion diseases like Creutzfeldt Jacob Disease (CJD) (Creutzfeldt, 1920), Gerstmann Straussler scheinker syndrome (GSS), Fatal Familial Insomnia (FFI), Kuru and Alper's syndrome. Alper's syndrome is the name given to Prion diseases in infants. Prions related diseases are characterized by long incubation and a symptomatic period of illness followed by loss of motor control, dementia, paralysis and ultimately death (Huang *et al.* 2004).

Epidemiology

More than one etiological factor might be responsible for the diseases. Humans might be infected by prions in following ways:

1. Acquired infection

Diet and medical procedures such as surgery, growth hormones, corneal transplants etc. *i.e.* infectious agent implicated. Recently it has been shown that pathological prion proteins are deposited in the neuroepithelium of the olfactory mucosa in patients with sporadic CJD, *i.e.* even nasal secretions could be infectious (Monaco *et al.* 2003).

2. Inherited genetic mutations

Sometimes rogue prions are produced by genetic mutation that is why some cases of CJD, GSS and FFI in humans are inherited in an autosomal dominant fashion.

3. Sporadic prion diseases

Majority of the cases seem not to result from infection, but occur as a spontaneous disease with or without a genetic predisposition (Mastrianni, 2004).

4. RNA mutation

Researchers have suggested RNA mutation as a possible initial event in the etiology of sporadic prion diseases (Garcion *et al.* 2004).

A new variant of CJD has been identified which is designated vCJD, is transmitted to humans by indigestion of BSE-contaminated food products in Great Britain (Scott *et al.* 1999). The disease affects approximately one person per million per year, worldwide (Mckintosh *et al.* 2003). The disease has no preference for a particular gender (Brown *et al.* 1987) and nor is there a predilection for a particular ethnic group although certain mutations within the PrPc gene that causes familial diseases may be more restricted to certain groups, based on the origin of mutation (Mastrianni, 2004).

PRIONS: ON THE TRAIL OF KILLER PROTEINS

The animal and human prion diseases are characterized by their ability to cause an invariably fatal degeneration of the nervous system. When the brain tissue is viewed under a microscope, the sections have a sponge-like appearance. The classic neuropathological features of human prion disease include spongiform degeneration, gliosis, and neuronal loss in the absence of an inflammatory response. Scrapie was the first example of this type of disease and has been known for many hundreds of years. Kuruî is the condition, which first brought prion diseases to prominence in 1950's. It was found in geographically isolated tribes in Papua New Guinea. It spread through cannibalism because the people used to ingest brain tissue of dead relatives for religious reasons.

The symptoms of Kuruî are congestion of blood vessels in brain, cortical atrophy, vacuolization in neuronal cytoplasm and dendrites, which gives the neuropil a spongy appearance. Ultimately the cells misfire and become non functional (Hadlow, 1959). Clusters of brain cells die and the brain stops looking like brain. In mad cow disease the cows become so uncomfortable and itchy that they frantically rub up against anything they can, finally scrapping off most of the wool and hair. Inherited prion diseases like GSS and FFI show the deposition of amyloid plaques in the cortical and sub cortical regions of the brain. Impairment of movement is observed when cerebellum gets involved where as decrease in memory and mental agility shows the involvement of cerebral cortex. Porter and Porter (Porter *et al.* 1973) have reported lack of specific immune response to the puzzling agent. Essentially, the brain cells fall apart, leading to symptoms such as a progressive sub-acute or chronic decline in cognitive or motor function.

(*a*) Different prions affect different regions of the brain

Symptom depend on which region of the brain is affected as in figs. : 2 (a), (b)

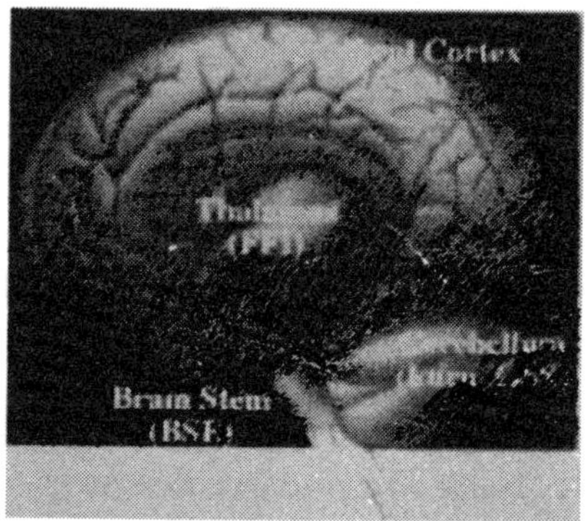

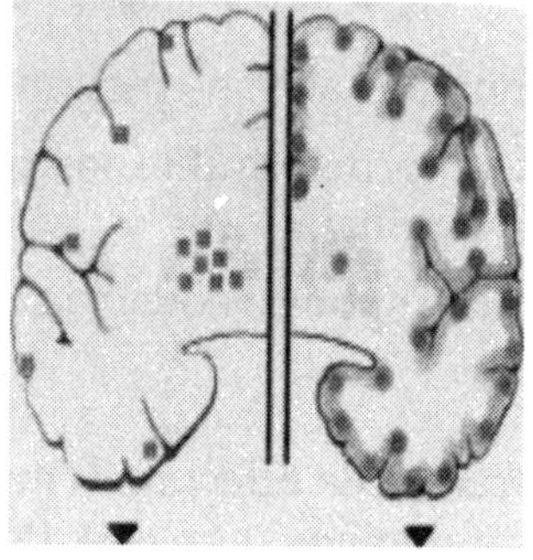

Fig 2. (a) and (b) Parts of brain being affected.

Cerebral cortex: When affected, the symptoms include loss of memory and mental acuity, and sometimes also visual impairment (CJD).

Thalamus: Damage may result in insomnia (FFI).

Cerebellum: Damage results in problems to coordinate body movements and difficulties to walk (kuru, GSS).

Brain stem: In the mad cow disease (BSE), the brain stem is affected.

(*b*) Spongy appearance of brain tissue

Mutations can result in different shapes of the prion protein that accumulates in different regions of the brain: In familial insomnia (FFI), mutated prions (violet squares) accumulate in the thalamus, with the result that the patients are unable to sleep. In Creutzfeldt-Jakob disease, the prion protein accumulates primarily in the cerebral cortex (red dots area). A precise diagnosis of a prion disease can only be made upon autopsy. The fig. 3 shows thin sections of diseased brains. FFI, with typical proliferation of astrocytes, the support cells of the brain, is shown to the left (arrows). CJD, with the characteristic spongiform appearance with vacuoles (arrows) is shown to the right. Due to their characteristic appearance, the diseases are known as the transmissible spongiform encephalopathies (TSEs) as they can be transmitted both experimentally and naturally. The TSEs consist of several closely related prion diseases, of which four similar diseases affect humans as shown in Fig. 3

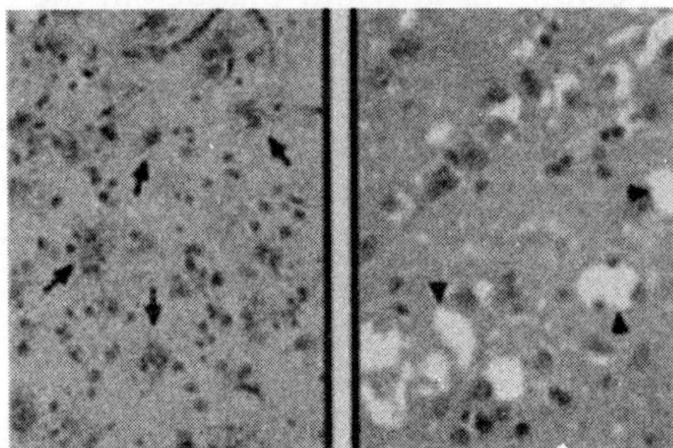

Fig 3. Different regions of brain affected by different prions.

1. Through horizontal transmission from e.g. a sheep to a cow (BSE).
2. In inherited forms, mutations in the prion gene are transmitted from parent to child *i.e.* familial.
3. They can arise spontaneously i.e sporadic (no known antecedent events).

Relevance

The prion diseases can affect both humans and animals, and it is invariably fatal due to the neurological damage it causes. Presently, there are no vaccines or medicines that will prevent the disease and, apart from nursing care, there are no treatments to halt or slow its inevitable course. The course of the disease is alarmingly rapid compared with other neurodegenerative disorders such as Alzheimer disease, Parkinson's disease and motor neuron disease.

Epidemiology and Clinical features of the different diseases

Background: The prion diseases are a large group of related neurodegenerative conditions, which affect both animals and humans. These diseases all have long incubation periods but are typically rapidly progressive once clinical symptoms begin. All prion diseases are fatal, with no effective form of treatment currently; however, increased understanding of their pathogenesis has recently led to the promise of effective therapeutic interventions in the near future.

Prion diseases are unique in that they can be inherited, they can occur sporadically, and they can be infectious. The first of these diseases to be described was Scrapie, a disease of sheep recognized for over 250 years. This illness, manifested by hyper excitability, itching, and ataxia, leads to paralysis and death. It is called Scrapie because of the tendency of affected animals to rub against the fences of their pens in order to stay upright, reflecting their cerebellar dysfunction.

Pathophysiology: A unifying feature of all the prionoses is their neuropathology.

These illnesses tend to affect the gray matter of the central nervous system (CNS), producing neuronal loss, gliosis, and characteristic spongiform change. The latter is a vacuolation of the neuropil, and to a variable degree, of the neurons. In addition, plaques with the typical staining properties of amyloid (eg, apple-green birefringence after Congo Red staining when viewed under polarized light) are observed in many of these conditions.

PRION RELATED DISEASES

Sporadic diseases include CJD, which occurs worldwide with incidence of about 1 in a million people. An example of an acquired disease is kuru, meaning shivering or trembling, which was discovered in Eastern Highlands of Papa New Guinea amongst members of the ForÈ tribe. It is also known as the laughing disease due to the facial grimaces it causes and is thought to be acquired through the ritual cannibalism of dead relatives who had caught the disease. There is also iatrogenic CJD and new variant CJD.

New variant CJD is commonly thought to result from eating food contaminated with the BSE agent.

Familial diseases are genetically inherited and include familial CJD, which accounts for about 15% of all CJD cases. Also inherited are GSS, with a global incidence of 1 in 10 million, and FFI, where insomnia is a major symptom, in addition to other symptoms including dementia. GSS diseases comprise dementing, ataxic and atypical GSS with neurofibrillary tangles. Included is the atypical prion disease, which does not easily fit the diagnostic criteria for prion disease. There are also several types of animal prion diseases affecting animals. This is explained by the fact that many are zoo animals possibly exposed to the agent in their feed.

HUMAN PRION DISEASES

In the care of patients dying of human prion disease, the precautions used for patients with AIDS or hepatitis are certainly adequate. In contrast to these viral illnesses, the human prion diseases are not communicable or contagious. There is no evidence of contact or aerosol transmission of prions from one human to another. However, they are infectious under some circumstances, such as ritualistic cannibalism in New Guinea causing Kuru, the administration of prion-contaminated growth hormone causing iatrogenic CJD, and the transplantation of prion-contaminated dura mater grafts. Familial CJD, GSS, and FFI are all dominantly inherited prion diseases; five different mutations of the PrP gene have been shown to be genetically linked to the development of inherited prion disease. Prions from many cases of inherited prion disease have been transmitted to apes, monkeys, and mice carrying human PrP transgenes.

Creutzfeldt-Jacob disease

Creutzfeldt-Jacob disease (CJD) occurs rarely throughout the world and typically presents as a dementia in either men or women at least 60 years old. Some cases are clustered in certain families, and the fact that some of these families also have an apparently higher incidence of Alzheimer's disease has led to the supposition that the two diseases may be related. Only CJD is readily transmissible to apes (and occasionally, rodents) however. CJD also appears to be transmissible to humans by contaminated products.

Coste E - Uro *et al.*(*PLos Pathog* , 2008) from the United Kingdom and France have identified four separate biochemical subgroups in a selection of cases of Creutzfeldt-Jakob disease. The study suggests that these subgroups could represent distinct prion strains in the most common human prion disease.

Prion diseases are transmissible neurodegenerative disorders characterized by accumulation of an abnormal isoform (PrPSc) of a host-encoded protein (PrPC) in affected tissues. Although considered a spontaneous disorder, the clinicopathological characteristics of sporadic Creutzfeldt-Jakob disease (sCJD) are variable and substantially influenced by a particular variation in the DNA of the prion protein gene (PRNP). Due to the strong influence of host factors on the characteristics of the disease, diversity of prion agents responsible for CJD remains extremely difficult to investigate.

In this study, using two new biochemical assays, the authors identified four distinct biochemical PrPSc subgroups in 41 sCJD cases. These subgroups correlate with the current sCJD sub classification. The subgroups were also found in 12 iatrogenic CJD (iCJD) cases from different countries. iatrogenic CJD occurs following human-to-human sCJD transmission. In contrast to the sCJD cases, however, there was no particular correlation with the PRNP codon in the iCJD cases, indicating that observed biochemical properties could be specific to the prion agent.

Creutzfeldt-Jacob disease (sporadic)

In a number of countries, there have been extensive epidemiological studies on the numbers of CJD cases. All of these countries obtained broadly the same results, approximating about 0.5-1 case per million. There is no significant case clustering other than in familial clusters. Cases are distributed apparently at random with a frequency related only to local population density. There is no evidence for case to case spread (other than with respect to iatrogenic CJD discussed below). There is now evidence of an association with local scrapie prevalence, since CJD is as common in scrapie free areas such as Australia and New Zealand, as it is in the UK, where scrapie is endemic. Despite the attention focused on links between BSE and CJD, the overall incidence figures do not appear to have risen more in the UK than in countries, which do not have BSE. Although an overall increase has been observed, it is likely to be due to increased accurate identification of cases due to increased attention, which is now focused on these diseases. This is especially relevant to the increased incidence noted in the elderly, where other dementing illnesses are common and could be confused with CJD. Occasional weak statistical associations with particular foodstuffs have been reported, although these have tended to disappear in subsequent years of the study and no consistent new risk factors have emerged. Despite this, concern has been expressed about a number of occurrences of CJD in UK dairy farmers.

Clinical features of CJD

According to Collinge & Palmer (1997) the core clinical syndrome of CJD is of rapidly progressive multifocal dementia usually with myoclonus. The onset is usually in the 45- 75 year age group with peak onset between 60-65 years. The clinical progression is typically over weeks progressing to akinetic mutism and death, often within two to three months. Around 70% of cases die in less than 6 months. Prodromal features include extrapyramidal signs, cerebellar ataxia, pyramidal signs and cortical blindness. About 10% of the cases present initially with cerebellar ataxia. Routine haematological and biochemical investigations are normal although occasional cases have been noted to have raised serum transaminases or alkaline phosphatase. There are no immunological markers and acute phase proteins are not elevated. They go on to mention that "Prospective epidemiological studies have demonstrated that cases with a progressive dementia and two or more of myoclonus, cortical blindness, pyramidal, cerebellar or extrapyramidal signs, or akinetic mutism with a typical EEG nearly always turn out to be confirmed as histological definite CJD if neuropathological examination is performed". This depends on neuropathological confirmation being "by demonstration of spongiform change, neuronal loss, and astrocytosis". PrP amyloid plaques are usually not seen with CJD, although the presence of protease resistant PrP can be demonstrated by immunoblotting of brain homogenates.

Iatrogenic CJD

Iatrogenic CJD is usually referred to as CJD despite cases arising from peripheral inoculation (not intracerebral) usually having a prominent cerebellar syndrome more reminiscent of Kuru (discussed later). Transmission may occur by any one of a number of routes involving accidental inoculation with human prions as a result of medical procedures; the routes involving the use of inadequately sterilised neurosurgical instruments due to the high heat resistance of the prion, dura mater and corneal grafting, as well as the use of gonadotrophin derived from the pituitary gland from cadavers. Iatrogenic CJD actively demonstrates the transmissible properties of the prion (described later). The first case was with a 55 year old woman who received a corneal transplant from a donor later found to have died from CJD. In such cases of neurosurgical transmission, the equipment was found to have been sterilised with 70% ethanol and formaldehyde vapour. This is normally enough for conventional agents, but is ineffective against prions. In view of this, the new sterilisation protocol involves immersion in 1 M NaOH for one hour to inactivate the prions.

New Variant CJD

Teenagers rarely get sporadic CJD, and when in 1995 two cases were reported in the UK, (Bateman, Britton) both were unusual in having Kuru-type plaques, which is only found in about 5% of CJD cases. Shortly after, a third young case emerged and a link was suggested with BSE. In the following months further cases emerged showing similar histological patterns which were unique from those observed with other prion diseases. These cases were named new variant CJD, despite being atypical in their clinical presentation with most cases not meeting the accepted clinical diagnostic criteria for probable CJD. Since the original reports, 124 cases of nvCJD have been diagnosed in Great Britain, 6 in France, 1 in Italy, and 1 in Ireland. These cases were extensively compared with other prion diseases and the lack of correlation seemed to confirm the arrival of a new strain of prion disease in the UK due to the incredibly small statistical probability of these similar cases occurring by chance. The symptoms of the new variant CJD start with behavioural and psychological disturbances, which include anxiety, depression, persistent pain in the limbs/face, withdrawal and behavioural changes which progress over neurological features later become apparent often after most patients have gone to see a psychiatrist due to the preliminary symptoms. Several weeks later, progressive cerebellar syndrome develops with gait and limb ataxia. Later dementia develops progressing to akinetic mutism; myoclonus was also seen in most patients and sometimes was preceded by choreoathetosis. The neuropathological appearances were very consistent with widespread spongiform change, gliosis and neuronal loss, most severe in the basal ganglia and thalamus, the most remarkable feature being the abundant PrP amyloid plaques in the cerebral and cerebellar cortex. The plaques consisted of Kuru like florid plaques (surrounded by spongiform vacuoles) and multicentric plaque types. The florid plaques were very unusual and seen previously only in scrapie, but were very consistent. There was also an extensive amount of PrP in the cerebral cortex, cerebellar cortex, and in the molecular layer of the cerebellum. The unusual young age of these patients and the link with BSE may point to a risk involving dietary exposure. The risk involved however is as yet undetermined, and will only be determined as time goes by and more or fewer cases are identified. This is due to the potentially long incubation time related to the species barrier.

Different (atypical) forms of CJD

These forms of the disease are well recognised, and around 10% of cases have a longer clinical course of over two years. These cases may be represented by people with heterozygous PrP polymorphisms. 10% of cases involve cerebellar ataxia rather than cognitive impairment, and are called ataxic CJD. Other types of CJD can include Heidenhain's variant of CJD, which refers to cases where cortical blindness predominates with severe involvement of the occipital lobes. There is also the panencephalopathic type of CJD where there is extensive degeneration of the grey matter; this was mainly reported in Japan. Amyotrophic variants of CJD have been described with prominent early muscle wasting and are usually seen in late disease when other features are established. However most cases of dementia with amyotrophy are not experimentally transmissible, and their relationship with CJD is not clear. It is possible that these are variants of motor neurone disease with associated dementia.

When evaluating a patient for possible sporadic CJD, the clinician should be guided by published case definitions; they are as follows:

Definite CJD

Characteristic neuropathology

Protease-resistant PrP by Western blot

Probable CJD

Progressive dementia

Typical findings on EEG

At least 2 of the following - Myoclonus, visual impairment, cerebellar signs, pyramidal or extrapyramidal signs, or akinetic mutism

Possible CJD

Progressive dementia

Atypical findings on EEG or EEG not available

At least 2 of the following-Myoclonus, visual impairment, cerebellar signs, pyramidal or extrapyramidal signs, or akinetic mutism

Duration less than 2 years

Kuru disease

The study of Kuru is immensely important for the understanding of these diseases, providing the most extensive clinical experience of acquired prion disease in humans, the epidemiology of Kuru clearly indicates the infectious, but non-contagious, properties of prions and provides extensive evidence that vertical transmission from pregnancies does not occur. Kuru reached epidemic levels in the ForÈ people in the Okapa district of the Eastern Highlands of Papua New Guinea (Gajdusek *et al.* 1957, Gajdusek, 1977) other areas affected included Keiagana, Kanite, Yate, Usurfa and Gima, which are in the close vicinity.

Work done by Gajdusek (1957) in an epidemiological survey, uncovered the link between the ritualistic cannibalism and the disease occurrence. This included the differential age and sex incidence which was due to the women and children preparing the food, and becoming exposed to the agent, as well as eating the less desirable brains and other internal organs, even the bones were pounded up to be eaten with greens later. The whole of the deceased sufferer was recycled by way of endo-cannibalism. This is a type of cannibalism. It could be termed 'transumption' as it is seen as disposing of the dead relative, recycling the remains back into the community, and most importantly as a spiritual and virtuous gift to the family members. These customs were thought to instill wisdom.

The ratio of occurrence was about 5 females to 1 male, with about 1% of the population affected. In the years after these practices were stopped, the only cases were from the older and not the younger people (who won't have been cannibals) indicating no evidence for vertical transmission, and due to the mortality rate much later, an incubation time of as much as 40 years or more.

The clinical duration had a mean of 12 months with a range of 3 months to three years, the course tending to be shorter in children. Cerebellar ataxia is the central feature with dementia usually absent (contrasting to CJD). Gross dementia can occur but is usually rare. There is a prodrome stage followed by three clinical stages which are:

1. *Prodromal stage:*

Kuru typically begins with prodromal symptoms consisting of headache, aching of limbs, and also joint pains which can last for several months.

2. *Ambulatory stage:*

At this stage there may be no objective signs of disease, but it can be self diagnosed as patients experience an onset of unsteadiness in standing or walking, or of dysarthria or diplopia. Gait ataxia worsens and patients develop a broad-based gait, truncal instability, and titubation. There is a postural tremor present which is coarse and accentuated by movement; patients characteristically hold their hands together in the midline to suppress this. Standing with feet together reveals clawing of toes to maintain posture. This clawing response is regarded as pathogonomic of Kuru. Therefore signs at this stage are of midline truncal ataxia, presumably due to paleocerebellar disease, which is reflected by the only clear macroscopic brain abnormality seen in Kuru, which is atrophy of the cerebellar vermis. As the gait ataxia and astasia worsens the patients require a stick, as ataxia progresses the patient passes to the next stage of the disease (Sedentary). The mean clinical duration of the first stage is around 8 months and correlates closely with total duration.

3. *Sedentary stage:*

At this stage patients cannot walk, but are able to sit. Walking attempts lead to a high steppage, wide based gait with reeling instability, and flinging arm movements in an attempt to maintain posture. Hyper-reflexia is seen, although usually plantar responses remain flexor and abdominal reflexes intact. Clonus is characteristically short lived. Athetoid and choreiform movements and parkinsonian tremors may occur. There is reduced muscle power but no paralysis, and obesity may be common at this stage but possibly linked to bulimia. Characteristically there is emotional lability and bizarre uncontrollable laughter, which has lead to the disease being referred to as 'laughing death'; one sufferer may start and this spreads throughout the group, although this is only a small aspect of the disease. Myoclonic jerking is rarely seen (unlike in CJD) in this stage which lasts about two to three months.

4. *Tertiary stage:*

Hypotonia and hyporeflexia develop and the terminal stage is marked by flaccid muscle weakness. Plantar responses remain flexor and abdominal reflexes intact. The patient is unable to sit unsupported, dysphagia occurs and the patients become incontinent. Inanition and emaciation develop and transient conjugate eye signs and dementia may occur with primitive reflexes developing in some cases. Brainstem involvement and both bulbar and pseudobulbar signs occur.

Respiratory failure and bronchopneumonia eventually lead to death. This final stage lasts one or two months. Often death may occur earlier due to the nature of the illness, for example the sufferers may role into a fire, and burn themselves when in the later stages of the disease. The sufferer is always very well cared for by their family. In the last five years, there have only been 18 deaths in the south ForÈ, with only one male death in 1998 and none yet in 1999, indicating the beginning of the end of the kuru epidemic.

INHERITED PRION DISEASES

Gerstmann-Straussler-Scheinker disease

This was first described by Gerstmann in the 1920s-30s as a peculiar heredo-familial disease of the central nervous system. The classical presentation is with a chronic cerebeller ataxia accompanied by pyramidal features, with dementia occurring later in a much more prolonged clinical course than that seen in CJD. The mean duration is around five years with onset usually in the third or fourth decades of life. The histological hallmark is the presence of the multicentric amyloid plaques. Spongiform change, neuronal loss, astrocytosis and white matter loss are also usually present. This is an autosomal dominant disorder and is caused by mutations of the PrP gene, which can now be classified within the spectrum of inherited prion disease. Interestingly, in several kindreds of GSS, extensive neurofibrillary tangle (NFT) formation is found (Ghetti, 1994). NFTs are an essential feature of Alzheimer disease, but are also observed in other neurodegenerative conditions.

Fatal Familial Insomnia

This disorder was first described in 1986 and is an autosomally dominant inherited sleep disorder. This presents itself as a progressive untreatable insomnia and autonomic dysfunction (i.e hyperthermia, hypertension, tachycardia, tachypnea, hyperhydrosis), dementia, and motor paralysis however, the phenotypic expression is very variable even within the same family. Thalamic atrophy is the main histological finding, with the anterior-ventral and medial dorsal thalamic nuclei being consistently affected, while the cortex shows moderate asterocytosis.

The age of onset is also variable, ranging from 18-60 years. Once symptoms begin, the course ranges from 6 months to 3 years. Because of the diversity of clinical presentations of this disorder, genotyping is very important for definitive diagnosis. Unlike other prionoses, spongiform change can be a minor feature or can be absent altogether. All patients with FFI have a missense mutation at codon 178 of the PrP gene where Asn is replaced by Asp, coupled with a Met at the polymorphic codon 129. The somewhat divergent clinical and neuropathological features of FFI, in comparison to other human prionoses, highlight the wide spectrum of disease associated with PrP dysfunction and suggest that other human illnesses have yet to be recognized as prionoses.

Familial CJD

This inherited disease occurs later on in life at a similar time to GSS with similar symptoms except for a more rapidly advancing dementia and an abnormal EEG.

CAUSES OF PRION DISEASES

Prion-related diseases are unique in that they can be related to infectious, sporadic, or familial causes:

Infectious causes

Kuru, a form of prion disease, occurred among the Fore people of the Eastern Highlands of New Guinea and was related to ritualistic cannibalism. The disease is believed to have started with the chance ingestion of body parts of a patient with sporadic CJD, followed by a serial passage of the disease Many cases of iatrogenic CJD have been reported. In clinical practice, CJD has been transmitted by surgical instruments, EEG electrodes, corneal transplants, dura mater grafts, human pituitary-derived gonadotrophins, and human-derived growth hormone.

Concerns that nvCJD could be transmitted by blood transfusion have not yet been borne out by a documented case. nvCJD in humans is presumed to have been caused by ingestion of beef products contaminated with BSE. BSE is presumed to have started because of the practice of supplementing the diets of calves and dairy cows with meat and bone products. These meat and bone products are thought to have been contaminated with scrapie material (from sheep) and/or with material from cattle with a sporadic form of bovine prion disease.

Recently, a number of cases of apparent sporadic CJD have occurred in the United States among young individuals (<30 y). The incidence of sporadic CJD among such young individuals has historically been about 1 case per billion populations. In the years

1979-1996, 4 cases of sporadic CJD were reported in the United States among individuals younger than 30 years. In the years 1997-2000, 5 cases have occurred in the United States among young patients. Two of these individuals came from adjacent counties in Michigan (ages at onset were 26 and 28 y), and 3 cases occurred among individuals who were known hunters of deer and/or elk.

Over the same period, a major outbreak of CWD occurred among the deer and elk populations in many western states. CWD is a form of prion disease that occurs naturally in the deer and elk population; however, the pathology has many similarities to BSE, including the presence of florid plaques. Significantly, transmission studies of CWD PrPSc in the laboratory have shown that it can cross the species barrier from deer to human PrP at about the same efficiency as the BSE prion agent. These observations have led to the speculation that limited transmission of CWD to humans has occurred recently in the United States.

Familial causes

The cause of familial forms of prion disease is related to mutations in the PrP gene. A number of mutations in the PrP gene are linked to autosomal dominant forms of prion disease. The diseases can also be related to infectious causes, such as nvCJD, that are thought to be related to ingestion of cattle products contaminated with BSE. The disease can also be related to a spontaneous conformational change in the normal PrPC to a PrPSc form. The PrPSc form has higher beta-sheet content and is self-propagating, inducing more PrPC to convert to the PrPSc form.

Sporadic causes

Sporadic CJD is the most common form of prion disease.

It probably arises as a spontaneous conformational change in PrPC to a PrPSc form.

The PrPSc form is then self-propagating, inducing more PrPC to convert to the PrPSc form.

ANIMAL DISEASES

Scrapie Disease

The oldest known prion disease, was described in 1759 (Leopold) and is characterized by the abnormal walk and scratching of infected sheep and goats. It is still found in parts of world despite attempts to eradicate the agent by destroying infected flocks. The disease is characterized by incoordination and unsteady gait, which becomes progressively worse until death. The infectious agent, in the form of extracts from infected brains has been passed experimentally to mice, hamsters, ferrets, mink and monkeys but apparantly is not infectious for chimpanzees or rabbits. The disease induced by the Scrapie agent in Mink is analogous to that in sheep and is indistinguishable from a Transmissible MinkEncephalopathy (TME) found naturally in in Mink. Likewise, Chronic Wasting Disease of mule deer and elk is similar both in pathology and known physical properties to the Scrapie agent. A condition in cattle, the so called ìMad Cow Diseaseî is believed caused by a prion transmitted in contaminated feed supplements derived from infected sheep and goats. Concern has been raised that this agent might also cause disease in humans eating contaminated meat. Scrapie was the first example of this type of disease to be noticed and has been known about for many hundreds of years. There are two possible methods of transmission in sheep:

1. Infection of pasture with placental tissue carrying the agent followed by ingestion, or direct sheep-lamb transmission *i.e.* an acquired infection.
2. Parry showed considerable foresight by suggesting that it is not normally an infectious disease at all but a genetic disorder. He further suggested that selective breeding would get rid of the disease.

Scrapie is classified as a slow infection. It is characterized by a prolonged incubation period of months, years or possibly decades in which the patient or the host animal is without symptoms. Once the illness begins, however, it progresses steadily and always leads to death. The name scrapie comes from the tendency of sheep that have the disease to scrape off much of their wool. The disorder has been known for more than 200 years, and by 1935 French workers had shown convincingly that it can be transmitted from one sheep to another by inoculation. This demonstration of transmissibility implies that the infectious agent is capable of reproducing itself in the host animal. The clinical and pathological signs of Scrapie are similar to those of Kuru and CJD. The initial symptoms are difficulty in walking and loss of coordination, indicating a functional disorder of the cerebellum. Other symptoms include irritability, excitability, restlessness, scratching, biting, and rubbing of skin, patchy loss of wool, loss of weight, weakness of the hindquarters and impaired vision. There is no sign of an inflammatory process or fever, and cell counts in the spinal fluid remain normal. This indicates that the immune system does not respond to the disease agent. Pathological effects brought on by the diseases are confined to the central nervous system. In some of the disorders numerous vacuoles give the brain tissue a spongy appearance. The natural mode of transmission in sheep is unclear, but scientists do know that it is transmitted by a filterable particle that is resistant to heat and formalin fixation. As yet there are no biochemical or cell culture assay systems for either Scrapie agent or any other of these agents. Transmission of the disease to mice or hamsters allows endpoint titration of biological activity to be performed, albeit slowly. With the finding that the interval from inoculation to onset of disease is inversely proportional to the dose of Scrapie agent injected intracerebrally into hamsters, it has become possible to assay samples with the use of four animals in 60 to 70 days- provided that the titers of Scrapie agent in the samples are sufficiently high. This assay has been used to help purify the infectious principle and investigate its structure on the molecular level. Variants of the Scrapie agent have been isolated from standard stocks by passage in mice and hamsters at limiting dilutions. Some of these exhibit different neuropathological effects, including markedly different incubation times and tissue tropism. These ìstrainsî seem to be stable and do not revert or interconvert with detectable frequency. The existence of such stable variants is most easily explained by the operation of a genetic mechanism based on the presence of a nucleic acid genome in the infectious agent. However as noted above, no such genome has been found and no virus like particles has been observed in infectious samples. The Scrapie agent (like viruses) passes through filters of 22 to 100 nm. It can be inactivated by strong alkaline solutions (1 M NaOH) but is highly resistant to nonionic detergents. Its resistance to nucleases and erlative sensitivity to proteases led to its consideration as an infectious protein (prion).

Timm Konold *et al.* BMC Veterinary Press (2008) from the Veterinary Laboratories Agency in Weybridge, UK, investigated the transmission of scrapie by feeding milk from scrapie-affected ewes to lambs that are genetically susceptible to contracting scrapie. The researchers were looking for the presence of the prion protein, PrPd, which is associated with the disease. Scrapie can be transmitted to lambs through milk, according to new research. The study provides important information on the transmission of this prion-associated disease and the control of scrapie in affected flocks. Scrapie is a fatal neurodegenerative disease of sheep and goats. Clinical signs include itchiness, head tremor, wool loss and skin lesions as well as changes in behaviour and gait. Eighteen lambs were fed milk from scrapie-affected ewes. Three of these lambs were culled and two were found to have PrPd in intestinal tissues. The prion protein was also detected in lymphoid tissue of the gut of the surviving lambs and in some control lambs mixed with the scrapie milk recipients after weaning.

This suggested that scrapie milk recipients were able to shed the infectious agent and infect other lambs. There was no sign of PrPd in tissue samples from a control group of 10 lambs (one culled and the rest alive), which were housed in the same building but fed milk from healthy ewes. The research will continue, to see whether the lambs with PrPd develop the disease as they get older. This work raises the possibility that other prion diseases could be transmitted in sheep *via* milk although it should have no direct implications for human health. Scrapie has been found in sheep and has not been shown to be transmissible to humans. BSE has not been found naturally in sheep and occurrence in sheep in the UK is considered to be unlikely. This research adds to our understanding of the transmission of prion diseases in sheep and would help to inform measures needed to protect human health if BSE were ever to be found in sheep.

Transmissible mink encephalopathy

Transmissible mink encephalopathy (TME) is a disease of farmed mink, probably caused by feeding on Scrapie-contaminated meat.

Chronic Wasting disease

Chronic wasting disease (CWD) is a Scrapie-like disease of obscure origin found in wild and captive mule deer and Rocky Mountain elk.

Bovine Spongiform Encephalophathy/Mad Cow Disease

BSE is a progressive neurological disorder of cattle which results from infection by an unconventional transmissible agent. From November 1986 until December 1995, an estimated 155,600 head of cattle in almost 33,000 herds were diagnosed with BSE in Great Britain. The epidemic peaked in January 1993 with almost 1,000 new cases per week. The nature of the transmissible agent is unknown but currently the most accepted theory is that he agent is a modified form of a prion protein. Scientists considered herd type, size and age structure, whether the animals had been in contact with sheep, and is so, whether Scrapie was present in the flock. Sheep were not present on 20% of the farms; therefore it was believed that the cattle food which was being supplemented with sheep bone meal was responsible for the outbreak. So far, the disease has been confined to the United Kingdom with only a few cases showing up outside. Those cases resulted in cows being exported from the U.K. Britain may have been at greater risk than other countries when the Scrapie agent unexpectedly moved from sheep to cattle, apparently sometime in the 1980s. Feed practices in Britain differ from those in other countries, and British farmers raise a large number of sheep-and hence Scrapie-infected carcasses-relative to cattle. Most of the herds affected are dairy, milk producing herds. The reason for this may be that dairy cattle eat more commercially produced food, and thus are more likely to ingest sheep ruminants. Beef cattle depend on grazing and vegetation for food. In 1996, scientists believed that there was a clear possibility that BSE might be infecting humans as well.

Pathogenesis

BSE has an incubation period of 2 to 5 years. Infection can show up in the brain, spleen, thymus, tongue, and gut. Cows infected with BSE appear alert but agitated, anxious and apprehensive. The cow takes a wide base stance, abdomen is drawn up, gait becomes abnormal and exaggerated, and causes tumbling and skin wounds. There are fine muscle contractions over the surface of the neck and body with occasional larger muscle jerks. The animal begins to lose weight and there are frenzied movements including aimless head butting. Disease is fatal for cattle within weeks to months of onset.

Epidemiology

At its peak the epidemic affected approximately one percent of adult breeding cows (annual incidence). The peak was in the winter of 1992/93 when a maximum of just over 1000 suspected cases a week were being reported. The incidence has subsequently fallen year by year at a rate of around 40% per year. The within head incidence has been monitored closely, as this is a good indicator of whether or not there has been spread of disease from cow to cow in the herd. This suggests that there has been little or no spread of infection from animal to animal (horizontal transmission). It is still unclear as to whether cows may become infected directly from the food or from asymptomatic mothers that have died. Post-mortem tests are the only type of diagnostic test which are currently available. Scientists examine the characteristic appearance of the brain under the microscope and also use electron microscopy techniques.

Nano-sensor For Better Detection of Mad Cow Disease Agent

Craighead Harold *et al. Analytical Chemistry* (2008) in New York are reporting development of a nano-sized sensor that detects record low levels of the deadly prion proteins that cause Mad Cow Disease and other so-called prion diseases. The sensor, which detects binding of prion proteins by detecting frequency changes of a micromechanical oscillator, could lead to a reliable blood test for prion diseases in both animals and humans the researchers say.

Prions are infectious proteins that can cause deadly nerve-damaging diseases such as Mad Cow Disease in cattle, scrapie in sheep, and a human form of Mad Cow Disease called variant Creutzfeldt - Jakob disease. Conventional tests are designed to detect the proteins only upon autopsy and the tests are time-consuming and unreliable.

In the new study, Harold G. Craighead and colleagues describe a high-tech, nano-sized device called a nano mechanical resonator array. The device includes a silicon sensor, which resembles a tiny tuning fork, that changes vibrational resonant frequency when prions bind. Its vibration patterns are then measured by a special detector. In experimental trials, the sensor detected prions at concentrations as low as 2 nanograms per milliliter, the smallest levels measured to date.

Feline spongiform encephalopathy

Feline spongiform encephalopathy (FSE) is a disease seen in domestic cats, and in a few other members of the cat family. The spongiform encephalopathies (feline and non feline) are assumed to be caused by feeding BSE-infected material to the animals in question, often in zoos. Mice, hamsters and monkeys have proved experimentally to be the most useful models of prion diseases.

Iron Is Involved In Prion Disease-associated Neuronal Demise

Imbalance of iron homeostasis is a common feature of prion disease-affected human, mouse, and hamster brains, according to Singh *et al.* (*PLos Pathogens*, 2009).These findings provide new insight into the mechanism of neurotoxicity in prion disorders, and novel avenues for the development of therapeutic strategies.

Unlike other neurodegenerative conditions, prion disorders are sporadic, inherited, and infectious, and affect both humans and animals; common examples are mad cow disease in cattle, scrapie in sheep, and Creutzfeldt-Jakob disease in humans. The causative agent is a misfolded protein referred to as PrP-scrapie that replicates itself by changing the conformation of neighboring copies of the same protein, namely the prion protein. Aggregates of PrP-scrapie are toxic to brain cells and cause a spongy-like appearance in diseased brains. Singh suggests that accumulation of PrP-scrapie alters the metabolism of iron in diseased brains. The imbalance of brain iron homeostasis worsens with disease progression, and is not an outcome of end-stage disease. Since iron is highly toxic when mismanaged, this condition is likely to contribute significantly to prion-disease associated neurotoxicity. The likely cause of this condition is loss of normal function of the prion protein in cellular iron metabolism demonstrated recently by Singh *et al.* combined with gain of toxic function by the redox-active PrP-scrapie complex; it was surprised to find that prion disease-affected brains are iron deficient despite a significant increase in their overall iron content. The group concludes that ferritin, a major iron storage protein, co-aggregates with PrP-scrapie in diseased brains and sequesters bound iron in the complex, creating a state of apparent iron deficiency. The brain cells respond to this condition by increasing their level of iron uptake, thus creating a vicious cycle of increased iron uptake in the presence of increased iron.These observations contribute to our understanding of how the prion agent causes neurotoxicity, and may enable the development of novel therapeutic strategies targeted at restoring brain iron homeostasis in prion disorders.

Prion Infectivity Found In White and Brown Fat Tissues Of Mice

Race *et al.* (*PLos Pathogen,s* 2008) from the National Institutes of Health and the Scripps Research Institute have found novel prion infectivity in white and brown fat tissues of mice. Prion diseases affect humans as well as wild and domestic animals. Distribution of prion infectivity in organs and tissues is important in understanding prion disease pathogenesis and designing strategies to prevent prion infection in animals and humans. Studies in animals including sheep, goats, cattle, deer, mink, hamsters and mice, have found prion infectivity mostly in nervous system tissues such as the brain and spinal cord. The tissues studied here in a mouse model demonstrate a proof of principle that white and brown fat tissues are sites of prion agent deposition and therefore may play a previously unrecognized role in prion infectivity and transmission of prion disease. It will be important to extend studies to prion-infected large animals, such as cattle, sheep, deer, and elk where they may be potential sources of contamination of human and domestic animal food chains. Results of the current and future studies may merit additional consideration of steps to eliminate from the food chain any fat from ruminants suspected of exposure to or infection with prions as shown in Fig.6 .

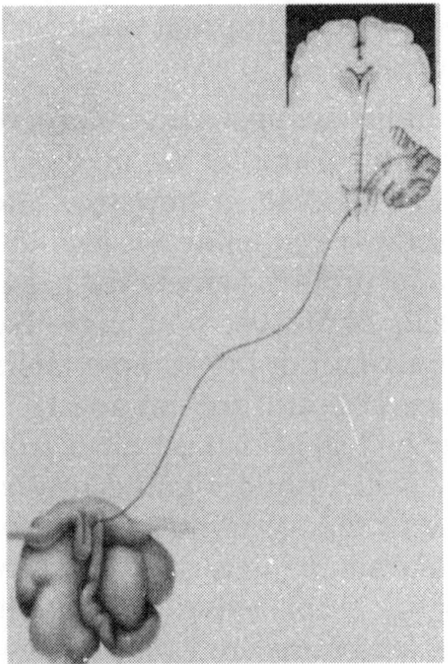

Fig. 5. Spread of infection across species barrier.

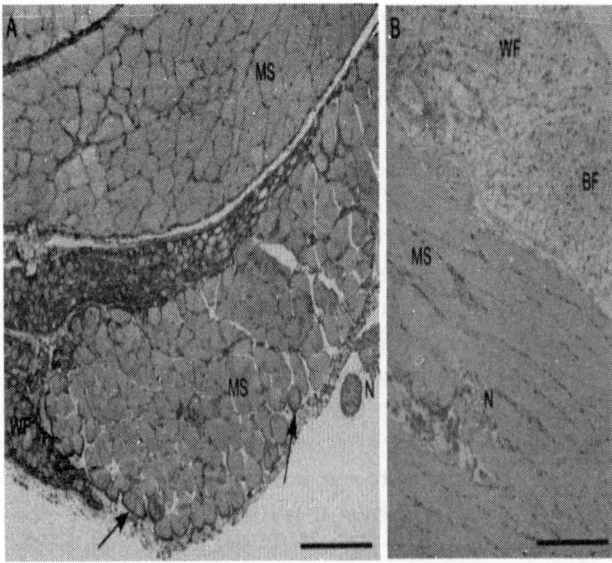

Fig. 6. PrP in white fat and adjacent skeletal muscle of Scrapie-Infected Mice.

Prion Discovery Gives Clue to Control of Mass Gene Expression

The common brewer's yeast of a new, infectious, misfolded protein -or prion -- by University of Illinois at Chicago molecular biologists raises questions about the roles played by these curious molecules, often associated with degenerative brain diseases like "mad cow" and its human counterpart, Creutzfeldt-Jakob.

Liebman and Patel propagated the new prion from a normal yeast protein called Cyc8. They note that like the Cyc8 protein, the prion of Cyc8 can affect the expression of a large number of yeast genes. Liebman *et al.* discovered that Cyc8 was a prion candidate using a genetic screen that looks for proteins that when overproduced can spur formation of new prions. To date, scientists have discovered only seven prions, six of which are only in fungi, including yeast. The latest two discovered, Cyc8 and another, identified as Swi1, The normal Cyc8 protein shuts down expression of more than 300 genes in yeast, including some genes that are involved in stress tolerance. Once Cyc8 is converted to a prion, it loses that function; This might provide some advantages under stressful conditions. Since the protein represses more than 300 genes, it's possible the prion form can activate the genes on a mass level." If an organism wanted to activate all the genes in a cell that the protein repressed, he said, "Converting the protein into a prion would be an easy way to do it."

Patel and Liebman's are testing the protein to see if this molecular mechanism does in fact take place naturally. They're also studying the interaction of prions to determine if pre-existing prions facilitate or destabilize new prions.

Whether the actions of prions in yeast are analogous to mammalian models is not yet fully known, "There could be prions in humans that are not causing disease but have important effects on the cell or organism," said Liebman. "They may even be related to the ones we find in yeast. The more prions we learn about and study, the more information we learn from them -- how they arrive, what proteins are needed to maintain them. As we study other models, we have a better idea."

THE CJD-BSE CONNECTION

The prion disease of humans that causes the most concern is variant CJD (vCJD). The cause of this disease has not yet been determined. One theory is that humans may become infected with aberrant prions after ingesting meat from cows infected with BSE. This would require that aberrant prions from one species (cows) could serve as a template for refolding normal prions of another species (humans). This jump from one species to another involves crossing what is referred to as the species barrier. In a laboratory setting, prions have been shown to cross the species barrier. The onset of disease in these cases, however, is delayed compared to cases in which the infecting (aberrant) prions are from the same species as the new host. It is believed that the origin of the BSE outbreak in Europe (and especially Britain) was due to aberrant sheep prions crossing the species barrier to infect cows. During the 1970's there were changes made to the process used to render offal for a livestock feed additive. These changes created conditions that allowed prions to survive the rendering process. The sheep disease known as Scrapie is prion disease. It is believed that prions from Scrapie-infected sheep survived the rendering process and were passed on to cows that ingested infected feed. Once the prions were established in the livestock population the feed could have acted as a reservoir of infective prions. The practice of using rendered offal as an additive to livestock feed has been discontinued. Whether or not bovine prions can be passed to humans is unclear. Human prion diseases like vCJD have long incubation times (typically more than 12 years). The details of the transmission of prions from one species to another have not been fully elucidated.

Route of infection

When cows are fed with offals prepared from infected sheep, prions are taken up from the gut and transported along nerve fibers to the brain stem. Here prions accumulate and convert normal prion proteins to the disease-causing form, PrPSc. Years later, BSE results when a sufficient number of nerve cells have become damaged, affecting the behavior of the cows as shown in Fig.5

Evidence that proteins are infectious

Proteins are not normally termed infectious agents and unlike viruses, bacteria and other infectious organisms, the prions have no obvious nucleic acid. It has been proven that these are present with the disease symptoms, but do these actually cause the disease from step one- this was to be found out. The prion diseases themselves are known to be transmissible, from the early studies of CuillÈ and Chelle, when they set out to determine the infectivity of Scrapie in 1936, to the likely cannibalistic transmission of Kuru. Parallels were drawn to the possible infectivity of similar diseases such as CJD. The infectivity of these diseases was demonstrated by intracerebral inoculation of Kuru into chimpanzees by Gajdusek *et al.* (1966) and by similar ways with CJD by Gibbs *et al.* (1968). With this evidence it can only be said that the disease is transmissible and not strictly due to the prions. The strongest evidence was by Prusiner *et al.* (1982) where he further purified the prion by demonstration using 3000-fold enrichment for infectivity, and it was the only macromolecule that could be associated with infectivity. However some investigators think that the agent is merely the pathologic product which coincidentally purifies with the "Scrapie virus".

Despite these ideas, there is little evidence to support this view as fractions containing

<1 prion molecule per ID 50 have not been found, if it were to be found, it would indicate that prions are not required for infectivity. There is now a lot of evidence that the discovered 27-30 prion protein is a major and necessary component of the prion disease, this evidence includes:

- That during the co purification of PrP 27-30 and Scrapie infectivity using biochemical methods, the PrP concentration obtained is proportional to the prion titre.
- The kinetics of proteolytic digestion of PrP 27-30 and infectivity are similar.
- Co purification of the infectious prion and infectivity by immuno affinity.

- Chromatography is achieved, and? -Prion antisera neutralises infectivity.
- The novel prion protein is detected only in clones of cultured cells producing infectivity.
- The prion amyloid plaques are specific for prion diseases of animals and humans, and the deposition of the prion amyloid plaque is controlled at least partially by the prion protein sequence.
- There is correlation between the infectious prion and disease in humans and animals.
- Incubation time in mice is affected by base substitutions in the mouse prion gene-Mo-PrP.
- The existence of the species barrier (discussed later) due to amino acid differences in the protein causes a longer incubation time.
- There is genetic linkage between mutations in the prion gene (also discussed later) and the onset of disease, with a mutation at codon 102 linked with GSS development and a mutation in codon 200, or insertions of six additional octarepeats linked with familial CJD.
- Mice expressing prion genes with a point mutation similar to that of GSS develop neurological dysfunction, spongiform brain degeneration and gliosis.

From the evidence discussed at this point most scientists will believe that there is some involvement of prions in the infectivity and pathology of Scrapie, BSE and CJD, but the nature of the involvement was still to be found. Investigators suggest an answer to this with the prion concept.

THE PRION CONCEPT

It took quite a while for this idea of the prion protein to be accepted because it at first sight appeared to challenge some of the most fundamental ideas about molecular biology. Of course, people are used to the idea that infections are caused by bacteria, which are large and can be seen, and viruses, which can be sometimes rather more obscure. Now, here was a situation where there was an infection but apparently no virus, and viruses contain nucleic acid and this was just a protein. And people probably misunderstood in the first instance, and thought that a protein was making more copies of itself, whereas the fundamental dogma of molecular biology says that the information required to make proteins is contained in nucleic acid which then has to be converted into the production of proteins. Of course, it later became obvious that the prion hypothesis doesn't actually challenge the fundamental laws of molecular biology at all, because the DNA that is responsible for making the protein is in the host - the animal that was infected - and indeed it's in all animals and all people. What happens is the abnormal protein directs the conversion which produces more abnormal protein from normal protein, but not from nothing, it doesn't make it itself, it still has to be made originally under the instructions of nucleic acid. It's just the conversion of its shape, which is directed by the protein to another protein molecule. It wasn't as heretical as some people first thought, but it is wholly new, because it's a new way in which information can be transferred from one molecule to another without going through nucleic acid, which is the normal repository of information in biology.

THE PRION STRUCTURE

Proteins consist of stretches of amino acids. Some fold into alpha helices; others fold into beta pleated sheets. These shapes are maintained by hydrogen bonds. The sheets and helices then fold together further to form the protein's tertiary structure. A protein, such as the one illustrated here, can include both alpha helices and beta sheets and can consist of several subunits that combine to form the quaternary structure. Prions have one characteristic that makes them unique: they can exist in two different conformations at the level of secondary structure. There is an alpha-helical portion of secondary structure in a normal prion that is refolded into beta-sheet formation in an aberrant prion. The primary structure of the aberrant prion remains the same, but its secondary structure is different. As a result, its tertiary structure will be different as well. The isolated prion protein was found to be a product of a cellular gene on the short arm of chromosome 20 in humans and in a systemic region of chromosome 2 in mice. Studies then on the hamster and mouse models, after uncovering a protein designated PrP 27-30 (27-30K daltons) led to the eventual molecular cloning and complete sequence of prion protein cDNA (Oesch *et al.* 1985, Locht *et al.* 1986, Collinge & Palmer, 1997). The normal product (PrPc) of the prion gene (called PRNP in humans) is found to be expressed in most cell types, with expression at its highest predominantly in the brain. In 1987 investigations involving PrPc and PrP 27-30 started where an experiment involving bacterial-derived stearic acid treatment of the prion enabling it to react with antisera indicated that PrPc is anchored to the cell surface by a glycolipid, components of which include ethanolamine, *myo*-inositol, phosphate, and stearic acid as shown in fig. 9.

Later experiments involving PrPc indicated that PrPc is expressed as a glycosylphosphatidyl inositol-anchored glycoprotein found on the outer cell membrane (Stahl *et al.* 1987, 1990, 1992). However, this normal form can be distinguished from the infective prion found in diseased brain samples by its non-resistance to proteolytic hydrolysis (Borchelt *et al.* 1990). The infectious prion protein will now be termed as PrPsc and the normal prion protein, its function later to be discussed is now termed as PrPc (the "sc" stands for Scrapie and the "c" stands for cellular).

Different terminology of the two forms of the protein has also come into light. This is the PrPsens and PrPres; these names are given to prions, which are not structurally identical to PrPc or PrPsc. They may be truncated or recombinant forms, and are not exactly the same as the scrapie (sc) and cellular (c) forms, the 'res' indicates resistance to proteolytic degradation, and the 'sens' is for sensitivity to proteolytic degradation. There are therefore two different isoforms of the prion protein, one is expressed normally and one is present aberrantly. Initial studies on the different forms of the prion (Borchelt *et al.* 1990) found there were differences in the proteolytic resistance. Also demonstrated was that PrPc can be released from normal, and Scrapie infected neuroblastoma cells which were treated with phosphatidylinositol-specific phospholipase C (PIPLC), and also be labelled by a sulpho-NHS-biotin. Using pulse chase experiments [35S] methionine was found to incorporate almost immediately into PrPc. PrPsc however was incorporated at a much slower rate. PrPsc was not released by PIPLC, which suggested structural differences. It also enabled to monitor the amounts of both PrPc and PrPsc in the cell. Although this gave no real clues to the mechanisms involved, it further established the presence of two different forms of the protein. Mass spectrometry and amino acid sequencing studies were undertaken on the Scrapie prion protein, this was done using denatured PrPsc that was digested with endoproteases and isolated by HPLC. Analysis was done using mass spectrometry.

Edman sequencing and the primary structure of PrPsc was found to be identical to that of the PrP gene sequence, arguing that neither RNA editing nor protein splicing feature in the synthesis of PrPsc. Mass spectrometry was also used to identify any posttranslational modification-differences between the two isoforms, and none were found contending that PrPsc molecules do now differ from PrPc at the level of an amino acid substitution or a post-translational chemical modification (Stahl *et al.* 1993). However, the possibility of a small fraction of PrPsc being modified by a yet unidentified post-translational process could not be eliminated, or that PrPc carries a modification that is removed in the formation of PrPsc. It is probable then, that the differences are in the secondary andalso enabled to tertiary structures as shown in Fig.7

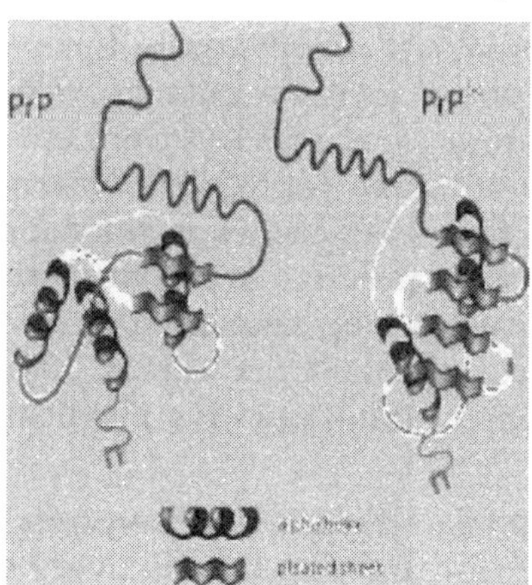

Fig. 7. A normal prion (left), compared to an aberrant, disease-causing prion (right).

The same amino acid sequence but two different structures

Work was then done on investigating the structure of the prion comparing PrPc to the structure of PrPc was found to have mainly a-helices and the structure of PrPsc was found to comprise mainly of b-pleated sheets. This was achieved using a pattern-matching approach (Cohen *et al.* 1985) and Fourier analysis of the primary sequence hydrophobocities to detect amphipathic regions, interestingly the secondarystructure showed some homology to that of the serum amyloid a proteins. The prion was linked with having a post-translational modification to yield amyloid A, which is more evidence for the prion involvement in amyloid plaque formation (Bazan *et al.* 1987). Later Pan *et al.* (1993) wrote a paper which described the conversion of a-helices to b-sheets as part of the formation of the scrapie prion proteins. Using Fourier Transform Infra-Red spectroscopy (FTIR) it was demonstrated that PrPc has high a-helix content (42%) and no -sheet content (3%), and PrPsc has high b-sheet content (43%) and low ahelix content (30%). Even higher b-sheet content is observed (54%) as compared to (ahelix content 21%) when using limited proteolysis to created N-terminally truncated PrPsc. This was confirmed by circular dichroism measurements. Pan *et al.* (1993) then remarked on the likeliness of a conformational transition being a fundamental event in the propagation of prions. Baldwin *et al.* (1994) produced a report which identified and confirmedthe differences between PrPc and PrPsc using spectroscopy to be conformational and even possibly transitional. For this FTIR spectroscopic methods were used, and also suggested the possibility of the transition of alpha to beta transition, they found that the N-terminal truncation of PrPsc by limited proteolysis did not destroy the prion infectivity. Instead, it increased the b-sheet content shifting the FTIR absorption to lower frequencies, which was typical of the cross b-pleated sheets of amyloids. Thus, the formation of PrPsc from PrPc involves a conformational transition in which one or more a-helical regions of the protein are converted to b-sheets. This transition was mimicked by synthetic peptides, allowing predictions of domains of PrP involved in prion diseases. Further evidence for the PrPsc formation from PrPc was shown in experiments involving the synthesis of a-helix peptides from regions of the prion protein, and their subsequent conversion to form amyloid peptides composed largely of b-sheets. Although this evidence is quite good, harder evidence was to follow (Gasset *et al.* 1992, 1993). Muramoto *et al.* (1996) performed studies on recombinant forms of PrPc. It is known that some truncation of the PrPc N terminus still permits conversion into the PrPsc isoform and so to assess whether additional segments of the PrP molecule can be deleted, the regions of the putative secondary structure were removed. It was found that removing a 36-residue loop between the second and third helix did not prevent formation of protease resistant PrP (PrPsc 106) that was also found to be soluble. Also, all the deletions of the 4 putative a-helical structures prevented PrPsc formation.

The results suggested that all of the proposed secondary structures are necessary for the conversion of PrPc into PrPsc and that the discovery of PrPsc 106 should facilitate structural studies of the prions and possible development of antibodies against PrPsc. A year later in 1997, studies investigating the possible posttranslational modifications involved with the PrPc conversion to PrPsc focussed on Scrapie infected ScN2a cells, where the metabolism of both PrP isolates involved cholesterol dependant pathways. It was demonstrated that both PrPc and PrPsc are attached to Trition X-100-insoluble, low-density complexes or "rafts". These "rafts" probably contain cholesterol due to their sensitivity to saponin. However, when extracted from solubility into insolubility using Triton X-100 at 37°C it was found that two different peaks of different densities were there, suggesting two populations of PrP-containing "rafts". This may permit the isolation of PrPc specific rafts. This suggested the possibility of "rafts" hosting the proposed PrPc to PrPsc formation, or if not some other aspect in the "life-cycle" of prions (Naslavsky *et al.* 1997).

The three dimensional structure of PrP

The three dimensional-structures for the cellular prion protein were predicted using was used, consisting of the prediction of secondary structures and of an evaluation of the packing of secondary structures. By using experimental and theoretical constraints four theoretical structures consisting of a-helix bundles were achieved. A group of amino acids was identified as being important for tertiary structural arrangement. Although the predicted structures gave a testable model, PrPc is thought now to have only three a-helix bundles (shown later). Possible inaccuracies could be in the computer programme or rather the experimental, theoretical constraints or data applied to the programme. Despite this, the structure of PrPc seems to be far more predictable than that of PrPsc (Baldwin *et al.* 1998). This is partially due to the very small amount of the natural protein isoforms isolated from the brain, and the solubility of PrP 27-30 in detergents that reduces infectivity even though the b-sheets can be regenerated. Despite this, the NMR (Nuclear Magnetic Resonance) experiments on recombinant protein can lead to a much better characterisation of the 3D structure of the prion proteins. In the studies involving the use of hamster and mouse models to elucidate the threedimensional structure, the hamster and mouse prion protein structure was compared, and is shown below. James *et al.* (1997) shows the hamster and mouse prion proteins superimposed, the mouse structure is more refined but shorter in length than the hamster prion, showing a close relation and therefore is a good model for transgenic studies (described later).

THE CONVERSION FROM PRPC TO PRPSC

The prion protein exists in two forms. The normal, innocuous protein (PrPc) can change its shape to a harmful, disease-causing form (PrPSc). The conversion from PrPc to PrPSc then proceeds via a chain-reaction. When enough PrPSc proteins have been made they form long filamentous aggregates that gradually damage neuronal tissue. The harmful PrPSc form is very resistant to high temperatures, UV-irradiation and strong degradative enzymes.

The wild type prion (PrPc) is found in the secretory pathway of cells expressing the protein (1) and moves to the plasma membrane where it is anchored by its GPI tail (2). There it may bind to an extracellular ligand (possibly copper) (3) before being cycledfrom the membrane into endocytic vesicles (4). At some point its cargo is released and the protein either passes to the lysozome for degradation or back to surface for another round of ligand binding. In this respect it resembles many other membrane-resident proteins.

The pathogenic form (PrPSc) also finds its way to endocytic vesicles where it coopts some of the wild type form to become pathogenic. PrPSc is resistant to degradation, a hallmark of the infectious form, so accumulates. Neurotoxicity is probably linked to the conversion event itself, perhaps through its interference with normal PrPc turnover, because there is considerable evidence to show that the accrued PrPSc is not inherently toxic. As shown in Fig.8

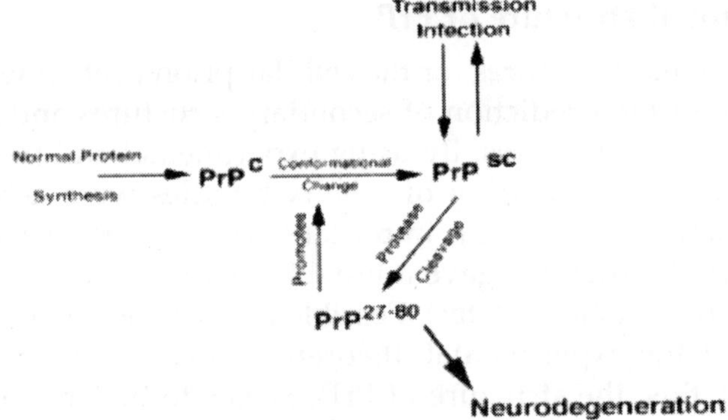

Fig. 8. Conversion of PrPc to PrPsc.

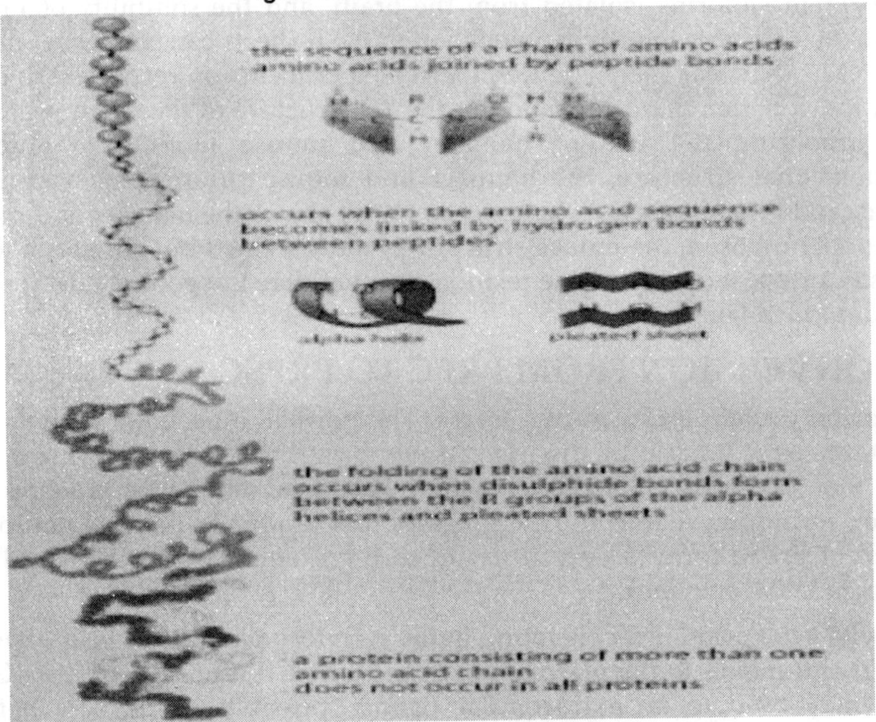

Fig. 9. Normal protein structure.

Infection of normal cells may occur when an aberrant prion acts as a template for the refolding of a normal prion into a new aberrant prion. It is thought that there is at least one more protein involved: the as-yet-unidentified Protein X. This protein is believed to mediate the folding from a normal into an abnormal prion.

When proteins are synthesized inside of a cell there are other special proteins (known as chaperones) that help in this process. Chaperones are proteins that bind to the newly synthesized protein or protein subunit, in order to ensure that the protein is properly folded into its secondary or tertiary structure. It has been hypothesized that Protein X is a type of chaperone.

The concept that a protein can transmit information (*i.e.* protein structure) is novel.

At the time that this idea was proposed, it was truly radical; it conflicted with many years of evidence showing that only DNA was capable of transmitting such information.

This does not mean that prion diseases cannot be genetic. If an individual happens to carry a gene that codes for a mutant prion, then all of the prions that their cells produce will be the abnormal, disease-causing prions. One such disease is familial CJD.

THE FUNCTION OF PRPc

Studies on transgenic mice to deduce the function of PrPc. From all the evidence put forward above, it seems certain that the production of the aberrant form of the PRNP gene product is responsible in some way for the pathogenesis of the disease. We do know that the PrPc protein is expressed as a glycosyl phosphatidyl inositol-anchored glycoprotein found on the outer cell membrane of neurons and to a lesser extent of lymphocytes and other cells (Stahl *et al.* 1987, 1990, 1992). However, the role of the normal cellular form of the prion is unclear and unproven. B eler *et al.* (1992) made studies on mice with inactive Prn-p genes (homozygously disrupted) and showed that the mice developed normally. The genes were disrupted by homologous recombination with a 4.8 kilobase DNA fragments in which codons 4 to 187 of the 254-codon open reading frame were replaced by a neomycin phosphotransferase (neo) gene under the control of the herpes simplex virus thymidine kinase promoter. Then were obtained mice heterozygous (Prn-p0/+) for the disrupted gene, the presence of which was confirmed by Southern analysis. Heterozygous mice were then mated to other heterozygotes to give a new population. Surprisingly, this population contained 24% Prn-p (0/0), 26% Prn-p(+/+) and 50% Prn-p(0/+) mice, surprising because they were all phenotypically the same, and Southern analysis confirmed these results. The Prn-p(0/0) mice were tested for many abnormalitiesin their behaviour, fitness and fertility and were all confirmed normal for at least 7 months, with no immunological defects. The reason that these results are surprising is that the Prn-p gene is highly conserved through evolution and so it might be thought that the function of this gene is vital to the organism. In contrast to B,eler *et al.* (1992), more recent studies have reported that the prion protein is necessary for normal synaptic function (Collinge *et al.* 1994) and that altered circadian activity rhythms and sleep has been observed in mice also devoid of the prion protein (Tobler *et al.* 1996). One might ascertain from PrP knockout mice, that although there appears to be no symptoms, by looking closely at the behaviour of the mice they might be seen to differ phenotypically from the PrP-normal mice. It seems that the studies in 1992 may have missed the subtle phenotypic differences observed in later more focused studies. They suggest that because of the intriguing similarities to rhythm and sleep alterations in fatal familial insomnia, where there is a profound alteration in sleep and the daily rhythms of many hormones, the loss of function PrP may be related to the normal function of the prion protein. Essentially, it can be seen that the probable function of PrPc, being a well-conserved protein present in many mammalian species, may be involved in maintaining sleep continuity or regulating sleep intensity. Another study by Sakaguchi *et al.* (1996) looked at the long-term effects of the disrupted gene on mice. At about 70 weeks of age, the PrP-null mice all began to show progressive symptoms of ataxia. A rotorod test evidently showed up a lack of co-ordination in these mice. As they got older, their coordination became progressively worse and theyfrequently collapsed. Closer examination of the afflicted mice gave a good indication of what Prion Proteins would normally be doing when there is plenty of it in the brain. The cerebellum, the part of the brain responsible for movement and coordination, had shrunk by almost one third in the specifically bred Prion free mice and upon pathological investigation, there was an extensive loss of Purkinje cells in the vast majority of cerebellar folia. This suggests that PrP plays a role in the long-term survival of Purkinje neurons (Moore *et al.* 1999). It was also reported that neuroprotective functions of PrP include its antioxidative and anti-apoptotic properties (Brown *et al.* 2001, Kim *et al.* 2004, Wong *et al.* 2001). In addition to this a role in copper binding and delivery and a signally function has been supported with experimental evidence (Brown *et al.* 1997, Stockel *et al.* 1998, Brown *et al.* 1999, Mouillet-Richard *et al.* 2000). Recent studies on yeast have shown that the infectious prions linked to the disease, could help organisms adapt to tough situations. In yeast, some cells carry the normal type of proteins, whereas others harbor the infectious form, which accumulates in clumps and is passed on from one cell to another. Scientist has shown that the Prions in their infectious form help the yeast to adapt, changing their rates of survival when they were grown in various nutrients or temperatures. Prions may therefore offer a speedy way for yeast to evolve (Osherovich *et al.* 2002). In addition to this, the rouge Prion fibres are being analyzed to be used as templates for nanowires (Nanobiotechnology). Scientists have used a Prion protein from yeast to template the formation of metal wires approximately 100nm wide. While other biomolecules have been used in nanofabrication strategies, the researchers argue that the physical and chemical study of Prion proteins make them more likely to be able to withstand industrial metallization processes (Lindquist *et al.* 2003). Another study by Mabbott *et al.* (1997) has looked at the possibility of the involvement of T-lymphocyte activation as a role of PrPc, this is due to the fact that PrPc is a GPI anchored protein (membrane attached via a glycosylphosphatidyl-inositol anchor).Certain GPI anchored proteins on lymphocytes have previously been shown to play an important role in an early step of cell activation, and as the expression of PrPc has also been identified in lymphoid tissues, they hypothesized that PrPc like other GPI-anchored proteins may play a role in lymphocyte activation. In this study it was demonstrated that the PrPc is expressed on the surface of murine lymphocytes, and following mitogen stimulation the lymphocyte PrPc expression was significantly enhanced suggesting a role of PrPc in lymphocyte activation. Also Also the mitogen responsiveness in PrPc-deficient lymphocytes was significantly reduced in comparison to wild-type controls. Therefore, the results imply that PrPc is expressed on the extra cellular leaflet of the lymphocyte membrane and involved in lymphocyte activation.

Although this report does suggest a role for PrPc being involved in immune function it must be noted that the PrPc molecule is not the only GPI-anchored protein expressed on the surface of murine lymphocytes. Other examples include the Thy-1, Ly-6, CD58, CD- 59 and Oa-2 proteins.

Previous work demonstrates that the non-specific removal of all GPI-anchored proteins with a phospholipase C enzyme abrogates the ability of murine lymphocytes to respond to stimulation by the mitogen Con A, so there is evidence that the GPI-anchored proteins are involved with activation of lymphocytes. However, PrPc is predominantly produced in the brain and not predominantly in the peripheral organs and those expressing lymphocytes (spleen, thymus gland, lymph nodes, gut wall and bone marrow).

But, this does not necessarily mean the lack of lymphocyte involvement of PrPc. PrPc may be involved in a multiple role, for example it may have a role in the brain and a role related to lymphocytes, this would explain the highly conserved nature of the gene, as a mutation in the gene would be more likely to give a loss of function or there is the possibility that PrPc can be involved but in a coincidental way, as many other GPI-anchored proteins are known to have the same effect.

Work done by De Armond *et al.* (1997) has uncovered the possibility that glycosylation can modify the conformation of PrPc. Glycosylation could also affects the affinity of PrPc for a particular conformer of PrPsc, and there by determining the rate of nascent PrPsc formation and the specific patterns of PrPsc deposition. They postulated that variations in the complex carbohydrate structure in the CNS could account for selective targeting (*i.e.* the formation of PrPsc in particular areas of the brain being due to differences in PrPc/PrPsc affinity). The experiments were done using deletions of Asn-linked oligosaccharides of PrPc in transgenic mice.

Recently Ikeda *et al.* (1999) have performed experiments on cultured hepatic stellate cells, and by using immuno electron microscopy, have revealed that PrPc resides on the plasma membrane of stellate cells. Their results indicate that PrP expression is closely related to stellate cell activation associated with fibrogenic stimuli. This has helped elucidate the pathogenesis of liver fibrosis, and develop possible therapeutic strategies to prevent liver cirrhosis. PrPc is not expressed in the normal healthy liver, but it was found to increase dramatically in diseased livers associated with stellate cell activation.

Applications

Armed with a plausible idea of how pathogenesis could arise, the way in which copper is bound to the surrounding sequence becomes highly significant. The role of copper binding by the octarepeats is important as well as the significance of repeats and their conservation. More intriguing still is the idea that copper (or other metal ion) binding it may, in part, modulate the conversion event. The high affinity site is in the region of the protein that appears to determine some of the disease characteristics; it has to be found whether mutations known to affect these disease traits influence the affinity or range of metal ions that can be bound. Equally, if perturbation of copper metabolism is at the heart of the pathology, then intervention strategies that target copper directly may offer prospects for disease management.

The search for sensitive and robust diagnostics should also take account of the fact that metal ions may modulate the conformation of the protein; antibodies that can distinguish the metal dependent conformation of prion protein may be useful diagnostic reagents. Lastly, a role for a redox active metal in the TSEs may draw them into a larger group of neurological disorders, including Alzheimer's disease that may have a shared basis of action. In that case there is every hope that advances in any one field will be beneficial to a wide range of related diseases.

Alzheimer's Prevention Role Discovered For Prions

Another role for prion proteins, the much debated agents of mad cow disease and vCJD, has been identified. It appears that the normal prions produced by the body help to prevent the plaques that build up in the brain to cause Alzheimer's disease. The possible function for the mysterious proteins was discovered by a team of scientists led by Medical Research Council funded scientist Professor Nigel Hooper of the University of Leeds. Alzheimer's and diseases like variant Creutzfeldt - Jakob disease follow similar patterns of disease progression and in some forms of prion disease share genetic features. These parallels prompted Professor Hooper's team to look for a link between the different conditions. They found an apparent role for normal prion proteins in preventing Alzheimer's disease. Their experiments have shown that the normal prion proteins found in brain cells reduce the formation of beta-amyloid, a protein that binds with others to build plaques in the brain that are found in Alzheimer's disease, explains Professor Hooper. He continues: In vCJD, the normal version of prion protein, PrPc, found naturally in the brain is corrupted by infectious prions to cause disease. The normal function of PrPc has been unclear. Using cells grown in the lab, the team looked at the effect of high and low levels of normal prion protein on the successful formation of beta amyloid, the source of Alzheimer's plaques. They found that beta amyloid did not form in cells with higher than usual levels of PrPc. In comparison, when the level of PrPc was low or absent, beta amyloid formation was found to go back up again. Mice genetically engineered to lack PrPc were also studied. Again, this revealed that in its absence, the harmful beta-amyloid proteins were able to form. It appears that PrPc, the normal prion protein, exerts its beneficial effect by stopping an enzyme called beta secretes from cutting up amyloid protein into the smaller beta-amyloid fragments needed to build plaques. Further evidence for the protective role of normal prion proteins is provided by mutated versions that are linked to genetic forms of prion disease because beta-amyloid fragments are able to form when the normal prion protein is corrupted by genetic mutation. According to Professor Hooper their findings have clearly identified a role for normal prion proteins in regulating the production of beta-amyloid and in doing so preventing formation of Alzheimer's plaques. Whether this function is lost as a result of the normal ageing process, or if some people are more susceptible to it than others this is not known yet. The next step for this research will be to look in more detail at how the prion protein controls beta amyloid, knowledge that could be used to design anti-Alzheimer's drugs. Theoretically, if a way of mimicking the prion's function is found it would be possible to halt the progress of Alzheimer's. However, there's still a lot of work to be done in looking at levels of prions in the human system and how these alter with age.

INTO THE FUTURE, AND THE UNKNOWN INCUBATION FACTOR

It is certain that due to the attention received, and the dangers well illustrated in the media, and by scientists, prion research will progress extremely quickly in the next five years. Attention will be focused on the function of PrPc, the development of effective rapid diagnostic techniques, and the molecular studies on the mechanisms of replication to provide targets for the development of effective chemotherapeutic agents. Perhaps most importantly and perhaps the most elusive fact to find out, however, is the possible incubation time for vCJD in humans.

The only model for comparison is the rise and fall of the kuru epidemic in Papa New Guinea. The maximum incubation time for that disease (not transmitted across the species barrier) has shown to be maximally about 40 years, with the peak onset after a few years after successful inoculation with the agent, as children as young as four have been reported with the disease. The big question is, what will be the effect of species barrier between cows and humans (and possibly even sheep and humans in the case of BSE fed sheep) in the case of vCJD. No one seems to know, some people think the 20-30 or so cases so far are the tip of the iceberg, others think it is mostly over.

However, it may be encouraging to think that the BSE agent's ability to traverse the species barrier also may mean that it is also faster at crossing this barrier than other prion diseases and so expression would be sooner (years) rather than later (decades). One way or another, time will tell. At present there is no cure for Prion diseases and the kill rate is 100%. More research is necessary to understand these devastating infectious agents and create opportunities to evaluate therapeutic strategies for these unusual diseases.

REFERENCES

Baldwin M James T Cohen F Prusiner S 1998, The three-dimensional structure of prion protein: implications for prion disease *Biochemical Society Transactions* **26** 481-486

Baldwin M Pan K Nguyen J Huang Z Groth D Serban A et al 1994, Spectroscopic Characterisation of Conformational Differences Between PrPc and PrPsc - an a-Helix to b-Sheet Transition Philosophical Transactions of the *Royal Society of London Series B-Biological Sciences* **343** 435-441

Bazan F Fletterick R McKinley M Prusiner S 1987, Predicted secondary structure and membrane topology of the scrapie prion protein *Protein Engineering* **1** 125-135

Belay E.D and L B Schonberger 2005, The public health impact of prion diseases *Annu Rev Public Health* **26** 191-212

Beler H Fischer M Lang Y Bluethmann H Lipp H DeArmond S *et al* 1992, Normal evelopment and behaviour of mice lacking the neuronal cell-surface PrP protein *Nature* **356** 577- 582

Bessen RA and Marsh RF 1994 Distinct PrP properties suggest the molecular basis of strain variation in transmissible mink encephalopathy *J. Virol.* **68** 7859-7868

Borchelt D Scott M Taraboulos A Stahl N Prusiner S 1990, Scrapie and Cellular Prion Proteins Differ in Their Kinetics of Synthesis and Topology in Cultured Cells The *Journal of Cell Biology* **110** 743-752

Brown D 1999, senior scientists dismiss fears of CJD epidemic The Daily Telegraph March 27th

Brown D R 1999, Prion Protein expression aids cellular uptake and veratridine-induced release of copper *Jourl of Neuroscience Research* **58** 717- 725

Brown D R Clive C Haswell S J 2001, Antioxidant activity related to copper binding of native prion Protein *Jourl of Neurochemistry* **76** 69- 76

Brown P Cathala F Raubertan R F et al 1987, The epidemiology of Creutzfeldt-Jacob disease Conclusion of a 15 year investigation in France and review of the old *Literature Neurology* **37** 895- 904

Caughey B Chesebro B 1997, Prion protein and the transmissible spongiform cephalopathies Trendsin *Cell Biology* **7** 56- 62

Chakraborty C S Nandi S Jana 2005, Prion disease: a deadly disease for protein Misfolding *Curr Pharm Biotechnology* **6** 167-77

Chesebro B 1998, BSE and Prions: Uncertainities about the *Agent Science* **279** 42- 43

Christian Wasmer Alice Soragni Raimon Sabaté Adam Lange Roland Riek Beat H. Meier 2008, Infectious and Noninfectious Amyloids of the HET-s (218-289) Prion Have Different NMR Spectra. *Angewandte Chemie*

Christina J Sigurdson K Peter R Nilsson Simone Hornemann Guiseppe Manco Magdalini Polymenidou Petra Schwartz Mario Leclerc, Per Hammarström, Kurt Wüthrich, and Adriano Aguzzi 2007, "Prion strain discrimination using luminescent conjugated polymers" *Nature*

Cohen F Abarbanel R Kuntz D Fletterick R 1985, Turn Prediction in Proteins Using a attern *Matching Approach Biochemistry* **25** 266- 275

Collinge J Palmer M 1997, 'Prion Diseases' Oxford University Press

Collinge J Whittington M Sidle K Smith C Palmer M Clarke A Jefferys J 1994, Prion protein is necessary for normal synaptic function *Nature* **370** 295- 277

Craighead Harold G. 2008, "Prion Protein Detection Using Nanomechanical Resonator Arrays and Secondary Mass Labeling" *Analytical Chemistry*

De Armond S S_nchez H Yehiely Qiu Y Ninchak-Casey A Daggett V et al 1997, Selective Neuronal Targeting in Prion Disease *Neuron* **19** 1337-1348

Gajdusek D C 1977, Unconventional viruses, the origin, disappearance of Kuru *Science* 943-960

Gajdusek D C Zigas V 1957, Degenerative disease of the Central Nervous system in New Guinea: The endemic occurrence of ìKuruî in the native population N Engl *Journal of Medicine* **257** 974-978

Gajdusek D Gibbs C Alpers M 1966, Experimental transmission of a kuru-like syndrome to chimpanzees *Nature* **209** 794- 796

Garcion E Wallace B Pelletier L Wion D 2004, RNA mutagenesis and sporadic prion diseases *Journal of Theoretical Biology* **230** 271-274

Gasset M Baldwin M Lloyd D Gabriel J Holtzman D Cohen F et al 1992, Predicted a-helical regions of the prion protein when synthesised as peptides form amyloid *Proceedings of the National Academy of Sciences of the United States* of Americ **89** 10940- 10944

Gibbs T Gajdusek D Asher D Alpers M Beck E Daniel P et al 1968, Creutzfeldt-Jacob disease (Spongiform encephalopathy): transmission to the chimpanzee *Science* **161** 388- 389

Hadlow W J 1959, Scrapie and Kuru *Lancet* **2** 289- 290

Hsiao K Prusiner S 1990, Inherited human Prion diseases *Neurology* **40** 1820 1827

Huang F P Mac Pherson G G 2004, Dendritic cells and Oral transmission of Prion disease *Adv Drug Del reviews* **56** 901- 913

Ikeda K Kawada N Wang Y Kadoya H Nakatani K Sato M Kaneda K 1999, Expression of Cellular Prion Protein in Activated Hepatic Stellate Cells American *Journal of Pathology* **153** 1695-1700

James T Liu H Ulyanov N FarrJones S Zhang H Donne D et al 1997, Solution structure of a 142-residue recombinant prion protein corresponding to the infectious fragment of the scrapie isoform *Proceedings of the National Academy of Sciences of the United States of America* **94** 10086- 10091

Karapetyan *et al.* 2009, Prion Strain Discrimination Based on Rapid In Vivo Amplification and Analysis by the Cell Panel Assay. *PLoS ONE* **5**

Khosravani *et al* 2008, Prion protein attenuates excitotoxicity by inhibiting NMDA receptors. *J. Cell Biol.* **181** 551-565 [link]

Kim B H Lee H G Choi J K Kim J I Choi E K Carp R I Kim Y S 2004, The cellular Prion protein (PrPc) prevents apoptotic neuronal cell death and mitochondrial dysfunction induced by serum deprivation Brain Research *Molecular Brain Research* **124** 40-50

Klein Alec. 2007, "Hospitals Rethink Reuse of Brain Surgery Devices." Chicago Tribun Chicagoland Final *ed* sec News **12** Access World News.

Lasmezas CI Deslys J-P Robain O Jaegly A Beringue V Peyrin J-M Fournier J-G Hauw J-J Rossier J and Dormont D 1997 Transmission of the BSE agent to mice in the absence of detectable abnormal prion protein. *Science* **275** 402–405

Lindguist S L *et al* 2003, Prion templates Proceedings of Natural Academy of Sciences USA **100** 4527

Locht C Chesebro B Race R Keith J 1986, Molecular Cloning and complete sequence of prion protein cDNA from mouse brain infected with the scrapie agent *Proceedings of the National Academy of Sciences of the United States of America* **83** 6372- 6376

Manuelidis L Fritch W 1996, Infectivity and host responses in Creutzfeldt-Jakob disease *Virology* **216** 46- 59

Mastrianni J A 2004, Prion diseases *Clinical Neurosciences Research* **3** 469-480

Mckintosh E Tabrizi S J Collinge J 2003, Prion diseases *Journal of Neurovirology* **9** 183-193

Monaco S *et al* 2003, Prions: They really do get up your nose N Engl *Jour Med* **348** 711- 720

Moore R C Lee I Y Silverman G L Harrison P M Storme R Heinrich C Kanunaratne A Pasternak S H Chishti M A Liang Y Mastrangelo P Wang K Smit A F Katamine S Carlson G A Cohen

F E Prusiner S B Melton D W Tremblay P Hood L E Westaway D 1999, Ataxia in Prion Protein (PrP) deficient mice is associated with up regulation of the novel PrP like Protein doppel *Journal of Molecular Biology* **292** 797-817

Mouillet-richard S Ermonval M Chebassier C *et al* 2000, signal transduction through Prion Protein *Science* **289** 1923-1928

Muramoto T Scott M Cohen F Prusiner S 1996, Recombinant scrapie-like protein of 106 amino acids is soluble *Proceedings of the National Academy of Sciences of the United States of America* **93** 15457-15462

Naslavsky N Stein R Yanai A Friedlander G Taraboulos 1997, Characterization of Detergent-insoluble Complexes Containing the Cellular Prion Protein and Its Scrapie Isoform *Journal Biological Chemistry* **272** 6324-6331

Oesch B Westaway D W‰lchli M McKinley M Kent S Aebersold R 1985, A cellular gene encodes scrapie PrP 27-30 protein *Cell* **40** 735-746

Osherovich L Z Weissman J S 2002 The utility of Prions Developmental *Cell* **2** 143- 151

Pan K.-M Baldwin M Nguyen J Gasset M Serban A Groth D Mehlhorn I Huang Z Fletterick RJ Cohen FE and Prusiner SB 1993.Conversion of-helices into-sheets features in the formation of the scrapie prion proteins. *Proc. Natl. Acad. Sci. USA* **90** 10962–10966

Porter D D Porter H G Cox N A 1973, Failure to demonstrate a humoral immune response to scrapie infection in mice *Journal Immunology* **111** 1407-1410

Prusiner S 1982, Novel Proteinaceous Infectious Particles Cause Scrapie *Science* **216** 136-144

Prusiner S B 1991, Molecular Biology of Prion Diseases Science **252** 1515-1522

Prusiner SB 1996, Molecular biology and pathogenesis of prion diseases. *Trend Biochem Sci.* **21** 482-7

Prusiner SB 1997, Prion diseases and the BSE crisis. *Science* **278** 245–251

Prusiner, SB 1994, Biology and genetics of prion diseases. *Annu Rev Microbiol.* **48** 655-86

Race B Meade-White K Oldstone MBA Race R Chesebro B 2008,Detection of Prion Infectivity in Fat Tissues of Scrapie-Infected Mice. *PLoS Pathog*, **4**

Sakaguchi S Katamine S Nishida N Moriuchi R Shigematsu K Sugimoto T *et al* 1996, Loss of cerebellar Purkinje cells in aged mice homozygous for a disrupted PrP gene *Nature* **380** 528 -531

Sciences USA 100 4527 Mabbott N Brown K Manson J Bruce M 1997, T-lymphocyte activation and the cellular form of the prion protein *Immunology* **92** 161-165

Scott M R Will R Ironsids J *et al* 1999, Compelling transgenic evidence for transmission of bovine spongiform encephalopathy Prions to humans *Proceedings of Natural Academy of Sciences* USA **96** 15137-15142

Singh *et al* 2009, Abnormal Brain Iron Homeostasis in Human and Animal Prion Disorders. *PLoS Pathogens* **3**

Stahl N Baldwin M Teplow D Hood L Gibson B Burlingame A Prusiner S 1993, Structural Studies of the Scrapie Protein Using Mass Spectrometry and Amino Acid Sequencing *Biochemistry* **32** 1991- 2002

Stahl N Borchelt D Hsiao K Prusiner S 1987, Scrapie Prion Protein Contains a Phoshatidylinositol Glycolipid *Cell* **51** 229-240

Stockel J Safar J Wallace A C *et al* 1998, Prion Protein selectively binds copper (II) ions *Biochem* **37** 7185-7193

Telling G Scott M Mastrianni J Gabizon R Torchia M Cohen F *et al* 1995, Prion Propagation in Mice Expressing Human and Chimeric PrP Transgenes Implicates the Interaction of Cellular PrP with Another Protein *Cell* **83** 79- 90

Timm Konold S. Jo Moore Susan Bellworthy and Hugh A. Simmons. 2008, Evidence of scrapie transmission via milk. *BMC Veterinary Research*

Tobler I Gaus S Deboer T Achermann P Fischer M R¸licke T *et al* 1996, Altered circadian activity rhythms and sleep in mice devoid of prion protein *Nature* **380** 639- 642

Uro-Coste E Cassard H Simon S Lugan S Bilheude J-M *et al.* 2008 Beyond PrPres Type 1/Type 2 Dichotomy in Creutzfeldt-Jakob Disease. *PLoS Pathog* **4**

Uro-Coste E Cassard H Simon S Lugan S Bilheude J-M *et al.* 2008, Beyond PrPres Type 1/Type 2 Dichotomy in Creutzfeldt - Jakob disease. *PLoS Pathog* **43**

Wadhwani C and Malik C.P 2007, Prions-the lethal proteins. In *Current advancres in Biotechnology.* Ed Malik C P Kaur B and Wadhwani C M D Publications New Delhi

Wong B S Liu T Li R Pan T Peterson R B Smith M A Gambetti P Perry G Manson J C Brown D R Sy MS 2001, Increased levels of oxidative stress markers detected in the brains of mice devoid of Prion proteins *Jour Neurochem* **76** 565- 572

12

NONI (*MORINDA CITRIFOLIA* L.) -A HOPE IN A BOTTLE

A. R. MALIK, S. H. WANI AND B. N. DAR

INTRODUCTION

Morinda citrifolia commercially known as 'Noni' grows widely throughout the pacific and is one of the most important traditional Polynesian medicinal plant. Noni is noted for its extremely wide range of environmental tolerance. It is found naturally in relatively dry to mesic sites or lowland areas in close proximity to shorelines, or as an important forest understory species in low elevation pacific island forests and rainforests. Noni has extensive range of environmental tolerances including exposure to wind, fire, flooding and saline conditions. Although not considered to be invasive to a degree that threatens ecosystems noni is treated as a weed in some settings, is very persistent and difficult to hill, and is one of the first plants to colonize harsh waste areas or lava flows. All parts of the plant have traditional and/or modern uses, including roots and bark (dyes, medicine), trunks (firewood, tools), and leaves and fruits (food, medicine). Noni is well suited for intercropping within traditional agroforestry subsistence farming systems or as a monocrop in full sun. The tree has attained significant economic importance worldwide in recent years through a variety of health and cosmetic products made from its leaves and fruits.

MAJOR COMPONENTS

A number of major components have been identified in the Noni plant such as scopoletin, octoanoic acid, potassium, vitamin C, terpenoids, alkaloids, anthraquinones (such as nordamnacanthal, morindone, rubiadin, and rubiadin- 1-methyl ether, anthraquinone glycoside), b-sitosterol, carotene, vitamin A, flavone glycosides, linoleic acid, Alizarin, amino acids, acubin, *L*-asperuloside, caproic acid, caprylic acid, ursolic acid, rutin, and a putative proxeronine (Levand and Larson ,1979 and Heinicke, 1985).

Noni fruit powder is high in carbohydrates and dietary fiber. According to the College of Tropical Agriculture and Human Resources at the University of Hawaii at Mānoa, a 100g sample of the powder contains 71% carbohydrate and 36% fiber. The sample also contained 5.2% protein and 1.2% fat (Nelson, 2006).These macronutrients evidently reside in the fruit pulp, as noni juice has sparse amounts of macronutrients. The main micronutrients of noni pulp powder include 9.81 mg of vitamin C per 1200mg sample, as well as 0.048 mg niacin (vitamin B3), 0.02 mg iron and 32.0 mg potassium (Nelson, 2006). Vitamin A, calcium and sodium are present in moderate amounts. When noni juice alone is analyzed and compared to pulp powder, only vitamin C is retained at a high level, 33.65 mg per 100g of juice (Nelson, 2006). Although the most significant nutrient feature of noni pulp powder or juice is its high vitamin C content, noni fruit juice provides only about half the vitamin C of a raw navel orange. Sodium levels in the noni juice blend (about 3% of DRI) (Nelson, 2006) are multiples of those in an orange. Although the potassium content appears relatively high for noni, this total is only about 3% of the recommended dietary allowance and so would not be considered excessive. Noni juice is otherwise similar in micronutrient content to a raw orange. Noni fruit contains a number of phytochemicals, including lignans, oligo- and polysaccharides, flavonoids, iridoids, fatty acids, scopoletin, catechin, beta-sitosterol, damnacanthal, and alkaloids. Although these substances have been studied for bioactivity, current research does not conclude anything about their effects on human health (Deng *et al.* 2007 and Mohd *et al.* 2007). Laboratory experiments demonstrated that dietary noni juice increased physical endurance in mice (Ma *et al.* 2007).

DISTRIBUTION

M. citrifolia is commonly assumed to have originated in Southern Asia and subsequently distributed by humans or other means into the islands of the Western pacific. The distribution of Noni is pantropical at latitudes of 19°W or S. The Indo-pacific distribution includes eastern Polynesia (Hawaii, lime Island, Marquesas, cook Islands), Melanesia (Fiji, New Guinea, Solomon Islands), western Polynesia (Samoa, Tonga, Rothmas and Tuvalu) and Micronesia (Pohnpei, guam, Palau the Marshall Islands and Northern Marianas), Indonesia, Australia and SouthEast Asia. Noni has also become naturalised on the open, shores of central and South America and on many islands of West Indies, the Bahamas, Bermuda, the Florida and parts of Africa.

Botanical description

The scientific name of Noni is *Morinda citrifolia* L. The botanical name for the genus was derived from the two latin words *morus*, mulberry, and *indicus*, Indian, in reference to the similarity of the fruit of Noni to that of true mulberry (*Morus alba*). The species name indicates the resemblance of the plant foliage to that of some citrus species. It belongs to family Rubiaceae (coffee family) and subfamily Rubioideae. Noni is the common name for *Morinda citrifolia* L and is also called great morinda, Indian mulberry,

Mengkudu (Malaysia), beach mulberry, Tahitian noni, cheese fruit or noni (from Hawaiian) in various cultures throughout the world.

The Noni plant is a small evergreen tree found growing in open coastal regions at sea level and in forest areas up to about 1300 feet above sea level. The plant is often found growing along lava flows. It's identifiable by its straight trunk, large, bright green and elliptical leaves, white tubular flowers, and its distinctive, ovoid, "grenade-like" yellow fruit. The fruit can grow in size up to 12 cm or more and has a lumpy surface covered by polygonal-shaped sections. The seeds, which are triangular shaped and reddish brown, have an air sac attached at one end, which makes the seeds buoyant. This could explain, in part, the wide distribution of the plant throughout the Polynesian islands. The mature Noni fruit has a foul taste and odour (Swanholm *et al.* 1959) and is not considered to be at risk in the wild.

GROWING HABITATS

Noni grows in shady forests as well as on open rocky or sandy shores. It reaches maturity in about 18 months and then yields between 4–8 kilograms (8.8–18 lb) of fruit every month throughout the year. It is tolerant of saline soils, drought conditions, and secondary soils. It is therefore found in a wide variety of habitats: volcanic terrains, lava-strewn coasts, and clearings or limestone outcrops. It can grow up to 9 metres (30 ft) tall, and has large, simple, dark green, shiny and deeply veined leaves.

The plant flowers and fruits all year round and produces a small white flower. The fruit is a multiple fruit that has a pungent odor when ripening, and is hence also known as *cheese fruit* or even *vomit fruit*. It is oval and reaches 4–7 centimetres (1.6–2.8 in) in size. At first green, the fruit turns yellow then almost white as it ripens. It contains many seeds. It is sometimes called starvation fruit. Despite its strong smell and bitter taste, the fruit is nevertheless eaten as a famine food (Krauss, 1993) and in some Pacific islands, even a staple food, either raw or cooked (Morton, 1992). Southeast Asians and Australian Aborigines consume the fruit raw with salt or cook it with curry. The seeds are edible when roasted. Noni is especially attractive to weaver ants, which make nests out of the leaves of the tree. These ants protect the plant from some plant-parasitic insects. The smell of the fruit also attracts fruit bats, which aid in dispersing the seeds.

ASSOCIATED PLANT SPECIES

The associated plant species include breadfruit (*Artocarpus altilis*), banana (*Musa* spp.), papaya (*Carica papaya*), plalms (e.g. betel nut palm, *Arecacatechii* and coconut, *Cocos nucifera*), pandanus (*Pandanus* spp.), beach hibiscus (*Hibiscus tiliaceus*), til (*Cordylinie fruticosa*) and piper species (e.g. kava, *Piper methysticum*). Some of these associated species are understory and some are overstory for Noni. It grows as a recent introduction around villages or in homegardens, in backyards and along streams and gulches.

USES AND PRODUCTS

Captain James Cook of the British Navy noted in the late 1700's that the fruit was eaten in Tahiti (Cheeseman, 1903). An 1866 publication in London explained that *Morinda citrifolia* fruit was consumed as a food in the Fiji Islands (Seemann and Flora, 1866). Later publications describe the use of this fruit as a food throughout the Pacific Islands, SouthEast Asia, Australia, and India. In Roratonga "the fruit was often eaten by the natives" (Cheeseman, 1903). Australian Aborigines were reported to be "very fond" of the fruit (Maiden, 1889). In Samoa, Noni fruit was common fare, and in Burma, the fruit was cooked in curries or eaten raw with salt (Morton, 1992). Merrill (1943) described *Morinda citrifolia* L as an edible plant in a technical manual of edible and poisonous plants of the Pacific Islands, in which the leaves and fruits could be used as emergency food. Abbott (1992) also reported that Noni had been used as a food, drink, medicine, and colorful dye. Very young leaves are cooked as vegetables and eaten with rice in Java and Thailand; mature leaves are wrapped around fish before cooking and then eaten with the cooked fish. The terminal bud is used as food. Dried leaves or fruits are used to make infusions and teas for medicinal use.

The medicinal history and accumulated scientific studies, to date, have revealed and confirmed the Polynesian's claim of the health benefits of Noni. The medical knowledge and pharmacopoeia of the Polynesians is now believed to have been fairly complex and modern scientific and medical communities are beginning to study the plants compiled from this knowledge base. The Polynesians utilized the whole Noni plant in various combinations for herbal remedies. The fruit juice is in high demand in alternative medicine for different kinds of illnesses such as arthritis, diabetes, high blood pressure, muscle aches and pains, menstrual difficulties, headaches, heart disease, AIDS, cancers, gastric ulcers, sprains, mental depression, senility, poor digestion, atherosclerosis, blood vessel problems, and drug addiction. Scientific evidence of the benefits of the Noni fruit Juice is limited but there is some anecdotal evidence for successful treatment of colds and influenza (Solomon, 1999). Allen (1873) reported some information on the ethnobotanical properties of Noni. He said that the fruit is used as deobstruent and emmenagogue. This is one of the earliest articles on the medicinal benefits of Noni. Isabel Abbott (1985), a former botanical chemist at the University of Hawaii, stated that, "People are crazy about this plant. They use it for diabetes, high blood pressure, cancer, and many other illnesses". Bushnell (1950) reported that Noni was a traditional remedy used to treat broken bones, deep cuts, bruises, sores, and wounds. Morton (1992) gave numerous references for medicinal uses of Noni. In addition, Polynesians are reported to have successfully used Noni to treat breast cancer and eye problems. Betz (1997), stated that "*Morinda citrifolia* has been tested for a number of biological activities in animal and anti-microbial studies." He reports that the dried fruit has smooth muscle stimulatory activity and histaminergic effects. In Hawaii, ripe fruits are applied to draw out pus from an infected boil. The green fruit, leaves and the root/rhizome have traditionally been used to treat menstrual cramps and irregularities, among other symptoms, while the root has also been used to treat urinary difficulties (McClatchey, 2002).

BIOLOGICAL ACTIVITIES OF NONI PRODUCTS

Antibacterial activity:

Acubin, *L*-asperuloside, and alizarin in the Noni fruit, as well as some other anthraquinone compounds in Noni roots, are all proven antibacterial agents. These compounds have been shown to fight against infectious bacteria strains such as *Pseudomonas aeruginosa, Proteus morgaii, Staphylococcus aureus, Baciillis subtilis, Escherichia coli, Salmonella,* and *Shigela*. These antibacterial elements within Noni are responsible for the treatment of skin infections, colds, fevers, and other bacterial-caused health problems. Extracts from the ripe Noni fruit exhibited moderate antibacterial properties against *Ps aeruginosa, M. pyrogenes and E. coli,* and were also shown to have moderate antibacterial properties against *Salmonella typhosa, Salmonella montevideo, Salmonella schottmuelleri, Shigella paradys,* and *Shigella paradys* (Bushnell, 1950).

Antiviral activity:

Umezawa (1992) found a compound isolated from Noni roots named 1-methoxy- 2-formyl-3-hydroxyanthraquinone suppressed the cytopathic effect of HIV infected MT-4 cells, without inhibiting cell growth.

Anti-tubercular effects:

Noni has been found to kill Mycobacterium tuberculosis. A concentration of extracts from Noni leaves killed 89 per cent of the bacteria in a test tube, almost as effective as a leading anti-TB drug, Rifampcin, which has an inhibition rate of 97 per cent at the same concentration.

Antitumor activity:

In 1994, Hirazumi, a researcher at the University of Hawaii, reported anticancer activity from the alcohol-precipitate of Noni fruit juice (noni-ppt) on lung cancer in C57 Bl/6 mice at the 83rd Annual Meeting of American Association for Cancer Research. The noni-ppt was shown to significantly prolong the life of mice up to 75 % with implanted Lewis lung carcinoma compared with the control group (Hirazumi *et al.* 1994). It was concluded that the noni-ppt seems to suppress tumor growth indirectly by stimulating the immune system (Hirazumi *et al.* 1996). Improved survival time and curative effects occurred when noni-ppt was combined with sub-optimal doses of the standard chemotherapeutic agents such as adriamycin (Adria), cisplatin (CDDP), 5 fluorouracil (5-FU), and vincristine (VCR), suggesting important clinical applications of noni-ppt as a supplemental agent in cancer treatment (Hirazumi and Furus, 1999). These results indicate that noni-ppt may enhance the therapeutic effect of anticancer drugs. Therefore, it may be of benefit to cancer patients by enabling them to use lower doses of anticancer drugs to achieve the same or even better results.

Anthelmintic activity:

An ethanol extract of the tender Noni leaves induced paralysis and death of the human parasitic nematode worm, *Ascaris Lumbricoides*, within a day (Raj , 1975). A botanist via Morton reported that Noni has been used in the Philippines and Hawaii as an effective insecticide (Morton, 1992).

Analgesic activity:

Betz (1997) reported that the Noni fruits posses analgesic and tranquilizing activites. A French research team led by Younos, tested the analgesic and sedative effects of extracts from the *Morinda citrifolia* plant. The extract did "show a significant, dose-related, central analgesic activity in the treated mice." They stated that "these findings validate the traditional analgesic properties of this plant." The analgesic efficacy of the Noni extract is 75 % as strong as morphine, yet non-addictive and side effect free (Younos *et al.* 1990).

Anti-Inflammation:

There have been recent applications for the use of oil from noni seeds (West, 2008). Noni seed oil is abundant in linoleic acid that may have useful properties when applied topically on skin, e.g., anti-inflammation, acne reduction, moisture retention (Letawe *et al.* 1998 and Darmstadt, 2002).

Hypotensive activity:

Dang Van Ho of Vietnam demonstrated that a total extract of the Noni roots has a hypotensive effect (Youngken, 1996). Moorthy and Reddy (1970) found that an ethanol extract of the Noni roots lowered the blood pressure in an anesthetized dog. Youngken's research team determined that a hot water extract of Noni roots lowered the blood pressure of an anesthetized dog (Davison,1927). A Hawaiian physician reported that Noni fruit juice had a diuretic effect (Asahina *et al.* 1994).

Immunological activity:

Asahina *et al.* (1994) found that an alcohol extract of Noni fruit at various concentrations inhibited the production of tumor necrosis factor-alpha (TNF- α, which is an endogenous tumor promoter. Therefore the alcohol extract may inhibit the tumor promoting effect of TNF-α.

Mental health and improved high frequency hearing:

A small human clinical trial of the effect of TNJ on auditory function and quality of life in the patients with decreased bone mineral density and auditory function has been conducted in UIC College of Medicine, Rockford, IL. This study showed that TNJ provided a positive benefit on mental health and improved high frequency hearing. The data suggests that increased amounts or extended duration of TNJ intake may be required to affect this disorder (Langford *et al.* 2002).

COMMERCIAL PRODUCTS

The primary commercial products from Noni include beverages (fruit juice, juice drinks), fruit powders (for manufacture of reconstituted juice or juice drink products made from dried ripe or unripe fruits), toiletries (lotions, soaps etc.), oil (from seeds), and leaf powders (for encapsulation or pills).The bark of the great morinda produces a brownish-purplish dye for batik making; on the Indonesian island of Java, the trees are cultivated for this purpose. In Hawaii, yellowish dye is extracted from its root in order to dye cloth (Thompson, 1971).

Noni as nutraceutical and functional food (NFF):

Noni can be used as a raw material for nutraceutical and functional food products. In Thailand, "Yor" leaves cooked with curry give a wonderful and nutritious dish. "Yor" fruits can be prepared as a Thai traditional salad "Somtumyor". Dried fruits or leaves powder boiled with water produces a tea, which is used to relief blood pressure, muscle pains and vomiting. There are more than 120 nutraceutical compounds identified in Noni (Solomon, 1999). Noni or Yor juice extract which is obtained from fermented Noni fruits is the most effective product that has helped relieved people ($n \geq 10{,}000$) from the suffering of about 22 conditions, such as arthritis, heart disease, diabetes, headache and muscle pain, high blood pressure, cancer, etc. (Solomon, 1999). Recently 'Noni' juice extract has been commercially processed and distributed internationally as a dietary supplement. Relatively unheard of 5 years ago, *M. citrifolia* has now exploded into nearly a billion-dollar industry.

ENVIRONMENTAL PREFERENCES AND TOLERANCES

Climate

Elevation range : 1-800 m, depending on latitude and environment.

Mean annual rainfall : 250-4000 mm.

Rainfall pattern : Noni can tolerate, a wide range of precipitation patterns, including summer, winter, bimodal and uniform.

Dry season duration (consecutive months with <40 mm rainfall) : At least 3-4 months depending on age, size of tree, temperature, relative humidity and soils.

Mean annual temperature: 20-35°C.

Minimum temperature tolerated: 12°C

Soils

Noni grows in a very wide range of soils and environments, with a notable ability to survive in harsh environments, such as those found on coral atolls or basaltic lava flows. It can also be found in solution pits or brackish tide pools near the coast, in limestone soils or outercroppings, on coral atolls, as a colonizing species of basaltic lava flows, as well as in native forests.

Soil drainage :

Noni tolerates a wide range of drainage conditions including seasonal waterlogging, but it prefers free, well drained soils.

Soil acidity :

It can grow in a wide range of acidity levels, from acidic to alkaline.

Special soil tolerances :

Noni tolerates shallow, saline, sodic and infertile soils.

StressTolerance

Mature, cultivated Noni can easily withstand drought for 6 moths or more. Wild noni plants growing in arid conditions can spend their entire lives in conditions of perpetual drought. The plant grows well in full sun. Noni can grow in a wide range of light intensities, from full sun to over 80 per cent shade. It can regenerate after fire by sprouting new foliage from roots or stems. Noni withstands and even thrives in brackish tide pools.It can also tolerate flooded conditions for long periods of time. It is very salt-resistant and tolerant of ocean salt spray. Noni is tolerant to extreme salinity in general and is thought to possibly gain nutritional benefit from the minerals contained in seawater. Although windy areas are not advised for commercial production, noni can grow in windswept locations however, yields and overall plant growth of noni in such areas are diminished.

Abilities

It has the ability to regenerate from shoots or root suckers rather than from seed, producing small thickets or groves. Noni is not considered to be self-pruning, although the woody branches of this plant are brittle and may be relatively easily broken during overly heavy fruiting loads or during high winds. Noni plants regenerate well, even after severe pruning, Noni may be cut back to the trunk (stumping) to promote the growth of a dense head of foliage.

GROWTH AND DEVELOPMENT

The growth rate is moderate, generally 0.75-1.5 m/yr, slowing as the tree reached maturity. Noni flowering and fruiting is continuous throughout year. Fluctuations in flowering and fruiting may occur due to seasonal effects (temperature, rainfall, sunlight intensity and duration). Noni does not compete well with grasses or with grassy weeds in deep soils as an agricultural monocrop. However, it is a good forest understory plant that can tolerate very harsh conditions and plant competition from forest trees, including allelopathic species. In fact, noni is one of the few plants that can thrive beneath the canopy of iron-wood (*Cstuarina equisetifolia*) trees.

PROPAGATION

Noni is relatively easy to propagate from seeds, stem or root cuttings and air-layering. The preferred methods of propagation are by seed and by cuttings made from stem verticals.

PROPAGATION FROM SEED

Noni seeds are reddish-brown, oblong-triangular, and have a conspicuous air chamber. They are buoyant and hydrophobic due to this air chamber and their durable, water-repellent, fibrous seedcoat. The seedcoat is very tough, relatively thick, and covered with cellophane-like parchment layers. A single large noni fruit can contain well over 100 seeds. Only soft, ripened noni fruits should be chosen for seed collection. The seeds must be separated from the fibrous, clinging fruit flesh. First, split the fruit by hand into smaller pieces. Separate the seeds from the flesh using a strong spray of water and a firm screen or colander, washing the pulp through the screen while retaining the cleaned seeds. Rubbing the fruit fragments on the screen by hand or with a blunt object can help force the fruit flesh through the screen. It may take 15 minutes or more of vigorous washing and rubbing to detach most of the flesh from the seeds.

Scarifying the hard seedcoat by nicking or puncturing it significantly reduces germination time, improves germination percentage, and promotes uniform sprouting. Whereas in nature the seedcoat must gradually decompose before water can enter, scarification overcomes this natural seed dormancy. Using a household blender to separate seeds from the ripened flesh can also result in nicking the seedcoats, or the seeds can be suspended in water and sub-seedcoats, or the seeds can be suspended in water and subjected to short pulses of blending. Noni seeds can be dried and stored, but the length of time they will remain viable is not known. After cleaning, spread the seeds out on newspaper and dry them in the shade or indoors for 2 or 3 days. Store the seeds in an airtight container at room temperature. Fresh Noni seeds can be planted immediately after extraction from the fruit. Some growers soak the seeds until they start to germinate, then plant them in containers, while others plant fresh seeds without pre-soaking treatments. Some growers just plant fruit fragments containing seeds directly into the field soil. Noni seeds require hot, wet conditions for optimum germination. Unscarified seeds need several months to a year before natural germination takes place, but this period can be reduced to a month or so using heat. The seeds can tolerate temperature of 100°F (38°C) perhaps even higher. Select the warmest spot in the nursery of greenhouse to germinate Noni seeds or heat can be supplied using nursery heating pads.

Generally Noni seedlings are grown in pots in full sun for a minimum of 9-12 months before they are transferred to the field. Seedlings upto a year old or more may also be planted. If Noni plants are transplanted too young they are most susceptible to weed competition, mechanical damage, and slug attack. Seedling and young plants grown from cuttings can be given liquid fertilizer once a month, or a controlled release fertilizer less often (depending on the formulation's release period). Balanced formulations such as 14-14-14 that also contain micronutrients (minor elements) are advised. Young plants also respond well to applications of dilute, liquid foliar fertilizers. As plants become established, granular, rapidly soluble formulations can be used. Noni is relatively salt-tolerant, and fertilizer burn is uncommon under normal conditions.

VEGETATIVE PROPAGATION

Cultivation of noni plants from stem cuttings (verticals or laterals) reduces the time required to obtain plants that are ready for transplanting. Cuttings from stems and branches will sprout roots readily under the proper conditions.

Select vigorous plants for propagation. Remove a branch or stem and check for fresh sap flow from the wound. If the sap flows readily, cuttings could be made from these materials. If sap does not ooze from the cut end, discard the material and select another plant, another location, or perhaps wait for a better time of year. Sap flow indicates a vigorous, actively growing plant with relatively high reserves of energy. Insert the cut end of the freshly cut Noni stem into a pot containing a general-purpose growth medium. Rooting hormone should improver or accelerate rooting of vegetative cuttings, but it is not necessary. As with seed germination, bottom heat enhances rooting, and an artificial, pathogen-free medium is preferred to untreated agricultural field soil. Select a location with partial shade and keep the cuttings well watered until rooting occurs. After rooting, move the plants into full sun and begin fertilizer applications. Noni plants may also be produced by air layering or by digging up plants that have sprouted from the root system of a mature plant.

SITE SELECTION

Avoid locations where other crops have been planted recently, due to the susceptibility of Noni to root-knot nematodes. Select a site in full of partial sun with well drained, well aerated soil. Avoid heavy soils, compacted areas, and flood-prone sites. Prepare a hole about the size of the pot and transplant carefully. In rocky location, "rip" the land (disturb or plow the subsoil) before grading to prepare a flat or gently sloping field.

PRUNING

Less than 3 years old should be pruned back after or during their first production of fruits. In the flowering years, the pruned plants will become bushy and because Noni trees can reach a height of approximately 20 ft, growers may wish to prune the plants to facilitate picking.

NUTRITION AND FERTILIZER

Young seedlings and transplants are given controlled-release fertilize formulations, while older, mature plants are given rapidly available granular formulations. Fertilizer should be applied away from the trunk at the "drip line" of the plant, the area where water drips from the edge of the leaf canopy

Noni plants of all ages respond well to sprays of foliar fertilizer. Noni flower and fruit production is very responsive to sprays of high-phosphorous foliar fertilizers (e.g., 10-45-10) and products (e.g., seaweed emulsions) containing nitrogen and minor elements.

Noni should be fertilized frequently using smaller amounts of fertilizer, rather than infrequently suing larger amounts. In high-rainfall areas, young plants up to a year old can be given $\frac{1}{2}$ pound per month of balanced fertilizer (14-14-14), and more mature plants can be given up to 1 pound per month. Effective organic fertilizers for noni cultivation include crushed coral, dolomite, K-mag 7-7-7, and composted chicken manure and macadamia nut husks. Some locations will benefit from yearly applications of lime, about 1 pound per plant.

IRRIGATION

Noni thrives with moderate irrigation and can survive extended drought once established and mature. When plants are less than 2-3 years old condition are dry, irrigate once or more a week, applying up 10 gallons per plant; for older plants, irrigate less frequently. Overwatering can accelerate damage from root-knot nematodes, cause root rot, and leach fertilize nutrients beyond the root zone.

CROP IMPROVEMENT

Among all the new innovations and activities, information technology and biotechnology ruled the world for the past two to three decades. In the field of Plant biotechnology, the technical advances have been tremendous and unique in their application as commercial tools to improve agriculture, sylviculture, horticulture and agro industries. Noni plants were taken for rapid multiplication using tissue culture techniques at the tissue culture lab in Chennai. Axillary buds were taken from young seedlings and after proper sterilization these buds were inoculated to Murashige and Skoog's medium containing various hormone combinations. After successful culture initiation at *in vitro* buds were subjected to rapid multiplication. An excellent rate of multiplication with enough elongation of the multiple shoots was achieved in the medium. After multiplication, shoots of 4cm in length and above were separated individually and inoculated to rooting media to produce roots. Enough roots were produced within 18 days *in vitro* (Subramani, 2006).

Other than micropropogation, Noni biotech division will also enter into new research area like:

- Cell culture to produce saline tolerant plants
- Micropropogation of elite clones
- Isolation and identification of cells *in vitro*
- Cell immobilization for production of nutraceuticals
- Bioreactor technology

PLANT DISEASES

Black flag of noni

A severe leaf and fruit blight occurs in the Puna district of the island of Hawaii. The disease is caused by the pathogen *Phytophtora botryose*. Symptoms include a black leaf and stem blight, stem or branch dieback, and a chocolate-colored or black fruit rot. The disease is favored by prolonged periods of wet weather. Control measures include pruning, weed control, and field sanitation.

Sooty mold

Sooty mold is a black, superficial growth of a nonparasitic fungus that utilizes the sugary exudates produced by soft-fungus insects such as scales and aphids. Sooty mold can easily be wiped off leaves by hand. Heavy infestation of the fungus can reduce photosynthesis, resulting in poor plant growth and reduced fruit size and quality. A soapy water spray can control sooty mold.

Noni root-knot disease

Noni root-knot is a disease of the roots caused by root-knot nematodes, *Meloidogyne* species. Above-ground symptoms include reduced vigor, plant stunting, and yellowing. Affected roots can be galled, swollen, cracked, and rotten. This disease can be minimized by using nematode-free transplants and by adding organic soil amendments such as composts containing chicken manure. Moderate irrigation and fertilizer use will also minimize the damage and reduce the magnitude of secondary root rot by saprophytic fungi and bacteria.

Stem rot

In stem rot, tissues at the base of the plant decay, resulting in stunted plants. The initial plant symptom is wilt, including leaf flagging, and drop. The disease arises from a complex of multiple causes including predisposing stresses (root-knot nematodes, flooding, stem injury) and stem-girdling decay by a soilborne fungus, *Sclerotium rolfsii.* Planting pathogen-free plants, selecting appropriate sites, and avoiding plant stress can minimize the occurrence and severity of stem rot.

INSECT PESTS

Pest attacks of cultivated noni can cause significant damage. Pests known to attack noni in Hawaii include aphids (*Aphis gossypii*), ants, scales (the green scale), mites (eriophyid mites), whiteflies (fringe guava whitefly), and slugs. Pest outbreaks are favored by monocultures of noni. Thus, intercropping with other species of nonhost plants can minimize the overall severity and frequency of pest attacks.

Eliminate weeds that are alternate hosts of noni pests to prevent outbreaks. Significant weed hosts of noni pest are identified by periodic inspections of both noni pests nearby weeds for the presence of pests such as aphids, scale, and whitefly. These and other soft-bodied insects can be effectively controlled with sprays of insecticidal soaps and oils. Mites can be controlled by pruning affected leaves and by applications of approved pesticides, such as sulphur. For slugs and ants, apply a temporary barrier at the base of the noni plant, such as copper tape (for slugs), or a sticky substance such as pine tar (for ants).

WEEDS

Noni transplants are susceptible to competition from weeds such a Guinea grass and sensitive grass (*Mimosa pudica*). Young noni plants should be weeded by hand until the stem is sufficiently large to withstand potential mechanical injury from mowers or string trimmers. Although weed mats can suppress weeds, they also inhibits granular fertilizers from penetrating to the roots zone, and heat can accumulate under the mat, which may favour nematode reproduction. Once noni plant are well established, competition from weeds is less damaging than when the plants are young. Care should be taken to eradicate weeds that are hosts of root-knot nematodes may be identified by inspecting their roots for the presence of the characteristic galls and swellings. Noni is susceptible to attack by dodder (*Cuscuta sandwichensis*), a parasitic weed. Dodder should be manually removed from infested Noni plants.

HARVESTING AND YIELD

Noni plants can begin to bear fruit about 9 months after planting. Fruits can be harvested at this early stage, although they generally are small and few. Some farmers choose to forgo harvest during the first and second years in favour of pruning back the branches instead. This pruning results in a bushy plant with more vertical and lateral branches and ultimately produces greater fruit yield.

In Hawaii, noni fruits may be harvested year round, although there are seasonal trends in fruit production that may be affected or modified by the weather and fertilizer applications. Fruit production may diminish somewhat during winter months in Hawaii. The expected yield from mature, healthy plants is 250-500 pounds per plant per year, depending on nutrition and plant spacing. However, yields at some locations can exceed 500 pounds per plant per year with good crop management. Fruits may be picked just before they begin to ripen fully and turn completely whitish-yellow on the tree. These fruits are suitable for shipping and will continue to ripen en route unless measures are taken to retard ripening. For processing locally, fresh fruits are picked when ripe, just before they fall naturally from the tree. Juice can be squeezed directly from the ripened fruits or allowed to drip from the fruits and ferment as in the traditional method.

AGROFORESTRY / ENVIRONMENTAL PRACTICES

Although Noni regrows well after pruning. Noni plants are generally not managed for mulch production in agroforestry situations. It is well suited for home gardens. A single plant is sufficient to meet the needs of one or more families. Noni can be used for boundary markers due to its persistence and ability to survive harsh conditions and extended periods of drought. The fruits and leaves are useful as animal feed or fodder (pets and livestock). Ripe fruits are a natural source of food for birds, rodents and insects. The flower nectarines are very attractive to honeybees. Noni is tenacious enough help to stabilize lands in harsh or unstable coastal environments. Although the naturalized *M. citrifolia* (the wild and cultivated Noni types) is considered by many to be a beautiful plant with shiny green foliage, some object to its use as an ornamental plant due to the strong and sometimes offensive odour of ripened fruits and because the fallen fruits attract many flies and other insects. The cultivated *M. citrifolia* variety 'Potteri' is a beautiful and functional ornamental plant with small fruits and green and white variegated leaves.

CONCLUSION

Morinda citrifolia assumed to have originated in Southern Asia has been used in folk remedies by Polynesians for over 2000 years, and is reported to have broad range of antibacterial, antiviral,antifungal, antitumor, antihelminthic, analgesic, hypotensive and immune enhancing effects. Noni plant has the ability to survive in harsh environments and wide range of acidity levels. It has rapid regeneration ability and can be propagated from seed,stem or root cuttings and air layering. Owing to its therapeutic and nutritional value, it can be considered as one of the most important medicinal plants.

REFERENCES

Abbott I A 1992 La'au Hawaii' traditional Hawaiian uses of plants v 3 Honolulu Hawaii: Bishop Museum Press; p 97- 100

Abbott A 1985 The geographic origin of the plants most commonly used for medicine by Hawaiians J Ethnopharmacol; **14**: 213-22

Allen WH 1873 London C Some information on the ethnobotanical properties of Noni (*Mor inda citr ifolia*) In: The useful plants of India

Asahina A Y Ebesu J S M Ichinotsubo D Tongson J and Hokama Y 1994 Effect of okadaic acid (OA) and Noni fruit extraction in the synthesis of tumor necrosis factor-a (TNF-a) by peripheral blood mononuclear (PBN) cells *in vitro The Procedings of the International Symposium of Ciguatera and Marine Natural Products* p 197-205

Betz J 1997 Pride Publishing Noni: Polynesia's natural pharmacy p 13

Bushnell O A Fukuda M Makinodian T1950 The antibacterial properties of some plants found in Hawaii *Pacific Sci* **4**: 167-83

Cheeseman T F 1903 The flora of raratonga the chief island of the cook group v 6 London: *Linnean Soc* p 261-313

Darmstadt G L Mao-Qiang M Chi E Saha S K Ziboh V A Black R E Santosham M Elias P M 2002 "Impact of topical oils on the skin barrier: possible implications for neonatal health in developing countries" *Acta Paediatrica* **91** (5): 546–554

Davison C 1927 Hawaiian medicine The Queen's Hospital Bulletin with Palama Clinic Section; **4**: 2-5

Deng S Palu A K West B J et al 2007 Lipoxygenase Inhibitory Constituents of the Fruits of Noni (*Morinda citrifolia*) Collected in Tahiti *J Nat Prod* **70**(5):859-862

Heinicke R1985 The pharmacologically active ingredient of Noni *Bulletin National Tropical Botanical Garden* **15**:10-14

Hirazumi A Furusawa E Chou S C and Hokama Y 1994 Anticancer activity of Morinda citrifolia (noni) on intraperitoneally implanted Lewis lung carcinoma in syngeneic mice *Proc West Pharmacol Soc* **37**: 145-6

Hirazumi A Furusawa E Chou S C and Hokama Y1996 Immunomodulation contributes to the anticancer activity of *Morinda citrifolia* (noni) fruit juice *Proc West Pharmacol Soc* **39**: 7-9

Hirazumi A and Furusawa E 1999 An immunomodulatory polysaccharide- rich substance from the fruit juice of *Morinda citrifolia* (noni) with antitumour activity *Phytother Res* **13**: 380-7

Krauss B H 1993 *Plants in Hawaiian Culture* Honolulu: University of Hawaii Press Honolulu

Langford J Doughty A Wang M Y Clayton L and Babich M 2002 Effects of *Morinda citrifolia* on auditory function and quality of life in patients with decreased bone mineral density and auditory function *J Complementary & Alternative Med* **15**:10-16

Letawe C Boone M and Pierard G E1998 "Digital image analysis of the effect of topically applied linoleic acid on acne microcomedones" *Clinical & Experimental Dermatology* **23** (2): 56–58

Levand O and Larson HO1979 Some chemical cons tituents of *Morinda citrifolia Planta Med* **36**: 186-7

Ma D L West B J Su C X Gao J H Liu T Z and Liu Y W 2007 Evaluation of the ergogenic potential of noni juice *Phytotherapy Res* **21**(11):1100-1101

Maiden J H 1889 Useful native plants of Australia including Tasmania Sydney: Tuner and Henderson Publisher p 45

McClatchey W 2002 "From Polynesian Healers to Health Food Stores: Changing Perspectives of *Morinda citrifolia* (Rubiaceae)" *Integrative Cancer Therapies* **1** (2): 110–120

Merrill E D 1943 Noni (*Morinda citrifolia*) as an edible plant In: Technical manual: emergency food plants and poisonous plants of the islands of the pacific Washington DC: US Government Printing Office

Mohd Z Z Abdul-Hamid A Osman A *et al* (2007) Isolation and identification of antioxidative compound from fruit of mengkudu (*Morinda citrifolia* L) *Inter J Food Properties* **10**(2):363-373

Moorthy N K and Reddy G S1970 Preliminary phytochemical and pharmacological study of *Morinda citrifolia* Linn Antiseptic **67**: 167-71

Morton J F 1992 The ocean-going Noni or Indian mulberry (*Morinda citrifolia* Rubiaceae) and some of its 'colorful' relatives *Economic Bot* **46**: 241-56

Nelson S C 2006 "Nutritional Analysis of Hawaiian Noni (Noni Fruit Powder)" *The Noni Website* Retrieved 15-06-2009

Raj R K 1975 Screening of indigenous plants for anthelmintic action against humanAscaris Lumbricoides: Part-II *Indian J Physiol Pharmacol* **19**: 47-9

Seemann B and Flora V 1866 A description of the plants of the Viti or Fiji is lands with an account of their history uses and properties London: L Reeve and Co p 1865-73

Solomon N 1999 The Noni Phenomenon Direct Source Publishing USA 43-47

Subramani J 2006 Crop Improvement in Morinda citrifolia Proc First National Symposium on Noni Research October 7-8 2006

Swanholm C E St John H and Scheuer P J 1959 A survey of alkaloids in Hawaiian plants *Pacific Sci* **13**: 295-305

Thompson R H 1971 *Naturally Occurring Anthraquinones* New York: Academic Press

Umezawa K1992 Isolation of 1-methoxy-2-foremyl-3-hydroxyanthraquinone from *M citrifolia* and neoplasm inhibitors containing the same Japan Kokai Tokyo *Appl*; 92/264: 311-17

West B J Jensen C J Westendorf J 2008 A new vegetable oil from noni (*Morinda citrifolia*) seeds *Inter J Food SciTech* **43**(11):1988-1992

Youngken H W Jenkins H J and Butler C L 1996 Studies on *Morinda citrifolia* L II *J Am Pharm Assoc* **49**: 271-3 57

Younos C Rolland A Fleurentin J Lanhers M C Miss L R and Mortier F 1990 Analgesic and behavioural effects of *Morinda citrifolia Planta Med* **56**: 430-4

INDEX

A

B

C

D

E

F

G

H

I

L

M

N

O

P

Q

R

S